J. Stewart Stein
CONSTRUCTION GLOSSARY: AN ENCYCLOPEDIC REFERENCE AND MANUAL

James E. Clyde
CONSTRUCTION INSPECTION: A FIELD GUIDE TO PRACTICE

Harold J. Rosen and Philip M. Bennett
CONSTRUCTION MATERIALS EVALUATION AND SELECTION: A SYSTEMATIC APPROACH

C.R. Tumblin
CONSTRUCTION COST ESTIMATES

Harvey V. Debo and Leo Diamant
CONSTRUCTION SUPERINTENDENTS JOB GUIDE

Oktay Ural, Editor
CONSTRUCTION OF LOWER-COST HOUSING

Robert M. Koerner and Joseph P. Welsh
CONSTRUCTION AND GEOTECHNICAL ENGINEERING USING SYNTHETIC FABRICS

J. Patrick Powers
CONSTRUCTION DEWATERING: A GUIDE TO THEORY AND PRACTICE

Harold J. Rosen
CONSTRUCTION SPECIFICATIONS WRITING: PRINCIPLES AND PROCEDURES, 2nd edition

Walter Podolny, Jr., and Jean M. Muller
CONSTRUCTION AND DESIGN OF PRESTRESSED CONCRETE SEGMENTAL BRIDGES

Ben C. Gerwick, Jr., and John C. Woolery
CONSTRUCTION AND ENGINEERING MARKETING FOR MAJOR PRODUCT SERVICES

*Construction and Engineering
Marketing for
Major Project Services*

Construction and Engineering Marketing for Major Project Services

Ben C. Gerwick, Jr.
John C. Woolery

A Wiley-Interscience Publication
JOHN WILEY & SONS
New York • Chichester • Brisbane • Toronto • Singapore

Copyright © 1983 by John Wiley & Sons, Inc.

All rights reserved. Published simultaneously in Canada.

Reproduction or translation of any part of this work beyond that permitted by Section 107 or 108 of the 1976 United States Copyright Act without the permission of the copyright owner is unlawful. Requests for permission or further information should be addressed to the Permissions Department, John Wiley & Sons, Inc.

Library of Congress Cataloging in Publication Data:

Gerwick, Ben C.
   Construction and engineering marketing for major project services

  (Wiley series of practical construction guides, ISSN 0271-6011)
  "A Wiley-Interscience publication."
  Bibliography: p.
  Includes index.
  1. Construction industry. 2. Engineering. 3. Consulting engineers. 4. Business consultants. I. Woolery, John C., 1949- II. Title. III. Series.
HD9715.A2G465   1982   658.8   82-8383
ISBN 0-471-09886-8   AACR2

Printed in the United States of America

10 9 8 7 6 5 4 3 2

Series Preface

The Wiley Series of Practical Construction Guides provides the working constructor with up-to-date information that can help to increase the job profit margin. These guidebooks, which are scaled mainly for practice, but include the necessary theory and design, should aid a construction contractor in approaching work problems with more knowledgeable confidence. The guides should be useful also to engineers, architects, planners, specification writers, project managers, superintendents, materials and equipment manufacturers and, the source of all these callings, instructors and their students.

Construction in the United States alone will reach $250 billion a year in the early 1980s. In all nations, the business of building will continue to grow at a phenomenal rate, because the population proliferation demands new living, working, and recreational facilities. This construction will have to be more substantial, thus demanding a more professional performance from the contractor. Before science and technology had seriously affected the ideas, job plans, financing, and erection of structures, most contractors developed their know-how by field trial-and-error. Wheels, small and large, were constantly being reinvented in all sectors, because there was no interchange of knowledge. The current complexity of construction, even in more rural areas, has revealed a clear need for more proficient, professional methods and tools in both practice and learning.

Because construction is highly competitive, some practical technology is necessarily proprietary. But most practical day-to-day problems are common to the whole construction industry. These are the subjects for the Wiley Practical Construction Guides.

M. D. Morris, P.E.

Preface

During the many years in which I worked in construction and engineering, I came to realize what a major role is played by marketing. The name of the game was getting the contract—getting it on the most favorable terms possible, but, in any event, getting the contract. Without contracts, the enterprise would soon wither and die. Low bids were not enough, however, for if they weren't on the right jobs, on the right terms, and in the right places, they could soon drain away our resources. And once landed, the contracts had to be performed. Their financial success depended to a high degree on our ability to negotiate change orders as they arose, and to "sell" the completed job when we were finished. Our best business came from satisfied clients, and we soon realized that a repeat client was the most valuable client of all: a source of pride and a testimonial to our capability.

When I then had the opportunity to move my career to a graduate engineering program at the University of California, I felt that one of the most valuable services I could render was to introduce young engineers to the principles of marketing. Initially I had difficulty convincing my colleagues that this was an appropriate subject for a graduate program in engineering. Their doubts were assuaged, however, as we found that the majority of senior professional contractors' and engineers' time was spent in one form or another of marketing.

And so for ten years a course in marketing has been taught in our program. It has developed, expanded and evolved as we have directed our attention to the many facets of marketing. While the majority of our students have been engineers, we have also had architects, landscape architects, mechanical engineers, and naval architects, all of whom have constructively participated in the course.

I had the great good fortune to have an exceptional student, John Woolery, in one of my early classes. He later went on to take his doctorate and to engage in the practice of construction engineering. He developed an interest in marketing and has actively augmented our earlier syllabi with additional examples from practice and with data from an extensive search of related business, engineering, and architectural literature.

At the instigation of our former students, we have undertaken to write the present book, summarizing our knowledge and opinions on this challenging subject. Marketing is a dynamic activity and hence is continually moving. While the many examples quoted here will be continually superseded, it is hoped that the underlying principles will prove of value to both present and succeeding generations of constructors.

BEN C. GERWICK, JR.

May 1982
San Francisco, California

Contents

1. **Introduction,** *1*
   - 1.1 The Construction Industry, *1*
   - 1.2 Marketing, *2*
   - 1.3 The Marketing Effort, *3*
   - 1.4 A Dynamic Response to the Client's Needs, *4*
   - 1.5 The Skills Required for Marketing, *4*
   - 1.6 The Study of Marketing Techniques, *5*

2. **The Function of Marketing—A Matching of Services to Needs,** *7*
   - 2.1 Proper Size, *7*
   - 2.2 Proper Type of Work, *8*
   - 2.3 Proper Location, *9*
   - 2.4 Proper Scope, *9*
   - 2.5 Proper Contracts, *11*
   - 2.6 Proper Schedule, *12*
   - 2.7 Proper Quality, *13*
   - 2.8 Proper Timing, *13*
   - 2.9 Proper Relations with Third Parties, *14*
   - 2.10 Proper Volume, *15*
   - 2.11 Proper Client, *15*
   - 2.12 Proper Price, *17*

3. **The Establishment of a Marketing Plan,** *19*
   - 3.1 Market Survey, *19*
   - 3.2 Increasing Contract Volume, *26*
   - 3.3 Bid Volume versus Contract Volume, *28*
   - 3.4 Game Theory, *28*
   - 3.5 Profitability, *29*
   - 3.6 Private Work versus Public Work, *31*
   - 3.7 Design of Your Sales Force, *33*
   - 3.8 Establishing Personal Contact, *34*
   - 3.9 Multiheaded Clients, *35*
   - 3.10 Look for Something Different, *37*
   - 3.11 Establishing a Marketing Plan, *39*

Contents

4. **The Role of the Contract as a Marketing Tool, 40**
    - 4.1 Types and Scopes of Contracts, 40
    - 4.2 The Lump-Sum Contract, 41
    - 4.3 The Lump-Sum plus Units Contract, 47
    - 4.4 The Unit-Price Contract, 47
    - 4.5 The Cost plus Fixed-Fee Contract, 52
    - 4.6 The Day-Rate Contract, 57
    - 4.7 The Target Estimate Contract, 57
    - 4.8 Consulting Contracts, 59
    - 4.9 The Construction Management, or Project Management, Contract, 61
    - 4.10 "Turnkey" Contracts (Design and Construct), 69
    - 4.11 Subcontracts, 70
    - 4.12 Packaging a Contract, 70
    - 4.13 Risk Sharing, 71

5. **Meeting Competitive Pressures, 74**
    - 5.1 Meeting Competitive pressures, 74
    - 5.2 Use of Allies, 76
    - 5.3 Brain Picking, 78
    - 5.4 Monitoring Your Market, 81
    - 5.5 How to Fight Unethical Competition, 81
    - 5.6 "Opportunism", 85
    - 5.7 Reacting to Market Conditions, 86
    - 5.8 Price Chiseling, 86

6. **Brochures and Other Marketing Aids, 90**
    - 6.1 Three Categories of Brochures, 90
    - 6.2 Directing the Brochure to Reach the Right Audience, 91
    - 6.3 Designing the Brochure, 92
    - 6.4 Other Marketing Aids, 100
    - 6.5 Examples of the Well-Designed Brochure, 128

7. **Written Communication—Technical and Business Letters, 129**
    - 7.1 Basic Principles, 129
    - 7.2 Manner of Expression, 130
    - 7.3 Business Aspects, 132
    - 7.4 Technical Aspects, 133
    - 7.5 Examples, 137

8. **Verbal Communication—The Call and the Conference, 138**
    - 8.1 Basic Principles, 138
    - 8.2 The Call—How to Make an Appointment, 142
    - 8.3 The Call—The Initial Contact, 142
    - 8.4 The Call—The Follow-up, 146
    - 8.5 The Call—The Business Lunch, 146
    - 8.6 The Call—Preparation, 147
    - 8.7 The Conference—How to Set It Up, 149

Contents   xi

- 8.8 The Conference—Robert's Rules, 150
- 8.9 The Conference—Participation, 150
- 8.10 The Conference—Minutes, 151
- 8.11 The Conference—A Case History, 152

9. **Prequalification, 154**
   - 9.1 Case Histories, 154
   - 9.2 Factors Involved in Prequalification, 155
   - 9.3 Personnel and Equipment, 157
   - 9.4 Financial Data, 160
   - 9.5 Subcontractors, Major Suppliers, and References, 163
   - 9.6 Special Considerations in Prequalification, 164

10. **The Proposal, 183**
    - 10.1 The Basic Proposal Format, 183
    - 10.2 Exceptions and Stipulations, 184
    - 10.3 How to Avoid Late Payments, 185
    - 10.4 Technical and Contractual Alternatives, 186
    - 10.5 How Not to Sell a Proposal, 187
    - 10.6 Legal Requirements, 189
    - 10.7 After the Proposals Have Been Submitted, 198

11. **Salesmanship and Advertising, 202**
    - 11.1 What Influences an Owner's Decision to Call You or to Award You a Contract?, 202
    - 11.2 Client Relationships on a Social Basis, 203
    - 11.3 Projecting the Right Image through Dress, 204
    - 11.4 The Expense Account and Its Abuse, 205
    - 11.5 The Use of a Preliminary Design or a Budget Estimate as a Sales Tool, 207
    - 11.6 Use of Client Involvement to Sell Your Ideas, 207
    - 11.7 The Psychology of Closing Sales, 209
    - 11.8 Advertising: Its Use and Misuse, 209
    - 11.9 Community Involvement, 210

12. **Product Development, 212**
    - 12.1 Innovation, 212
    - 12.2 A Schedule for Product Development 213
    - 12.3 What Does Happen in Reality?, 214
    - 12.4 Forecasting, 215
    - 12.5 Some Case Histories, 216
    - 12.6 How Can an Innovative Idea Be Protected from Competitors?, 217
    - 12.7 Creativity, 219
    - 12.8 Franchise Extension, 219

13. **Pricing, 220**
    - 13.1 Proper Valuation of Construction and Engineering Services in Relation to the Market, 221
    - 13.2 Proper Pricing, 222

- 13.3 Purchasing Decisions on Factors Other than Price Alone, 224
- 13.4 Tailoring Services Offered to Prices Available, 224
- 13.5 Market Penetration, 225
- 13.6 Unintentional Underpricing, 225
- 13.7 Pricing to Anticipate Additional, Changed, or Future Work, 226
- 13.8 Protective Pricing, 227
- 13.9 Long-range Effects of Pricing Too High, 227
- 13.10 Markup, 228

14. Selling a Professional Service, 233
    - 14.1 Feasibility Study, 233
    - 14.2 Personal Contacts, 234
    - 14.3 Enthusiasm, 235
    - 14.4 Special Products or Services, 235
    - 14.5 Maintaining a Proper Client Relationship, 236
    - 14.6 Public and Professional Relations, 237
    - 14.7 Presentations, 238
    - 14.8 Proposals for Professional Services, 239
    - 14.9 Competition between Professional Services, 241

15. Product and Professional Liability, 243
    - 15.1 Consulting Contracts, 243
    - 15.2 Construction Contract, 245
    - 15.3 Design and Construct Contracts, 247
    - 15.4 Case Histories, 248

16. Negotiations, 249
    - 16.1 Recognizing the Right Opportunity to Negotiate, 249
    - 16.2 Establishing a Negotiation Strategy, 251
    - 16.3 Setting the Stage, 252
    - 16.4 Destroying the "Opponent" Concept and Replacing It with the "Mutual Endeavor" Concept, 254
    - 16.5 Various Appeals to Reason, 255
    - 16.6 Avoid Arguments, 255
    - 16.7 Self-Discipline, Self-Control, and the Proper Use of Temper, 256
    - 16.8 Proper Use of Legal Talent, 257
    - 16.9 Involvement of Subordinates, 258
    - 16.10 Trades—But Not Too Soon: The Value of Taking Time, 259
    - 16.11 Concluding a Deal, 260
    - 16.12 Finalizing It in Detail: Writing It Up, 261

17. Changes, Change Orders, and Claims, 264
    - 17.1 Claims Caused by Changed Conditions, 265
    - 17.2 Claims Caused by Changed Orders, 268
    - 17.3 Claims Caused by Impact or Ripple, 269
    - 17.4 Claims Caused by Owner-Furnished Items, 269
    - 17.5 Claims Caused by Inspection, 270
    - 17.6 Claims Caused by Differences in the Interpretation of the Plans and Specifications, 271

- 17.7 Claim Letters, 272
- 17.8 Documentation, 277
- 17.9 Using The Owner's Records to the Contractor's Advantage, 279
- 17.10 Legal Assistance, 281
- 17.11 Advantages of Deductive Change Orders, 282
- 17.12 "Walking Off the Job", 282
- 17.13 Vindictive Owners, 283
- 17.14 Lost Causes, 284
- 17.15 How to Avoid Claims, 284

18. **International Marketing, 286**
    - 18.1 How to Sell to the International Market, 286
    - 18.2 Marketing Techniques, 288
    - 18.3 Short-Term Business Development, 291
    - 18.4 Long-Range Business Development, 294

19. **Ethical Considerations in Marketing, 299**
    - 19.1 Bribes and Payoffs—International Problems, 299
    - 19.2 Ethical Solutions to International Problems, 304
    - 19.3 Bribes and Payoffs—Domestic Problems, 307
    - 19.4 Legitimate Fees for Services, 308
    - 19.5 Ethical Considerations in the Negotiation Process, 309
    - 19.6 Boycotts, 309
    - 19.7 Legal Standards versus Ethical Standards, 310

20. **Business Development in Your Career, 311**
    - 20.1 A Changing Profession, 311
    - 20.2 Professional Responsibility, 311
    - 20.3 Education and Experience, 312
    - 20.4 Planning Your Career, 314
    - 20.5 Finding the Right Job, 317
    - 20.6 Office Politics and Diplomacy, 319
    - 20.7 Handling of an Emergency, 321
    - 20.8 Give Your Ideas or Die with Them, 322
    - 20.9 Changing Jobs, 322
    - 20.10 Accepting or Rejecting New Assignments, 323
    - 20.11 Career Growth, 323
    - 20.12 In Conclusion, 324

References, 326
Appendix 1. Sample Exercises, 327
Appendix 2. Sample Correspondence, 331
Appendix 3. Conference Problems, 340
Appendix 4. Presentation Exercise for Sales Calls, 347
Appendix 5. Sample Contracts, 350
Index, 417

*Construction and Engineering
Marketing for
Major Project Services*

1
Introduction

1.1. The Construction Industry

The construction industry—construction companies, engineering consultants, and architectural firms—viewed in all its aspects, represents perhaps the largest single segment of the nation's economy. It is the nation's investment in the future. On a worldwide basis, it represents humanity's dedication to civilization, the plow back of today's physical goods in order to provide for the needs of tomorrow. These needs are both material and cultural, and construction is the means by which the needs, real and perceived, are met. For the less developed nations, it is to a large extent the lack of facilities, of housing, of irrigation, of transportation, of utilities, of resource development and industrial plants which prevents the establishment of an acceptable standard of living. The construction industry involves the entire gamut of planning, feasibility studies, financing, engineering, construction proper, infrastructure, start-up, maintenance and repair.

The construction industry can also be viewed on another basis as involving heavy construction (dams, airports, canals, reclamation); transportation construction (highways, railroads, pipelines); engineering construction (power plants, water and sewage treatment plants); marine construction (harbor structures, dredging, breakwaters, and offshore platforms); industrial aspects (refineries, chemical plants, factories, warehouses); utility (water, sewage, communication and power distribution); housing (single-family and multi-family residences); and commercial structures (shopping centers, parking garages, schools, office buildings, community facilities, governmental buildings). Even the above list only partially covers the wide range of construction activities carried out by a technologically oriented society such as ours.

To fulfill these services, a wide variety of organizations have become established. These are divided by types: construction companies, engineering consultants, architectural firms. To integrate their activities, we have seen in recent years the growth of construction management organizations.

One unusual characteristic of the construction organizations in the United States is their size; they tend to be small to medium in size, as compared with the few large firms that dominate the manufacturing industry. Thus they share many of the personal characteristics of the agricultural segment. The relative uniqueness of each project apparently requires the flexibility and the individual attention that limit the effectiveness of large-scale organizations. At some level, varying according to the type of activity and location, the incremental costs of management appear to exceed the incremental benefits of size.

Thus we see throughout the nation and the world a tremendous demand for construction, a demand limited primarily only by available financial resources, coupled with a widely spread and heterogeneous group of organizations devoted to carrying out the construction activity.

1.2. Marketing

Marketing is the process by which these construction capabilities are matched to the needs. It encompasses a far wider scope than merely obtaining a contract; rather, it is the process by which mutually beneficial relations are established and maintained between the client, who needs the facility, and the construction organization, which has the abilities and resources to fulfill that need.

As such, marketing is a two-sided affair. The client needs to find a construction organization, the contractor a client. Some industries, such as the oil industry, have felt it necessary at times to deliberately establish a group of construction organizations to supply their planned future needs. Large industries with continued construction demand and some governments—for example, The Netherlands, whose country's very existence depends on an availability of construction capabilities—have established policies to encourage the growth of construction organizations. On the other side, once established, construction organizations need a continuity of activity, not only to keep their own resources fully employed but to generate a return on investment (profit) that will attract the necessary capital for continued existence and for growth.

Like all human activities, construction organizations cannot long exist at a steady state. It is an aphorism that they either grow or decay. Without growth, they sooner or later lose their position in the industry and succumb to the inroads of other more aggressive, growing organizations. Growth continues until the organization becomes unwieldy: then it either finds a better organizational pattern (such as decentralization), fragments involuntarily, merges into a supercompany, or slowly dies.

Marketing is thus seen as an essential element to the survival and growth of construction organizations.

Henceforth, marketing as used in this book will be defined as that activity carried out by the construction organization to establish relations, both contractual and supportive, that will enable it to supply its services on a basis that will be most favorable to its growth. This means the obtaining of contracts for the supply of its services of the right kind and scope, in the right place and time, and on the most favorable terms. It further means the establishment of the supportive relations that will enable the contract to be carried out and concluded favorably.

1.3. The Marketing Effort

In practical terms, construction organizations devote a large proportion of their efforts to marketing. Many smaller firms would question this, but that is because they do not comprehend the full scope of the marketing function. Business development activities include finding markets and evaluating job potentials, establishing contacts, gaining vital information, and prequalifying with the client. Estimating the costs furnishes the basis for establishment of the price. Because estimating per se is usually treated as a separate subject, and because of its specialized complexities, it will not be further treated in this book, except to note that it is primarily a marketing function. Submitting the proposal, entering into a contract, and negotiating changes and claims are all marketing functions. Finally, so is product development, whether the product be a construction technique, new equipment, or a new contract form.

During the actual construction operations, the client must be kept informed of progress; changes must be processed; claims, if any, developed. Then at the finish of the construction, the contract must be favorably settled and concluded, and an ongoing relationship must be established that will lead to future contracts with that client and favorable recommendations to other clients. Thus the marketing effort extends from the earliest phase to the last: it is a continuous, ongoing and vital part of construction activity.

Construction executives fully acknowledge that marketing consumes 50 to 70 percent of their time. They reward and promote juniors largely on their evaluation of the ability to "get business and to establish proper relationships with clients, in short, to market." Their greatest complaints are that "engineers can't write a letter" and that they cannot exercise judgment in their relationship with the client.

Marketing is not carried out by executives only: the project managers and engineers in the field are extremely important foci of the ongoing marketing effort and the superintendents and foremen have their part to play. The successful construction organization is the one which is "market oriented," from top to bottom.

In recent years major construction organizations, such as Raymond International, have, as a matter of policy, publicly announced reorienting their organizations to be market conscious. Certainly the supergiants of the construction industry, Bechtel, Fluor, Morrison-Knudsen, Parsons and others, demonstrate continuously the value they place on the marketing function and their concentration on it.

1.4. A Dynamic Response to the Client's Needs

Throughout this book, the emphasis will repeatedly be placed on the concept that successful marketing must be directed to fulfillment of a need.

Construction organizations should not try to "sell their services": when they do this, they become unconsciously locked into a rigid position. Too many organizations establish a specialized technological service, then take the position that this is what they have to sell and go out to sell it. This difference in outlook is subtle but terribly important. When one "sells his services," he no longer is psychologically adapted to fulfillment of a customer's needs. Many times the sequence develops as follows: A client has a need which the construction organization perceives. The contractor then develops the capabilities, sells himself to the client, fulfills the contract, and makes a profit. He then finds other clients with the same need. As the business grows, the contractor becomes locked to the product and technology and service. Then the clients' needs change, but he is unable to change with them. This sequence has led to the ultimate downfall of many organizations.

Throughout this book, therefore, the emphasis will be on "dynamic responses to the client's needs." For example, we will suggest the "marketing of foundations," not the "selling of piles." We will try to market the "mass movement of soils," not "dredging" (or "cat-scraper operations"). We will try to market "bridges," not just "sell factory-made prestressed concrete girders." Peter Drucker, the management consultant, has repeatedly urged companies to ask: "What is my business?" "Where are my markets?" "Who are my potential clients?" And he has then advised us to think in abstract terms, that is, categories, rather than in terms of specific products and names. There is another subtle but terribly important point to make. The "need" is that which is perceived or recognized by the client as a need. The contractor is therefore responding to a "perceived need" generated by the client himself or, in some cases, instigated and abetted by the contractor-marketer.

1.5. The Skills Required for Marketing

The skills required for marketing are:

1. The ability to develop ideas logically
2. Empathy or the sensing of a need
3. Arts of verbal and written communication

These skills should be implemented with sincerity, conviction, creativity, and persistence.

Marketing is carried out by means of communication, both written and verbal. These communications must have a sound technical basis: engineering, architectural, and contractual. They must be based on sound construction ideas and practices. However, all this is useless unless it is conveyed with clarity and understanding to the recipient. One must possess a fully adequate command of business English: vocabulary, grammar, and style.

Beyond these two fundamentals lie the manner of presentation: the sensitivity of the writer or speaker to the recipient's needs, his prejudices, his constraints, even his "saving of face." Only then will the recipient really "receive" the information. Communications are made with an objective in mind: the success of a communication is judged by whether or not it did accomplish that goal.

It is similar with negotiations. One is not trying to show his skill as an orator or as an engineer but is attempting to achieve a favorable agreement.

The world has now become enmeshed in one great economic sphere or, rather, a series of overlapping economic spheres, in which construction operates on an international and multinational basis. To be effective in this worldwide arena, one needs an understanding of the geographical, political, social, and cultural environments. The effective marketer will be sensitive to the varying needs, laws, customs, and morals of each area and adapt his marketing to the particular areas in which he operates, seeking to integrate the differences within the framework of his home country's laws and policies.

Marketing is essentially a function of human relations. It is part of and draws its strength from humanity's cultural environment. Its motivation, rewards, and ethics are all bound up in human attributes and human values. Therefore, marketing of construction cannot be divorced from the value system that enwraps all human activities in the social-cultural-ethical system in which we live.

1.6. *The Study of Marketing Techniques*

Marketing techniques are essentially amoral: they may be used constructively or destructively. Since this is a competitive world in which we live, we need to understand all techniques. Whether or not we use them becomes a case for individual determination and decision. However, even the most ethical marketer will only be able to succeed and combat the unethical if he is aware of what may be done. He may properly and wisely refuse to adopt the unethical practices of others, but he should not be naively taken in by them. While the licentiousness of an often amoral marketplace may appear to dominate in the short term, the fundamental principles, such as integrity, eventually emerge as the overriding essences of an effective, long-term marketing program and construction program.

Why should an engineer (or architect) study marketing or any management science for that matter? In the past and sometimes today, engineers do reach top-management positions without having formally studied management science. However, today, we find business-educated individuals invading the construction industry and the engineering industry that were once havens for technically competent engineering managers with marginal management skills. There is a growing trend for the M.B.A. to replace the engineer in top-management positions. The person with just a business background and no technical background often seems, at least in the short term, to be winning out over the person with just a technical background and no business background. Of course, a person with both a business and a technical background will be by far the most effective. Therefore, a knowledge of marketing can be a valuable tool for the engineer who aspires for a top position in management.

In conclusion, marketing is a legitimate and necessary function designed to match the services which can be provided by the contractor to the perceived needs of the user. Without marketing, most construction companies and professional practices would simply die due to lack of work. Unfortunately, construction today is a disorderly business, with generally poor marketing practices by contractors, architects, and engineers. What is needed is an orderly flow of work that fits the services which the company can provide at a profit. The study of marketing can help bring this about. The industry needs reform, and the best reforms always come from within.

2

The Function of Marketing— A Matching of Services to Needs

The basic function of marketing is to match the right construction, architectural, or engineering organization to the correct project or client. This is a matching of service to needs. The successful marketer has the ability to appraise the situation to develop a proper program.

In the establishment of a marketing plan for a construction, an architectural, or engineering organization, the following aspects should be considered.

2.1. Proper Size

Contractors can and often do specialize in the size of contract sought. Chicago Bridge and Iron has been highly successful in taking a small part of a big job. It specialized in the site erection of steel tanks (both simple and complex); it developed the ability to prefabricate the tanks, move onto a job, erect the tanks, and move off.

McAlpine's of the United Kingdom directs its main marketing efforts toward large contracts—over £100 million—and small specialty jobs. On the large contracts, McAlpine's finds that there is little competition, that it can take full advantage of its large financial resources. On the small specialty jobs, it has found that it has a competitive edge because of its sophisticated technological capabilities. For McAlpine's, the middle-sized contracts attract too much competition from other very capable contractors. It has to bid on too many proposals before it gets a contract. When it does get a middle-sized contract, the profit margin is too low.

Back in the United States, on a much smaller scale, a West Coast marine contractor found that many small waterfront contracts in Alaska attracted few, if any, bidders, while large construction contracts were being bid much too low due to excessive competition. By concentrating on the smaller, more specialized contracts, it made good money.

The foundation division of a contractor we will call D & H Contractors analyzed the pattern of its particular business. Its medium-sized pile foundation projects always made money. If there were productivity or technical problems on these jobs, it had time to work them out. Conversely, the very small jobs, lasting only a few days, never had a chance to get going. These small jobs were draining resources and profits.

2.2. Proper Type of Work

Periodically, any corporation or individual should define what type of business it or he is in. Most markets are constantly changing—few are static. Consider the case of a large national piling company, which at one time had a 60 percent share of the U.S. market for piling. This company was in the business of selling a proprietary type of foundation piling for bridges and buildings. As the years went by, this pile was largely replaced in many areas of the country by the prestressed concrete pile for reasons of cost and soil-pile interactive performance. However, it continued stubbornly trying to sell its own piles. As a result, its own market share shrank while strong competitors emerged. This company should have recognized that it was in the business of selling "foundation structures" and adapted its work to new techniques such as prestressed piles. Similarly, today the drilled shafts (caissons) are being increasingly used. Foundation contractors should be alert to redirect their marketing efforts in response to technological developments.

Changing the type of business you are in without proper planning can be disastrous. Consider the case of a well-known major foundation contractor. This contractor dominated the New York foundation construction market; it had more foundation and underpinning expertise than anyone else in the business. Armed with confidence and foundation experience, it entered the heavy construction field of dam building on the Mississippi River. Unfortunately, it was not armed with large-scale earth-moving and heavy-concrete experience and, more important, with management capabilities for such large projects. It lost money in this new work and wisely returned to its primary business of foundation construction.

G Constructors was a highly competent and successful marine contractor. It was invited to participate in a very large military construction project as a minority partner in a joint venture. Problems developed; the original sponsor gave up; and G, which lacked expertise in this type of work, nevertheless had to finish the job, incurring a substantial direct loss. Even worse, this diversion of its management personnel prevented it from undertaking the marine work that was its proper field of construction.

2.3. Proper Location

Location affects marketing decisions. A soils engineering consultant would probably choose to locate in a city like San Francisco or New Orleans where the soil conditions require complex foundations. A custom-housing builder would want to locate in an area like Napa, California, where there is a housing boom, and homes sell for several hundred thousand dollars each; a prefab-building contractor would prefer to be located in a boom area like Houston or Atlanta. It is desirable to be in the area where the work is. Raymond International, along with some other major industrial contractors, has recently moved its headquarters from New York to Houston in order to be closer to its oil industry clientele.

In considering the location of work to bid on, distances in time and not distances in mileage should be considered. From San Francisco, Los Angeles or San Diego is just an hour by plane but parts of northern California can be 8 hours by car, with little or no air service. Back in the days when the Interstate Highway program was just starting, a small company with the improbable name of Transoceanic Constructors found a noncompetitive market in small bridges. Unfortunately, it simultaneously bid and won three contracts, each at very remote locations, with all the jobs to be performed at the same time. It was not able to provide proper supervision, and some jobs went bad. The company ended up bankrupt.

2.4. Proper Scope

Picking the proper scope is picking that part of the project that is the best for you; the scope can be too big, too small, or just right.

Atlanta architect John Portman offers a useful and attractive scope of services to his clients. He offers life-cycle building management, which includes management from start-up to intensive use, in addition to acting as developer, architect, and structural engineer. He has found that some clients, particularly first-time building investors, find his knowledge and management capability in maintenance, maintenance costs, real estate, and finance invaluable and a major reason for choosing his services.*

Kaiser Engineers, acting as a general contractor, specialized in fixed-price construction contracts in the heavy-engineering construction field. Unfortunately, it had four bad jobs in succession. First, it self-insured on an American dam project and suffered a big loss as a result of a 100-year flood. In the Mideast, on a mineral development project, it suffered another large loss, primarily due to poor engineering design by a European engineering firm.

---

* *Christian Science Monitor*, May 19, 1975, p. 16.

(Note: With inadequate and erroneous designs, the contractor usually suffers most.) At the same time, it incurred a large loss on a dam project in South America; this resulted from foundation problems that resulted in large overruns on material quantities that, in turn, depleted existing supplies and required material to be imported from greater distances than expected. Then the resultant delay threw Kaiser into a flood season that jammed the closure gates. Finally, on a relatively small job in southern Europe, it suffered a relatively large loss due reportedly to bad management. These losses were disatrous even for a large company like Kaiser. The top management made the decision to stop work for four days. It rented a lodge in the Sierra Nevada, assembled all its key people, and asked, "What is to be done?" Evaluating its capabilities and the services it offered, Kaiser decided that its expertise lay in design and construction management and that it would no longer seek fixed-price construction contracts but, rather, concentrate on offering turn-key "design and construction management packages." Its new marketing plan worked amazingly well. It was attractive to its clients, particularly the mining industry. As a result of this new marketing approach, its operations again became profitable.

Ben C. Gerwick, Inc., offered the individual services of pile driving, sheet pile bulkheading, and underpinning for building foundations in a very competitive market. Bid shopping and sometimes bid chiseling by unethical building contractors left very little profit. Ben C. Gerwick, Inc., decided to repackage these services into a complete foundation package and simultaneously changed the client from the building contractor to the owner. It offered to perform this work on a fixed-price basis that removed the risk of cost overrun to the owner. All below-ground work was bid with a guaranteed no-changed conditions clause. Without this risk of cost overrun, the owner was better able to negotiate financing; this was a big plus to the owner. Ben C. Gerwick, Inc., was able to accept this risk because of its detailed knowledge of the location and its specialized ability to deal effectively with changed conditions if they were encountered. This change of scope proved to be highly profitable to both Gerwick and the owners. However, in about three years, the competition, which had been closed out, also started offering similar packages. This illustrates the dynamic nature of marketing.

As an example of change of scope in the other direction, Ben C. Gerwick, Inc., found it was unable to be competitive in bidding on state highway bridges as a general contractor. For a typical job in that period, there would be 12 to 16 bidders. Despite Gerwick's specialized ability in cofferdams and piles, it always came in second or third.

Traditionally, bridge jobs do not make money because they are bid so competitively; however, contractors like to build bridges. For many contractors, it fulfills a long-term goal. As a result, the number of bidders on small- and medium-sized bridges tends to be excessive.

Gerwick decided to change its scope and submit subcontracts on the piling and piers only, bidding to many or all of the prospective general contractors. Gerwick found that its subcontract bids were accepted; the industry welcomed the specialized capability. Gerwick achieved a high success rate in getting jobs and made money on all of them. It was a far better use of its resources. A similar philosophy had been adopted with great success nationwide by Raymond Concrete Pile Company.

2.5. Proper Contracts

It is not enough to get contracts; they must be on terms which are fair and equitable to both parties and do not throw excessive risk on the contractor.

As a general remark, in addition to the usual considerations in the negotiation of contracts, the opportunities offered by risk sharing should be investigated when appropriate. Similarly, the proper financial contract terms—those that improve a contractor's cash flow so he will not require excessive financing—can often result in a substantial saving to both parties. They can make the proposal appear more competitive and more attractive while reducing the costs of doing business.

Even with public agencies, there can exist opportunities to improve contractual terms. Although the bid cannot be qualified, the contractor can bring objectionable clauses to the public agency's attention beforehand and try to get these clauses removed. This sometimes has been done effectively through an industry group like the Associated General Contractors. Also, soon after the contract award (while everyone is in a positive frame of mind), contract terms can often be modified or even radically changed under a value engineering clause. A prejob meeting to agree on how terms are going to be interpreted is very useful. Although this has no legal status, it often sets a pattern which will avoid adversary relationships and adverse interpretations. Such a meeting and frank exchange of opinions can be particularly useful if potential difficulties and delays are discussed before they have occurred and a general agreement is reached as to the best way to handle them. A contractor's most effective time for negotiations on a fixed-price contract is immediately after the award while the "honeymoon" is still on.

A very large U.S. contractor entered into some government contracts with adverse contract terms during the Vietnamese conflict. First, there was a non-cost reimbursable clause that excluded many items, such as travel expenses and home office expenses. Second was a clause that allowed the government to requisition key people by name from the company wherever they were working at the time and reassign them to the Southeast Asian job. The first clause alone cost the contractor almost the same amount as the fee: thus it made no profit on the job. The second clause caused the company to have serious problems on its other major jobs around the world, from which the key people had been requisitioned.

A major San Francisco building contractor developed a set of contract terms that proved very favorable to its organization. It would build a high-rise building for cost plus a small percentage fee—terms highly attractive to the owner. However, there was a clause in this contract that all interior work over the next 10 years must be performed by this same contractor on a cost plus 25 percent fee. It should be noted that the costs for the interior work are usually passed to the lessees.

2.6. Proper Schedule

Two major ports in Southern California were commencing long-term expansion. The G. F. Company did its market research well on this program. It began bidding with a very low markup so as to obtain each contract as it was let. Although each contract already had a generous schedule, G. F. would ask for and receive an extension of time as soon as each contract was awarded. This extended schedule enabled G. F. to run the group of small jobs as one big integrated job, moving progressively from one to the next. The rescheduling of the many small jobs proved very profitable for G. F. At the same time, the owners all saved money on their own project. With sad hindsight, the competition learned why G. F. could afford to bid so low on the individual small contracts.

A large shipping company needed quickly to build new terminal facilities in Oakland. They wanted to tie up the first ship in less than a year, and the Port District required more than a year for just engineering. So the shipping company approached several large private contractors including Ben C. Gerwick, Inc. Gerwick responded immediately to the request; an entire plan of operations was developed, complete with materials suppliers who could give firm commitments for delivery. Gerwick offered a lump-sum contract for design and construction with a guaranteed completion data. The shipping company accepted the proposal before the larger contractors even responded to the invitation. Gerwick correctly saw that the shipping company's main objective was meeting its schedule. By choosing Gerwick, it was able to tie up the first ship and unload containers on the scheduled date. Gerwick made a fine profit.

The State of California Highway Department put out bid proposals for a very long interstate highway bridge across the Yolo Bypass (the floodplain for the Sacramento River). The schedule in the bid proposal showed the project taking three years. The area is usually subject to flooding for four months a year, and the state highway construction planners and estimators calculated that 20 to 24 months of dry working conditions would be required—thus the three-year schedule. All but one of the contractors who submitted a bid automatically planned his work to the 36-month schedule. Thus these other contractors had 36 months of overhead, 36 months of equipment rental, three periods of mobilization and demobilization, and so forth, in their cost estimates. One bidder, Tom Polich, came up with the idea that by working fast—two shifts per day, six days per week—and using large crews, the project could be completed in a single eight-month-long dry season. The contract inflated to 1981 dollars was worth about $30 million. Tom Polich's bid was about $3 million below the second bidder. Using the eight months' crash schedule, Polich finished within one dry season, made a healthy profit, and the state acquired use of the bridge two years early. Everyone won on this.

2.7. Proper Quality

Proper quality can be used two ways as a marketing tool: some clients demand high quality and some clients want low price regardless of quality. A contractor can build a reputation and a market for his services on high-quality workmanship. A good example is a San Francisco building construction firm, which has a reputation for performing work of outstanding quality. It was chosen to build Saint Mary's Cathedral in San Francisco because of its reputation for quality. In this case, price was considered secondary.

In the private industrial sector, CNP is often chosen because of its reputation for high quality and its ability to meet the schedule. Conversely, the low-cost new home market is very often an example of where quality is a secondary consideration: the basic requirement is for homes that will just meet the minimum FHA standards. The average home buyer, given the choice between a home of average to good quality and a home of low quality—and marginal workmanship—but offered at a slightly lower price will almost always choose the one with the lower price. After all, he has probably already stretched his buying power to the utmost.

You should ask yourself, "What quality does my client really need?" You know what he has asked for or was told to ask for; but does he really need it? The city of H asked for public bids for a long sewer outfall contract. The contract specifications called for Monel metal manholes. Three hundred were required for the outfall. Now a painted steel manhole may cost about $200 but may corrode in time, while a manhole fabricated from Monel costs about $5000. The winning bidder, under a value engineering clause, proposed cadmium-plated cast-steel manholes, at a cost in between, which resulted in a substantial saving that was shared by both owner and contractor.

2.8. Proper Timing

Most markets are either increasing or decreasing; few are static. As a general comment, most markets fluctuate on a three- to five-year cycle. The number of years in a cycle may vary; what is important is recognition of the cycle. In marketing there is a tendency to look only one year ahead. You need to look five years ahead, for example, in order to properly plan a marketing program.

In Chinese, the words for risk and opportunity are the same. Entering a market in a period of decline in the cycle can be both a risk and an opportunity. In 1975, the port capacity for Los Angeles appeared more than adequate. The demand for new facilities was low. In 1979, the demand for new facilities was still low; however, it appeared that the increasing Asian and Chinese trade would create a surge in demand in the early to mid-1980s. To intensify activity in this market in 1979, with commitments of equipment and personnel, presented both a risk and an opportunity. Indeed, less than six months later, the port announced a huge long-range expansion problem.

2.9. Proper Relations with Third Parties

A contractor or engineer needs to have proper relations with third parties—for example, government and public interest and environmental groups. A contractor can accomplish more at work on the job than in court trying to fight the EPA. A contractor who can get along with the rules and regulations of, for example, the city of San Francisco, can make a profit, whereas the one who ignores or fights each rule has nothing but trouble.

Emory Construction was trying to get a permit to build a waterfront restaurant for a client. Emory had a contract, the owner had proper financing, but they were unable to obtain a permit to build. Finally, Emory called in a consultant who had established a sound relationship with the permit authority. The consultant contacted the permit authority who told him that this was the first time anyone had personally talked to them about the project on an engineering basis. Up until then, Emory would send a general letter asking for the permit, and the authority would send the routine form letter saying that the permit had been denied. Emory had never directly asked what the problem was! The consultant quickly found out that the problem was earthquake safety, and, with relatively small but important design changes, this problem was resolved to the satisfaction of both parties. The permit was granted, the contractor made a profit, and the client moved in on schedule.

The cliché "you can't fight city hall" can sometimes be amended to "you can't fight the Sierra Club," as in the next example. Many California land developers have been unsuccessful in getting the necessary permits for coastline development. One land developer, who was successful, worked with the Sierra Club on his design and proposal before applying for the permits. The proposed project—that received Sierra Club approval—is being built today. Society was also the winner because of the environmentally sound design.

Survey Engineering of Cincinnati is active in subdivision planning and design. The approval of a proposed subdivision by the county in many cases is anything but automatic. After Survey hired a retired county engineer, the interface with the county became much easier; the proper relations with the county had been established because he knew the procedures and the requirements.

As a footnote, marketing is an ongoing function. For example, if you have a construction job in a highly visible area, like downtown San Francisco, run the job neatly and efficiently—lots of people are watching. Many are very influential "third parties."

2.10. Proper Volume

When you make a discontinuous break in volume, you incur both risk and opportunity. For example, a big increase in volume could lead to a gain or loss in profit. A general rule of thumb for successful growth is, do not expand too fast—limit growth to about 15 percent a year. This is valid even if the market is increasing. Otherwise, your ability to perform efficiently is impaired. If, in spite of this warning, you see a unique opportunity and choose to expand quickly, recognize the dangers ahead of time and get the support of people whose capabilities you know. One way to hedge on risk when increasing volume quickly is to form a joint venture.

T. O. Contract Managers formed an organization of specialized steel construction engineers and successfully expanded rapidly. First, it recognized a great opportunity and need for high-technology inspection with the North Sea oil industry. What was needed was a quality-assurance organization to ensure the high-quality welding demanded by the designers and regulatory bodies. T. O. asked for and received proper contract terms—adequate fees, clear definition of scope and responsibility, and all support services provided by others. Most important, it used only skilled people it knew were highly competent in steel fabrication. In many cases, it hired them away from other organizations by offering them a major increase in their salary and by picking up all the benefits they would lose. This was expensive, but it got a lot of good workers fast. Therefore, it avoided the danger of expanding quickly with people it did not know.

Morrison-Knudson (M-K) in 1977 announced that it was shifting from emphasis on revenues (volume) to profitability as a bidding criteria.* The president stated that previous managers had opted for volume instead of seeking overall corporate profitability. What was implied was that M-K was willing to reduce volume in order to increase profitability.

2.11. Proper Client

Choosing the right or wrong client to whom to direct your marketing efforts can mean either profits or losses. Knowledge about clients can be gained through a market survey or experience. In construction and engineering

---

*ENR, Nov. 17, 1977, p. 30.

design, the oil industry has traditionally been considered a good client. It is a healthy industry with the resources to do the job right. For the contractor or consultant who has an established relationship with the railroad industry, it is an excellent client, even though it is not an expanding one. However, the railroad industry practically never gives a contract to a stranger unless it is a small test contract. It can be very difficult for the new firm to break into this market. A similar pattern was previously noted for the petroleum industry.

The State Department of Transportation of one state has the reputation of being a poor client. A contractor there states, "There is no way to come out without losing money on the job." Through the years, the contractors have learned that any profitable jobs were offset by losses. Prices are just too low due to excessive competition, and they prove inadequate to cover the costs imposed by strict, erratic, and even harsh interpretation and enforcement of specifications by inexperienced inspectors. Of course, there were exceptions in this state: for example, when the project was so large or so specialized that only knowledgeable and experienced inspectors were assigned by the state.

Small engineering research and design contracts for the federal government have a reputation for not making money for the design consultant. This is because of the elaborate proposals required, the policy of price negotiation after submittal of an offer (essentially a bid), and, after award, the administrative requirements and the huge amount of reports that must be submitted.

At a municipal utility district, contractors, equipment, and material suppliers would often direct their marketing efforts at the wrong person or client. In this organization, as in many other bureaucratic organizations, an outsider may face great difficulty in identifying the offices and individuals who make the decisions. It is obviously essential to dig this information out. Often, a district engineer would politely listen to the salesperson, let the salesperson take him out to lunch, and never tell the salesperson that he should be talking to someone else.

Marketing to committees or groups can be nonproductive. Be prepared to take lots of time. Committees tend to make decisions very, very slowly. Also, there will always be someone in the group who will not like your idea, and this can lead to a difficult job, particularly if you encounter some problems or delays, even if they are not your fault. Some contractors tend to avoid committees and groups altogether, considering them unreliable and sometimes financially irresponsible. However, in marketing, there are no fixed rules; there can always be an exception.

One such exception was a group of farmers whose dam was condemned as unsafe. They needed the water for irrigation in a hurry. Several contractors turned them down. "Farmers never have any money." Company Q went up to meet with them on a Sunday. They all turned out to be substantial, well-to-do farmers, who voluntarily put the money for the entire contract in the bank! They were about to lose their entire tomato crop, worth several million dollars!

Another such exception was a fisherman who came into G's office asking to have a small unloading facility built. He was in his fishing clothes. He said he had been to three competitors; they had all turned him down. G was curious; he followed up, did the job in a week, and was paid in cash. For decades after, the entire fishing community would deal with no one but G.

2.12. Proper Price

Proper price is discussed last to emphasize that giving a low price is not the only way to get the job and may be the least profitable. As a general comment, most contractors, engineers, and architects price their services too low. Unfortunately, most engineering consultants value their service at about one-half that which a good lawyer gets. This commonly acknowledged situation should by itself make the consultant ask why. Is it because the lawyer typically addresses himself to the client's overall problems, that is, his real needs, whereas the consultant tends to confine himself to technical expertise in his specialty, forcing the client to put it all together in proper proportion?

Conversely, the price can be too high even if you are getting the jobs: this usually occurs for contractors who enjoy a monopolistic position. Overpricing attracts unwanted competition or stimulates development of alternative solutions.

One manufacturer developed a new lightweight aggregate for concrete. He kept the price reasonable while he created his market. It was a very successful product, the volume grew, and for many years he had an extremely sound and profitable operation. New owners came in and, finding that they had 100 percent of the market, raised the price drastically. Within one year, three major competitors had sprung up and the price had dropped back. Now the original company had only one-fourth of the market!

Price margins should vary, depending on the actual work performed. Many contractors will bid piling for bridges and piling for wharves at the same price—after all, they will say, both jobs use the same piling. Piling for bridges is much more expensive to install; the structural engineers for the bridge will require that the piling be installed to exacting tolerances to prevent eccentricities. They will generally require deeper penetrations and higher resistance. Piling for wharves can usually be installed according to less rigid tolerances. If the wharf pile is off a little in position, it can be corrected by adding a small amount of reinforcing steel to the wharf deck. Contractors tend to bid too low in one case and too high in the other case.

In a similar case, a Japanese firm began producing hollow core floor slabs. The floor slabs were priced correctly to make an adequate profit. When floor slabs have been erected, they are typically covered with topping or tile, so minor surface defects are completely covered. The company then discovered a new market for this product—facing panels for the walls of cold storage warehouses. After erection, the panels are painted with a translucent plastic. The facing panel thus requires much higher quality control than the floor slabs. The slightest defect in finish, such as bleed or air pockets, will show up when painted. Joint tolerances had to be within 1 mm. About 50 percent of facing panels produced would not meet the quality requirements and had to be replaced, but the firm still priced its facing panels at the same price as floor slabs. They were losing money with this price structure. Ultimately, they recognized the differences and raised the price for facing panels.

In summary, then, a judicious reduction in price can be an effective tool in obtaining work, increasing volume, and discouraging competition. The increased volume can result in reduced costs, especially if greater utilization can be made of plant and equipment. Conversely, pricing too low can lead to ruin. Raising the price even slightly can often make a dramatic increase in profit if you can still get adequate volume. Since the net profit in the construction industry, after overhead, but before income taxes, averages about 2 percent, it is obvious what a difference a 1 percent raise in price level can make, provided, of course, the firm can still get an adequate number of jobs.

3

The Establishment of a Marketing Plan

Once a year, usually about October, corporate goals and objectives should be set for the next year. Your corporate abilities should be evaluated: you should determine and establish what you do best. Then you try to match your abilities with opportunities. For your abilities, who is the proper client, what is the proper scope of work, and so on.

In establishing your corporate goals and objectives, you need to evaluate your corporate financial health and status. Corporations seem to grow (increase in volume) or die. Few are able to remain at a constant level, no matter how hard they try. There is an inevitable need to expand, improve, or increase your share of the market.

3.1. Market Survey

A market survey is the appraisal of your corporate abilities coupled with research on the historical, existing, and projected market. In a market survey it is vitally important to ask the right questions. The following can be considered:

What Is My Business?

In the case of Kaiser Engineers, after a series of serious construction losses, Kaiser decided that its business was construction management, not general contracting. On the surface and especially in hindsight, this question appears straightforward to answer; however, at the time it was an extremely difficult decision. In the case of the national pile company discussed in

Chapter 2, it had defined its business as selling a proprietary pile, rather than a more proper definition as "pile foundations" or perhaps just "foundations." A business should be a function, not a skill. You need to keep reminding yourself that the same skills that erect a steel bridge can also be used to erect a concrete bridge. One international erector of steel tanks for the petroleum industry pioneered the development of offshore steel tanks and installed three such tanks in the Arabian Gulf. It concurrently set up a large research facility. Thus it offered the industry the unique combination of practical experience and technical leadership in a new and expanding field.

The concept of offshore storage was adopted by a number of oil companies for the North Sea. Here the much more severe wave climate and the economic conditions of the area made prestressed concrete a more economical solution.

Although some members of the steel tank company wanted to offer a concrete tank, the company decided erroneously that its business was "steel tanks," not "offshore oil storage." The result of this decision was that new competitors in concrete took all the subsequent jobs, and the steel company got none, so that over the years, its significant lead vanished.

What Is My Market?

The question can be asked for both the geographical area and for the type of client. When establishing or determining a geographical market, you need to evaluate your resources. The area must be one you know, one you can manage, and one you can serve effectively. It is best to concentrate and not spread beyond your resources and capabilities until you are really ready: then commit yourself to one new area at a time.

With regard to type of client, many companies find it useful to concentrate by industry or technology. Consider the petroleum industry as a potential client. For a company with an established relationship with one segment of the petroleum industry, a new segment of that industry is probably a good market. For another company with no past relationship with the petroleum industry, it can prove to be a fruitless approach.

One very successful contractor followed the principle of "going where the money is." He meant that he would select industries, such as the pharmaceutical industry (in past years), that were making large profits and were expanding.

How Big Is the Market?

The size of the market needs to be evaluated by categories. It is usually best to find a large market consisting of healthy industries, like the current petroleum industry. Working for industries which are not healthy and which are cutting corners themselves usually results in low profits for the contractor;

even though he may do an excellent job. The company just never has the funding for a properly scoped, designed, and administered project.

Contractors often fall into the trap of becoming intrigued with a small specialty; one which requires great effort yet does not generate enough total volume. Contractors must be keenly aware of their annual overhead costs, their break-even points, and the need for an adequate volume to justify the organization's existence.

There are, of course, two ways to approach the matter of size. One is to ensure individual contracts of sufficient scope that a substantial volume passes through the contractor's books. A few large contracts, which generate even a modest percentage profit, may develop an adequate gross return. The large engineering construction firms, such as Bechtel and Fluor, follow this pattern.

The other way is to increase the number of contracts, sticking to your specialty but spreading out geographically. A company such as Chicago Bridge and Iron, with its specialized work on steel tanks of all types, or Ryerson Steel, a prestressing subcontractor, exemplify this role.

The management demands are quite different for each of these patterns. The marketing approaches also differ greatly.

Whichever pattern is selected, there must be a gross demand that will ensure that the portion of the projects you obtain will generate enough return to support your organization and return an adequate profit.

How Continuous Is the Market?

Many markets tend to remain steady—like the oil industry. Other markets are very cyclic—like the mining industry. If your market will have four really good years, followed by six bad years, you will need to plan for this. Ideally, you should try to find two markets whose cycles are historically always out of phase. This is diversification at its best.

Who Now Serves the Market?

Most industrial design before 1940 was done in-house. If an industry built one new plant every 10 years, the in-house staff was busy for two years and lost touch with the field the next eight years. This practice resulted in the emergence of construction management (CM) services. The CM firm offered the latest management techniques, which allowed the highly paid in-house staff to be reduced.

Sometimes no one is serving the market. Bechtel Corporation discovered many years ago that the pharmaceutical industry was using only batch processes in manufacturing. Bechtel had extensive experience in the refinery and chemical industry which uses continuous process operations. It successfully transferred this technology to the drug industry. This created an entirely new market for its services.

Is the Market Increasing?

In general, it is best to enter a new market that is increasing—for example, construction of petrochemical plants or the sewage treatment-water purification industry. The overall field should be profitable. It is hard to move into a static or declining market. But there are exceptions such as initial establishment of a new service. A period of temporary decline is often the period when the most effective selling can be done. This is when the client has time to listen to you, and when he is most receptive to innovative proposals.

How Does This Market Fit with Other Markets My Corporation Serves?

A San Francisco-based construction company was originally involved with heavy construction—dams, canals, and power plants. It branched off into the mining business. It used its experience in the mining industry to build haul roads and to develop materials-handling processes in open-pit mining. Later it expanded its mining capability by adding process operations work. The mining business became very profitable, whereas the construction operations ran into a long drawn-out period of low profit. The company realized that construction operations were draining management time and talent and financial resources. So it sold the construction part of the operation and limited itself to mining, with very profitable results. Meanwhile, the buyer of the construction branch now devotes full attention to construction and is also making a profit.

As another example, many building contractors find that expanding into real estate development is highly compatible with their existing services.

Conversely, an engineering construction company once tried to enter the electronic communications field. It not only turned out to be a failure, but it diverted valuable management attention from the primary business of construction.

Who Are My Competitors?

In a marketing survey, it is necessary to correctly identify your competitors. An obvious direct competitor for a contractor would be another contractor who historically works in your area. Also, a competitor that is new to your area can move in. This type of competition can be difficult or impossible to predict. Also, there can be indirect competitors; for example, air freight competes with warehouse construction; pipelines compete with railroad construction. Be sure to include indirect competition in your market survey.

What Can I Offer That Is Better?

Before entering a new market, you need to determine what things or services you can provide which are better than existing services—for example, you may be offering an innovative technique, modern management skills, or lower prices due to better equipment. If you do not offer something better than the competition, you probably will not be successful.

Have I Picked the Right Client?

Assume that you have picked the industry; have you picked the right company or organization? Is it one which has adopted a program for construction? Is it one which has the money with which to pay? Have you picked the individuals who will make the decisions? Do you know their positions in the company? These matters will be discussed later.

How Do I Sell This Market?

Established markets usually have relatively fixed practices. There will be extreme resistance to change, unless you can offer something dramatically better. Expect this resistance; you will usually have to direct your efforts to top management, as only it can overrule established practice. At the same time, you have to elicit the enthusiasm of the lower-level decision makers and recommenders.

Try to determine what it is the client thinks he wants—and what he really needs. Whom does he have to convince? What constraints does your client's client impose? What constraints do governmental regulations or financial institutions impose on your client?

How Does This Market Establish Price?

Contractors historically have established price on the basis of their costs. On a $10 million job a contractor will add, let us say, 6 percent profit. An engineering consultant will similarly multiply his salary costs by a factor of, say, 2.4 to determine his price. This practice of establishing price on the basis of cost severely limits income; it is basically a poor business policy. The market price should be established by its utility—not your cost. Your services should be priced to meet the market; you should use the price which you believe will bring you the greatest return. This, of course, is classical economics.

A dredging contractor, bidding in a strong competitive market for a multimillion-cubic yard channel-dredging project which included the building of levees on each side, came up with a unique cantilevered disposal line which, with a booster pump, enabled him to place the hydraulically dredged spoil directly on the levee, without the use of a shore pipe, side boom cats, and crew. He figured his costs at 20 cents per cubic yard of borrow. The engineer's budget was 40 cents. The dredging contractor, desperate for the job, bid 23 cents. His lowest competitor was 36 cents, based on the traditional shoreline method of building levees. As so often happens, there were lots of bugs to work out, but when he got it working, it was very successful. His final costs were 22 cents.

Would he not have been wiser to have bid, say, 30 cents?

More sophisticated dredging contractors often estimate backward whenever they have an edge on a job; that is, they estimate their competitors' costs in order to determine the upper-bound pricing level.

A foundation contractor was bidding a job for a tied-back bulkhead in sand. The standard method involved progressive casing, a slow and tedious job. He developed an augur method which enabled the tendon to be placed and the grout pumped through a central hole in the augur stem. He figured his costs and added his conventional profit of 15 percent. He was far below the next bidder. When he designed the equipment, he had to increase the motor size and the support frame. Initially there were many small difficulties. Finally, it worked and well. The extra equipment cost, however, just equalled his profit!

Consider the case where an engineering consultant is preparing a report—based on years of specialized experience—for a new client. It might take the consultant just two weeks to prepare the report. His cost plus profits, based on time, might be $5000; however, any other consultant would have had to charge $25,000 because they would not have had the experience and it would take them 10 weeks to prepare the report. It would seem only reasonable that the engineering consultant should charge a lump-sum price somewhat higher than the $5000. However, for an old or continuing client, he will probably wisely charge only the $5000.

Sources of Data

In construction or engineering, there is a need for long-lead information or intelligence on new work. You cannot just wait for new work to be advertised; by the time it is advertised, it will usually be too late. For example, some big jobs like MARTA in Atlanta require that the bids be in two weeks from the time they are formally advertised. Two weeks does not allow enough time to prepare an intelligent estimate. With sufficient advanced knowledge, you can work with the architect or engineer to learn more about the project requirements. If it is private work, you may get the opportunity to negotiate the job for your company. By working with the architect or engineer, at least you will be on a selected list of bidders. On either public or private work, you might be able to influence the design to better fit your equipment or methods or, at least, to have a favorable alternative included. However, never do so at a sacrifice of good engineering. Unfortunately, too often proprietory services, processes, and materials are sold in a nonprofessional and sometimes even unethical manner, in that they do not represent sound engineering. Not only may a company incur liability in case of flagrant misrepresentation, but even in less serious cases, the company can lose valuable goodwill and contacts for future business if they allow their eagerness for a sale to overrule sound judgment.

3.1. Market Survey

With early information you have time to study the problem and maybe adapt your own innovative ideas to accomplish the work at less cost than your competitors. Sources of early information are shown in Table 3.1.

TABLE 3.1. Sources of Early Information

Source	Description
Soils engineers	When an owner starts to plan a project and takes out an option on a piece of real estate, he will normally engage a soils engineer to determine the foundation cost. Therefore, soils engineers are often the first to find out about a new project.
Architects and engineers	They can be a particularly good source of information for a contractor—if they will talk. Establishing relationships, whereby they feel free to call on you for early advice, will open up many leads, even confidential leads. Membership in professional and technical societies can be a way to get to know the architects and the engineers.
EIS or EIR firms	Firms that specialize in environmental impact studies or environmental impact reports often find out about new projects during the earliest planning phases.
Trade journals	There exist a number of trade journals: *Engineering News Record*, *Daily Pacific Builder*, *Daily Construction Service*, *Commerce Business Daily*, and so forth. Some are national, some regional, some local. *Engineering News Record* comes out every week and has a section called "Pulse." "Pulse" gives information on jobs that are in the planning and bidding stage. This is fairly long lead information. However, all your competitors are able to read it as early as you. *Daily Pacific Builder* has good detailed information and is quite current. *Daily Construction Service*, which has different editions for different areas, gives information on contracts, general contractors advertising for subbids, results of bids, and so on. *Commerce Business Daily* provides information on Federal Government contracts, both for construction and engineering services.
Owners and their in-house engineering staff	They can be a sound and early source of information. Sometimes the owner is very reluctant to talk because he does not want to stir up an interest in a particular area. Perhaps he may not yet have been able to conclude negotiations for the property he needs.
Building trade union officials	This group can be a very good source of information; they have a good ear to the ground about what is coming up because it is in their interest, too. Also, they are often more free in discussing advance information and rumors.
Business groups	In many, but not all, cities, membership in a business group, like Rotary, cannot only provide sources of information but also valuable business contacts. Also, membership in Rotary can be particularly valuable for a young firm trying to establish itself.
Local banker or realtor	Both of them know ahead of time about many long-term projects. In general, realtors are more likely to talk than bankers.

Moving Average

Moving averages can be useful in charting or forecasting sales trends. The formula for a moving average follows:

$\bar{x}$ = moving average
n = number of elements in the moving average

$$\bar{x} = \frac{x_1 + x_2 + x_3 + \ldots + x_n}{n}$$

$$\bar{x} = \frac{x_2 + x_3 + x_4 + \ldots + x_{n+1}}{n}$$

The smaller the n, the more sensitive the moving average is to trends.

For example, consider the Up and Coming Construction Company, given the following records:

Year	Contract Volume of Work on Hand
1976	$29,040,000
1977	$15,840,000
1978	$25,080,000
1979	$22,444,000
1980	$39,600,000
1981	$32,640,000
1982	$51,480,000

Calculate the three-year moving average beginning with 1978.

3-year average for 1978 = $23,320,000
3-year average for 1979 = $21,121,000
3-year average for 1981 = $31,561,000
3-year average for 1982 = $41,240,000

3.2. Increasing Contract Volume

One way to increase volume is geographical expansion. For example, a San Francisco Bay Area building contractor could extend his marketing efforts to the Sacramento area, southern Oregon, or Hawaii. Another way to increase volume is by better market penetration—increase your share or percent of the market. Prices can temporarily be lowered to increase your market share and then gradually raised to a more profitable level—a type of advertising. However, in a competitive industry, if your price is too low, you will go out of

3.2. Increasing Contract Volume

business. Better than low price is a good marketing plan (the emphasis of this book).

Diversification is another way to increase volume. One way to diversify is to buy a small specialty contractor. With this practice, you need to be sure that the firm you are buying is really sound. Do not just look at the financial sheet; look behind the figures. For example, what is the condition of the equipment? Has the firm you want to buy allowed its equipment to wear out or become obsolete in order to increase profits on last year's financial sheet? Also, who are the key people? Are they going to stay with the firm?

Another way to diversify is to start in a different type of work on your own. This takes careful preparation and planning. Also, it takes a commitment to go into this new field on a long-term basis. An average period of three to five years is often required for the new area to become profitable. As an example of the lack of planning, a local building contractor tried to get into heavy construction on a subway job. On his first contract he was low bidder, with a bid about 70 percent that of the next bidder. He went through with the contract but took a disastrous loss.

An even more catastrophic loss was incurred by a building contractor whose practice it was to estimate labor and materials and then add 25 percent to cover equipment, overhead, and profit. He bid a tunnel job, which required major equipment, with its heavy costs of fuel and maintenance and repairs. The 25 percent margin was adequate only for about half the equipment costs, leaving nothing for overhead and profit.

A more conservative approach is to "practice bid" first. That is, prepare the estimate and proposal, but do not submit it. This enables a company to test out the market and become familiar with its pricing levels, suppliers, and subcontractors. Visiting the job that was "practice bid" while the work is underway will reveal costs (and opportunities) not recognized during the practice bid.

Only then does the company go after the next job seriously.

A third way to increase volume is to increase scope. One example of this practice has already been presented in Chapter 2 with the building foundation example. Roe Construction, a mechanical contractor from San Francisco, found that for some building contracts, 40 percent of the contract was mechanical and electrical, for which it had the in-house ability. Therefore, Roe increased its scope from being a specialty mechanical subcontractor to that of being a general contractor. Roe would then subcontract the concrete and foundation work to its former clients.

One of these general contractors then decided that if he was going to maintain his position, he, too, would have to increase his scope. In his case, he started up a mechanical firm and an electrical firm as wholly owned subsidiaries.

While both these expansions accomplished their initial objectives, they brought up new problems over the long range: problems which proper foresight and study could have helped to mitigate.

28   The Establishment of a Marketing Plan

3.3. Bid Volume versus Contract Volume

In order to develop a growth plan, you need to determine bid volume versus contract volume—overall and by categories. The following would be typical for various areas of construction:

	Ratio of Number of Contracts Awarded to Number of Contracts Bid On
Building construction	1-10
Heavy construction	1-6
Specialty construction	1-3

Using your own area-specific experience, you can estimate how many total projects you need to bid on in order to obtain the desired increase in volume. Also, you can estimate average overhead and average profit. The following is typical:

	Corporate Overhead (Percent)	Profit before Overhead (Percent)
Building construction	3-4	4-5
Heavy construction	4-7	8-10
Specialty construction	10	15-20

From the above, it can be seen that a building contractor would need a large volume to generate a reasonable profit. A plus for building construction is that it is a low-risk business.

Be careful that increasing your volume does not lead to an excessive increase in overhead. In order to get more jobs, you have to bid more jobs, and that means more estimates. To run more jobs means more superintendants, more accountants, more personnel administrators. More jobs means more cash demands, hence more borrowing and more interest to pay. More jobs means more executive demands as well. Since the new hires are seldom as efficient as the old-time employees, your overall overhead percentages may remain about the same as before or even increase.

In construction, volume and sales can be 10 to 20 times the capital investment. Therefore, a 2 percent net profit on the volume can be a sizable return on investment (ROI).

3.4. Game Theory

Probability models can be used in bidding. These models are generally valid in a market situation which is steady both in demand and supply. Consider the case of the cross-country pipeline industry. Pipeline spreads consist of a fleet of equipment: side boom cats, wrapping machines, ditchers and dozers. Organizations are expensive to assemble and jobs are few. For any given year, there might be only eight jobs for nine pipeline spreads in the country. In this type of market, probability models are useful. Several bid strategies prevail. One is to bid low at first and get a job so that you have the total summer construction season to complete it, hence low field costs. Another is to wait until the end of the construction season and bid high. There will be, for example, just two jobs left and only three spreads available. So your chances are constantly improving. But you will need to work at an accelerated pace to finish by the end of the season.

Most construction markets tend to be dynamic. In dynamic situations, where the demand and supply are often out of balance, there may exist an outside factor that will invalidate the probability model. For example, a bidder will come in from outside, or a joint venture is formed that changes the probability for competition.

Psychological patterns need to be watched. The first job in a series tends to be best; maybe it is because there are no psychological pressures placed on the estimator. A threat of outside events can open an opportunity to get jobs. For example, a period of strikes can upset the psychology of bidding. By staying cool, you can frequently get good jobs during such a strike period. It may be that then the odds go from 1 in 10 to 1 in 3.

Do not become excessively worried about your failure to obtain a specific project. In the construction industry jobs seem to run in patterns. You can bid eight jobs in a row without success and then be low bidder on three jobs in a row. Patterns seem to occur; observe them, but do not panic. Do not put in a low bid just to break a pattern—it will often turn out to be the one you would have obtained at a reasonable price. Sometimes you will find a new pattern that can lead you to a new or better market. For example, you might have three successes in a certain geographical area, and this indicates that you should concentrate in that area. On the other hand, you might find a prolonged bad pattern that indicates you should pull out of a market.

A key to successful marketing is finding dynamic noncompetitive solutions in a highly competitive business. You can do this by taking each job as it comes along and treating it as your most important job, concentrating your marketing skills on this one job.

3.5. Profitability

One way to increase profitability is to be selective on the projects or clients you seek. Experience shows that some types of projects or clients prove to be better in the long run. For example, pile driving is a low-profit, competitive business. Ben C. Gerwick, Inc., found that very small pile-driving jobs that were competitively bid in the San Francisco area usually turned out to be bad jobs. By ceasing to bid these jobs, and instead concentrating on larger work, its profits increased. Some clients, like the governmental agencies previously mentioned, tend to be consistently bad clients and hurt your overall profitability. They are just too demanding, the work is too competitive, the inspection too difficult. On the other hand, some clients always seem to be just and fair; you always seem to have a good job when working for them. This type of client should be given "most favored status." You can also be selective as to when you should be the general contractor and when you should take the role of the subcontractor.

Increasing scope to increase volume has been discussed. Scope can, also, be decreased to increase profitability; scope can be narrowed to just that part of the job which is profitable. A good example of narrowing scope has been discussed in connection with the Kaiser Engineers example, where it narrowed the scope of its business to seek only project management and engineering contracts.

It is necessary to seek proper contractual terms, as previously discussed, in order to maintain profitability. There must be a reasonable sharing of responsibility and risk, especially for matters beyond the control of the contractor.

Proper price is a key to increasing profitability. A New York foundation contractor once made a complete evaluation of its pricing policy. Historically, its profit after taxes was about 4 percent on a volume of 90 jobs for a total of $30 million of work yearly. The study showed that by adding 5 percent to every bid, its volume would be cut in half by the price increase, but their profits would double. This illustrates, again, that net profit is not directly proportional to net volume.

Value engineering can be a tool to increase profits. "Value engineering" is a term used to connote any one of several methods by which each increment of the project is subjected to rigorous scrutiny to ensure that it is the most economical way in which the objective can be attained. As the term is used here, it is a contractual provision which allows a contractor, who has been successful bidder on a contract in exact accordance with the plans and specifications, to submit alternatives which he believes will accomplish the intended objectives at a savings in cost. Such contractual clauses prescribe that the savings will be shared between owner and contractor, usually on a fifty-fifty basis. With value engineering, you can be paid for your good ideas, provided the contract administrator will act promptly enough to enable you to place the ideas in practice in a timely manner.

A contractor who has trained his estimating staff to be alert to catch value engineering opportunities, although there is always the risk that the proposed alternative will not be accepted, has a means for enhancing the profitability of jobs.

In one case, contractor A recognized that the huge concrete intake tower in a reservoir was designed that way in order to withstand the design earthquake. It had been made stronger and stronger as the earthquake responses had been recalculated on the basis of an equivalent lateral force. Each increase in concrete quantity increased the actual mass and the hydrodynamic added mass and, hence, disproportionally increased the design lateral force.

After contractor A was awarded the contract, he proposed a slender independent tower, prestressed to the rock foundation, designed to respond dynamically in an acceptable fashion to the earthquake. The alternative was accepted, with a significant savings in cost and time to both contractor and owner.

However, value engineering is only really beneficial if it occurs sufficiently early in the contract period to enable complete and thorough engineering, and proper time for procurement and integration into an orderly construction schedule. Rushed or delayed implementation of value engineering changes is almost always unsatisfactory.

Frequently it is the contractor himself who overestimates the saving because he fails to see all the miscellaneous aspects needed to put the change into effect.

On occasion, the owner, acting on the advice of an overzealous architect-engineer, has demanded a larger saving than that estimated by the contractor for the value-engineering change. The contractor is caught in the bind of the schedule and does not know how to proceed. He either has to offer back too much or else give up the change altogether. This is a case where careful preparation in detail and good communication at all levels are needed: otherwise value engineering is self-defeating as a concept.

Concentration in geographical areas can increase profit. It has been previously noted that excessive geographical spread can stretch supervision beyond its capabilities. Concentration often provides the opposite results.

More than one contractor from the U.S. mainland has found a good market in Honolulu. By making an effort to get successive and concurrent jobs there, contractors achieved significant economies. Supervision and local overhead could be spread over several jobs. A larger total volume gave better buying power.

In conclusion, one economic function of profit is to assist in the matching of abilities to needs. For example, tax benefits for coal-fired power plants are intended to direct engineering and construction services from nuclear power plants to coal-fired plants by means of increased profit in the coal-fired area. (Please note: In your own marketing efforts, remember that profit has become a tainted word—"fee" or "markup" is more acceptable. If you tell a client you want a 4 percent profit, he will resist it. If you tell a client you need a five percent fee, he will say, "That sounds reasonable.")

Regardless of semantics, profit is essential for a business to live and grow. It is the reward that society pays for services.

3.6. Private Work versus Public Work

There are many advantages of doing business with a private owner. One reason is that with a private owner, it is possible to negotiate the terms of your contract. Further, the experience of a number of large companies over the years shows that contracts with private industry generally yield a substantially better gross profit compared to that obtained on public work. This does not mean to imply that the private owner gets less for his money; on the contrary, the greater efficiency and lack of red tape often make it possible for the contractor and the engineer to be much more productive. Also, once you have contracted and performed well with a private owner, he is likely to come to you again with his next package of work. This is generally not legally possible with public work, of course.

Private work is generally less burdensome regarding administrative details. Nevertheless, you must be meticulous regarding costs and records. For example, Ben C. Gerwick, Inc., had negotiated a long-term maintenance contract for the owners of a San Francisco Bay Area cargo company. It was cost plus a fee and a nice contract to have. One year an accountant for the owner noticed that a chain saw—which is used to cut down old timber wharves—was being charged at 75 percent of the equipment dealers' established rental rates. He calculated that the yearly rental would buy two chain saws. Now, the life expectancy of a chain saw in such an overwater operation is not too long; however, this particular chain saw just happened to survive longer than expected. The accountant complained because he thought Gerwick was overcharging. This one item almost cost Gerwick the next year's contract. A private owner must be assured that he is getting his money's worth.

Insurance and indemnification clauses are usually more reasonable for private work than public work. Indemnity and insurance programs in many public work documents are becoming quite difficult. As an example, a California city has a clause in its contract that can make the contractor liable for any accident or damage connected with the construction operations, regardless of whether or not a city employee was partially or wholly responsible. On a recent federal project, the contract required that the contractor indemnify and hold harmless the federal employees, the architect, and the engineer for all actions, direct and indirect, whether due in part or in whole to the negligence of the federal employee, architect, or engineer.

There are ways to get around this problem with public work. You can bid as a subcontractor and qualify your subcontract bid so as to exempt or modify the clause, provided of course that the general contractor will accept the risk. Or you may be able to get insurance (at a high premium), which will be endorsed to cover this liability.

A third method is to qualify your bid as to this clause, using carefully drawn phraseology that provides full protection yet does not inflame. One such clause that has been used is:

Stipulation. The contractor in submitting this proposal does so on the basis that mutually agreeable modifications will be made in the indemnity clause, Section _____ .

Note: All such clauses should be approved by your lawyer.

If you're not low bidder, the qualified bid doesn't matter. If you are low bidder, the owner may throw you out. Hopefully, he will class it as an informality and enter into negotiations.

Cases have occurred where the contractor using this approach was the only bidder: his competitors were afraid to take the risk.

By the time a public project survives the scrutiny and obtains the approval of the many and varied agencies, such as city, county, state, and federal government, not to mention the Sierra Club, the Save the Bay Group, and so on, it is so well known and so long anticipated that it is likely to generate 10 or more bids from various hungry contractors. In short, there is not much romance in bidding public work unless there are certain unique features about the project which either eliminate the average contractor or which lend themselves to the particular expertise or equipment your company possesses.

However, it is important to continue to bid public work which particularly suits your company. There are several reasons. First, this will keep you abreast of costs, prices, and the competitive situation. Costs can get away from you in a matter of months; they do not stand still. Second, you will establish contact with subcontractors and material suppliers and equipment dealers that you might otherwise not know of. If you are getting 10 percent of the work you bid, you are staying active in today's market. Third, the volume generated gives you the opportunity to innovate and adapt the special equipment and facilities in your company to fill the needs of a particular job. Fourth, it is proof to a private owner that you can hold your own in the competitive marketplace. In short, by bidding in the public field, you and your organization stay "sharp."

3.7. Design of Your Sales Force

The marketing of engineering services or technical services involves "technical representatives," "sales engineers," "business development representatives," all of whom have to be paid in some manner by their companies. Companies will often put these people on some type of a commission system. This could be anything from a direct commission to a large yearly bonus. Other companies will pay their sales engineers a direct salary, which has no direct relationship to the amount of contracts sold. People on a commission are sometimes perceived as "slick operators" by the clients. People on a direct salary are more often perceived as honest, sincere engineers who are not only trying to sell their company's product but also are interested in their client's best interest too. There are effective and noneffective salespersons working under both systems.

However, the most effective sales, in the construction and engineering field, are usually made by the persons who are actually involved in a line position. Thus the project manager on one job can speak with knowledge and conviction in attempting to obtain a new job. The engineer-manager can talk with knowledge and confidence as to his ability to meet the requirements on a desired schedule. Such "line" people usually require guidance by management as to pricing, for they often lack the objectivity that is necessary in hard negotiations over price. Sometimes they seek too high a price as a protective hedge against performance, tending to cover all risks and contingencies. Alternatively, they become too anxious and, hence, too optimistic about production rates and costs.

These remarks lead quite obviously to the concept of the sales team approach; the sales engineer to find the project and set up the meeting, the line construction manager or engineer-manager to talk authoritatively and convincingly, and the executive to guide the negotiations as to policy and business aspects (insurance, liability, contract provisions, and price).

3.8. *Establishing Personal Contact*

In a marketing program it is very important to establish personal contact with the prospective client in order to create an atmosphere of receptivity and confidence. Recently a large contractor wanted to enter the offshore oil industry construction market. The contractor placed expensive advertisements in trade journals, sent out brochures, and mailed business letters announcing its new services. Then it sat back and waited for the inquiries to come. Unfortunately, its prospective clients ignored its marketing campaign. The contractor did not even receive a single inquiry.

The problem with this approach was that the contractor failed to establish personal contact with the prospective client. The petroleum industry is actually a very closed fraternity (as are most other industries). They compete hard, buy hard, and demand performance. They tend to have more faith when dealing with former suppliers that they know will perform, that understand their needs, even if plans and specifications are not complete, and that understand their scheduling problems. Thus the industry decision makers inherently distrust newcomers, even if they really would like a wider source for the supplies and services they require.

It is extremely difficult to break into this closed fraternity. You have to establish a personal relationship based on acquaintanceship over a period of time and mutual interest. You must be seen with others in the industry, participate in their professional meetings, and be willing to furnish information freely and openly. Entertainment is one part of this process in many industries including the petroleum industry, but it is not the only part nor even an essential part.

Typically, one company will give the newcomers a very small contract, just to see how they perform. If they do the job well, on time, and within budget, they will give them a little larger slice with which to prove themselves. These should be viewed as marketing opportunities by the newcomer, who should be grateful for the opportunity to make contacts and recognize that these first jobs will undoubtedly be "loss leaders."

3.9. Multiheaded Clients

Who are the players? The key players are: the architect, the engineer, the construction manager, the contractor(s), the key subcontractors, the owner(s), the investor(s), and the user or principal lessee. The referees are governmental agencies, environmental groups, and city building departments. The roles of the players may change from job to job. You need to determine the objective of each player who makes the decisions. In building construction, the architect could be the key decision maker; in heavy construction, the CM could be the key decision maker.

What are the objectives of each player? The architect's main objective may be to see his design built. The owner's objective may be to get a functional building at a low cost. The engineer may be trying to get enough volume to meet his overhead costs, while the principal lessee is concerned about early completion.

Who can help you? Who can hurt you? The construction game is played with people, and you have to be able to interact with them. Most architects seem willing to accept change, except perhaps on aesthetic aspects. When a good idea comes along, they are usually willing to accept it. Conversely many engineers are unwilling to accept a change in their structural design. They do not like to be challenged on technical matters, and they are painfully aware of the interaction of many aspects. Therefore, to get an engineer to change, a nontechnical approach, perhaps presented by a business graduate, is most effective. Investors will often accept a higher price to eliminate risk of overrun in time or cost. Developers are normally looking for the cheapest price. You need to guess how the players may react before presenting your ideas.

In marketing, it is necessary to identify both those who make the decisions and those who approve the decisions at the various levels. In the old days, if your marketing efforts were directed to the Ford Car Company, the builder of the Model T, the answer to who makes the decisions and who approves the decisions would be simple—Henry Ford.

Today, things are a bit more complex. Consider the problem of marketing to a West Coast oil refinery. Assume the purchasing manager of the refinery calls you into his office and asks you to prepare a proposal to do some construction work in the existing refinery. He hands you the plans and specifications. The "multiheaded clients" might be identified as the following:

Purchasing manager
Plant manager
Plant engineer
Safety engineer
Treasurer, district office in San Francisco
Process equipment design office, Houston

Consulting structural engineer, San Francisco

The purchasing manager is the one to whom proposals are addressed and who will probably sign the contract. However, the others are of great importance and may be the actual decision makers. The plant manager is concerned with maintaining production during the period of construction and will need to be assured on this matter. The plant engineer knows the planning; he knows what lines can be shut down and how to bypass certain units. He probably will be the man who supervises your actual work. The plant safety engineer will undoubtedly have restrictions on operations that need to be investigated; for example, welding, burning, and pile driving. For example, he may require timber or bronze pads under a pile hammer. The treasurer needs an updated financial statement on your company, a letter from your bank, and perhaps a letter from your bonding company (even if the job is not bonded). In an actual case, one purchasing manager "thought" that the treasurer had this information when he did not, and consequently the low bidder was not awarded the job. The process equipment design office will have responsibility for the technical requirements and constraints. The consulting structural engineer may have prepared the actual plans and specifications on which you base your bid.

A multinational integrated oil company has a large number of subsidiary organizations. Let us look at a few of these.

The research and development subsidiary is often the one engineers feel most comfortable with, for this R&D branch initiates and funds development and feasibility studies. It is usually on a tight budget and limited scope. While it may introduce the contractor to important projects, his potential direct earnings from this subsidiary are limited.

Then comes exploration and production. In exploratory drilling and particularly when oil or gas is found, contractual and economic factors place a tremendous incentive on getting the job done, properly and on time, regardless of cost to some degree.

The same applies to pipeline and terminal construction: moving the crude to market. The emphasis is on performance, and here is where profits can be generated commensurate with skill, effort, and accomplishment.

If we then go to the oil company's refining subsidiary, we find a disciplined and rigorous control of costs. The work is usually thoroughly engineered and carefully controlled and administered. Risks are reduced, but percentage profits are also more limited. The generally large size of such projects may make them good business for the engineer-contractor but at a much lower rate of return on sales volume.

These three all carry the same name (of the oil company) but are in reality three different clients, or three heads of a multiheaded client. The sales approach, the relationships, the contractual aspects are all handled in completely different manners.

Multiheaded clients also exist on a broader scale, where, for example, a multinational firm operates through national companies and joint ventures, supporting them through other companies that provide technical, financial, and legal support. The success of multinational companies is their ability to draw upon the resources of these widely scattered units and to concentrate them on a particular problem. Thus marketing to one operating unit alone is not enough: several of the "heads" will play a part in making the decision.

3.10. Look for Something Different

One of the smartest men ever to go into the construction business was George Pollock—long since dead. First, he would watch the market and wait; he was able to do this because he had a very low overhead (three or four men). Pollock would wait for the large, unusual contracts and prepare an estimate; but the estimate would not be submitted. This was done to keep abreast of the market and find at what level the market was being priced. He was looking for something different. The first success came on a contract for a highway which was to be constructed along a section of the coast where the mountains ran down to the sea. On this contract Pollock determined that there was no statement as to where to spoil excess cut material. Since there were no environmental restrictions in those days, he just pushed the cut material over the bank with a bulldozer. Pollock had also foreseen that there could be a tremendous overrun in material to be moved because of the unstable cliffs above. That was the way the actual construction turned out, with the net effect that he got almost twice the expected income; yet his costs were half the estimated costs. After this great success, he sat back and waited again, watching for an opportunity. After several years, one arose.

A large tunnel was being built by a joint venture consisting of several major companies. Tunneling technology was not as advanced as it is today. There were problems with rock overbreak and squeezing soils. The owner, a regional district, was creating inspection difficulties, demanding stringent compliance with details, and holding up payments for failure to maintain schedule. The owner refused to discuss changed conditions, categorically denying them. It got so bad that the joint venture finally shut the job down. This created massive litigation. All parties involved were suing each other. The regional district terminated the contract and called for bids to finish the last quarter of the tunnel. No one would bid on "such an obviously bad contract" except George Pollock. Pollock determined that the biggest problem on the job was improper contract administration by the owner. He reasoned correctly that on this new contract the district would bend over backward to ensure the fairest possible inspections, payments, and schedules in order to "prove" that the original contractor was solely at fault. Also, Pollock, as well as the district, knew the actual rock and construction conditions by this time. Although he did bid high, he was the only bidder and was awarded the job. The high price, the reasonable inspections, and the thorough investigation resulted in a good profit. Pollock had the inherent ability to recognize an opportune situation.

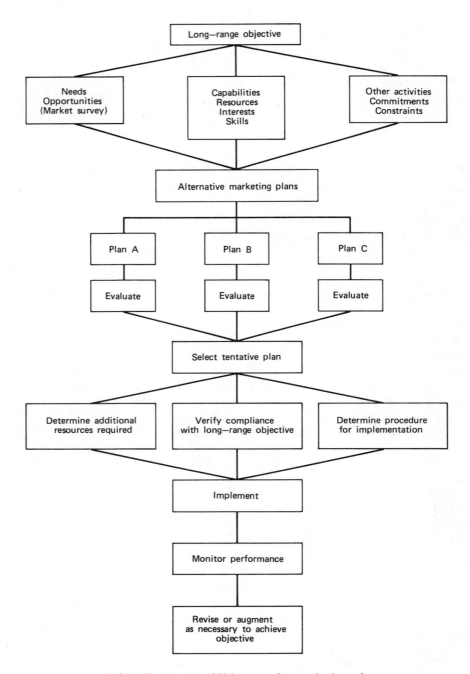

FIGURE 3.1. Establishment of a marketing plan.

3.11. Establishing a Marketing Plan

Based on the considerations presented earlier in this chapter, the results of the market survey, and a clear understanding of your company's long-range objective, a logical approach can be followed to the development of a marketing plan.

The logic diagram in Figure 3.1 is intended as a methodical means by which a firm can match its resources with the market needs and opportunities. This diagram is a quite standard approach, with a few exceptions. One is the inclusion of the box "Other activities, commitments, constraints" in order to ensure that the new endeavor will be compatible with your other objectives and markets.

The second is the emphasis on "implementation." Many good plans fail because the decision maker fails to allocate enough time, money, and technical skills to ensure its success. It is as though one had conceived a brilliant battle plan that could turn the tide of war and then committed only a few battalions to achieve it.

The third innovation is the item "Monitor performance." As the program proceeds, it is important to monitor it to see whether it is truly moving toward the objective. Does it require additional resources with which to carry it out?

Sometimes early action uncovers opportunities far beyond those felt possible in the decision stage. Do these new opportunities require a reassessment of the situation?

Monitoring should be periodic, not continuous. Neither individuals nor firms can effectively carry out the action phase of a program if they are constantly evaluating it. Commitment is all important to success.

There usually is a bird to be found in the bush if you know how to beat the bush properly.

Forbes Epigrams

4

The Role of the Contract as a Marketing Tool

4.1. Types and Scopes of Contracts

Contracts can be extremely effective marketing tools. The ideal contract is attractive to the client and ensures profitable performance for the contractor. No one contract is the perfect marketing tool for every case. The contract that is the most advantageous for each of the two parties depends on the case at hand. Tables 4.1 and 4.2 present an overview of the types and scopes of contracts covered in this chapter. Given all the different types of contracts in the tables, the opportunity exists to select one to best fit the case in hand. The marketer has to use the right one at the right time.

For example, consider a case where a private owner is asking for lump-sum contract bids from five competitors. Assume that none of the five competitors has any competitive edge. A smart competitor might then propose, as an alternate to his lump-sum bid, to perform the work on a cost plus fixed-fee basis, stating that he was willing to set a guaranteed maximum price. After discussing this with the owner, if this smart competitor felt that he was in a weak negotiating position, he could offer a guaranteed maximum price equal to his lump-sum bid, whereas, if he felt he was in a strong negotiating position, he could offer a guaranteed maximum price equal to his lump-sum bid plus 3 percent. The owner may be interested in the alternative contract because he feels he has a chance to save some money over the lump-sum price.

As a counterexample, Ben C. Gerwick, Inc., was asked for a cost plus fixed-fee bid for the foundation work for a large building in San Francisco. There was other competition for this job; therefore, Ben C. Gerwick, Inc., had to find some competitive edge. There were many subsurface risks at the site—including old ships' hulls from 1849. During the gold rush, the site was actually at the edge of the bay, and some of the ships that carried the miners had been abandoned and sunk as everyone left for the gold fields. There were some pictures showing the sunken ships in the client's office. Ben C. Gerwick, Inc., decided to submit a lump-sum contract, guaranteeing to complete the work on time despite whatever conditions were encountered, even if it ran into the sunken ships. This was attractive to the owner who had become fearful of potential changed conditions and delays if subsurface obstructions were encountered. Despite reportedly lower estimates submitted by other contractors (but without guarantees), Gerwick got the job. In this case it actually did encounter pre-earthquake building foundations which turned out to be a sizable but manageable problem. On this job, the owner wisely issued no change orders to Gerwick; it did not want to reopen the contract and thus present an opportunity to negotiate. So later the owner let a second small contract for the needed small internal changes so as to keep the first contract clean and simple. The bottom line was that Gerwick made its fullanticipated profit although it spent all its contingency. The owner got his job done on time and within his budget.

In negotiated work, an owner is rare indeed who will give the contractor carte blanche to proceed with a project, regardless of cost. Therefore, it is essential that the parties arrive at some mutually reasonable estimate of cost. There remains the matter of pinning down the basis of the contract. This could range from a cost plus percent fee to a lump-sum contract. Obviously, the contractor might prefer an open-ended cost plus fixed-fee arrangement, whereas the owner might want a cost plus fixed fee with a guaranteed maximum price. Frequently, the contractor will have to make some compromise on terms in order to conclude an agreement.

Do not forget that sometimes the best contract of all for both parties is a lump sum.

Note: Do not confuse the term "fixed price" with "lump sum." Fixed price means that the prices are quantitatively set ahead of time. However, the contract itself may be lump sum, unit price, even day rate, just as long as the numerical prices are fixed and not subject to adjustment on the basis of actual cost.

4.2. The Lump-Sum Contract

Many clients want a lump-sum bid. From the client's point of view, a lump-sum contract offers a fixed price that facilitates budgeting and financing. The owner has a fixed total that he can take to his board of directors or the investors who are going to finance the project. With a lump-sum bid, he feels that his risks are reduced.

A lump-sum contract is also attractive to the contractor, even though it involves risk, because it presents the possibility of making a better profit on the project. Generally speaking, the managers of most mature private industries and the managers of most public agencies want the contractor to make a

TABLE 4.1. Types of Contracts

Type of Contract	Contractor's Viewpoint		Owner's Viewpoint	
	Potential Advantages (+)	Potential Disadvantages (−)	Potential Advantages (+)	Potential Disadvantages (−)
Lump sum	Clean cut Easy bookkeeping Good profit	Risk Adversary status	Fixed price Can budget Easy to finance	Claims Adversary status High cost
Unit price	Easy to estimate Eliminate risks of quantity changes Unbalancing potential	Units may not reflect cost Unlisted items Disagreement over measurements Adversary status	Easy to adjust and determine final cost	Some claims Adversary status
Cost plus fixed fee	No risk on cost Both parties on same team	Low fee Risk on errors and accidents Extra bookkeeping	Quick start and finish Both on same team Changes are easy	Cost control Cost overrun Lack of incentive for contractor

Cost plus fixed fee with guaranteed maximum price	Easy to sell Can negotiate (if contract contains agreement to split savings, possible additional profit)	Risks all one way Limits scope	Ceiling on costs Both on same team? Possible savings	Some claims in event of overrun
Target estimate	Easy to sell Can negotiate Limits losses	Limits profit	Both on same team Potential for savings	Possible overrun No incentive to contractor after exceeding extreme limit
Lump sum plus units	Clean cut Eliminates risks of quantity overruns Unbalancing potential	Risk on costs Adversary status	Easy to administer Reduced claims	Adversary status Unbalancing potential
Day rate	Reflects actual costs Possible increased income	Disallowed days	Easy to administer Quick start Changes are easy	Cannot control productivity
Cost plus fixed fee converted to lump sum	Negotiation advantage in establishing price Good profit		Earlier start Fixed price to meet agency regulations	High cost

TABLE 4.2. Scopes of Contracts

Type of Contract	Contractor's Viewpoint		Owner's Viewpoint	
	Positive	Negative	Positive	Negative
Construction only	No design or ultimate performance liability	Adversary status with architect-engineer Impractical design	Competitive bid Review plans Select best engineer and contractor	Slow start Adversary status Lack of contractor's practical input
Phased contracts (multiple contracts)	Simple scope, hence easier to organize Competition usually more limited	Conflict with other contractors Minimizes his management capabilities and role	Earlier start and finish May reduce contingencies for contractors	Final cost unknown until all contracts are let Change orders and claims increased
Turnkey—design and construct	Integrated approach Practical design Competitive edge	Responsibility	Integrated approach Save time Both on same team	Usually less control on cost
Turnkey—design, construct, operate	Competitive edge	Responsibility Higher costs for same fee?	Complete package	High cost
Construction management (Project Management)	New No risk (?) Utilizes management skill to fullest Larger jobs Entry to markets	Lots of paperwork Low or moderate fee	Phased contracts Budget control Schedule control Value engineering One labor contract in many cases	High cost (?) Possibly many changes Owner "involved"

4.2. The Lump-Sum Contract

legitimate profit. These owners want the contractor to survive and be available for their next job.

The most serious problem with lump-sum contracts is that an adversary status can be created that leads to disagreements and divergence of interests. Each is trying to get the most they can from the lump-sum agreed price.

How can you as a contractor protect yourself against undue risk on this type of contract? What you can do on a lump-sum bid is to qualify your bid. On any private contract, this is perfectly legal; on public contracts, while this may present a legal problem, there still may be opportunities. Consider a private contract that contains the clause "will accept all risks and contingencies." You can come back by qualifying your bid with a "we accept all risks and contingencies with the following exceptions. . . ." In a foundation contract where there could be rock and water problems, the exception might be "will accept all rock up to 100 cu yd and all water up to 15,000 gal/day." In your marketing effort you can point out that it would be unlikely to encounter rock and water in excess of the amount listed. Also, when qualifying a bid, the word "stipulation" often sounds better to a client than the word "qualification." The word "qualification" has taken on a negative connotation in the construction business.

How can you qualify your bid when the contract states qualifications are not acceptable? In a private contract, this means that qualified bids "may" not be acceptable. However, if you are the low bidder, the client will surely look at your bid carefully. Maybe, on second thought, your stipulation will not seem all that bad to him. Of course, in many contracts there is no need to qualify your bid; the point is that qualification (stipulation) can be used as a marketing tool.

As a case in point, a West Coast contractor was trying to break into the offshore construction business against some excellent and experienced competitors. The first contract on which he bid was required to be a fixed-price contract. The West Coast contractor really bid low to try to get a first job. He had an advantage as to location—the contract was in California—and the competition was from the Gulf Coast. However, the California contractor was not awarded the contract. A year later he found out from an inside source that he had not been the apparent low bidder, even though he had bid the job at cost, without markup. The experienced Gulf Coast competitors had qualified their bid, limiting certain risks, which proved acceptable to the oil company (although the contract document said "qualifications were unacceptable"), and reduced their bid accordingly. For example, the specifications required that the contractor accept all weather delays. The experienced contractors said they would accept a weather delay up to five days (the average for that area), but they were to be paid on a "day-work basis" for weather delays over five days. The California contractor had bid higher to provide itself with contingency against a possible longer weather delay. Similarly, the specifications required the contractor to install the piles to a penetration of 200 ft in dense sands and silts. The Gulf Coast contractors stated that their bids included driving with the specified hammer to a blow count of 400 blows per ft and that all subsequent jetting, drilling, and driving would be paid as "extra work" on a day-rate basis. The naive West Coast contractor had included allowances for such contingencies in his bid.

With private bidding, you usually do not ever learn the other bidder's exact bid prices. In general practice, these bid prices are never revealed. Also, in private work, the low bidder is not always awarded the contract. Sometimes other factors influence the award. With public bids, the bids of all participants are made public at the bid opening. This information is useful. The second bidder should follow the contract award: sometimes the low bidder is disqualified or has made a mistake, and the contract is awarded to the second bidder. For example, the low bidder may have omitted inclusion of the bond, as required, or forgotten to sign the bid or to file his minority subcontractor plan as required. In such a case, the low bidder can legitimately file a protest.

For a lump-sum contract for a private company, often more than one bid can be submitted. The second or additional bids are called alternate bids. This useful marketing tool enables an unqualified bid, in exact accord with the plans and specifications, to be submitted, and also a second bid with qualifications. In many cases in the public sector, it is possible and legal to bid on two projects on the same day and to stipulate that if your corporation is low bidder on both contracts, only one contract will be accepted. This stipulation may be necessary to enable you to obtain a bond.

Many public bids (and some private bids) require that a bid bond be submitted with your bid, guaranteeing that you will sign the contract if you are low bidder. There is an agreement that if you are awarded the contract, the performance and payment bonds will be obtained from the company that issued the bid bond. When a private company requires a bond, many contractors state in their proposal that the bond will be furnished as required but the owner is to pay the premium.

More often than not, in order to close a deal, the contractor will be asked by a private owner to reduce his lump sum (or guaranteed maximum) by an amount "which will bring him within his budget." It is not always possible to anticipate such a request, but the contractor should be prepared for it and know ahead of time how much he is willing to give back. There are several possible ways to cope with such a request if one does not want to compromise on the established lump sum or guaranteed maximum price. First, the contractor can reduce the scope of the work, hopefully eliminating some feature of the project which he believes the owner can live without. Another way would be to see whether the owner can contribute something of value, which does not necessitate capital outlay for him, such as the use of a crane which he owns and operates or use of structural steel or timber he may have in stock. Although not necessarily recommended, a stalling tactic is to ask the owner to wait until the contractor can talk with management to see whether any compromise on price is possible. The ultimate way to cope with a request for price reduction is to tell the owner that he, the contractor, has gone as far as he can go. In this event, he must be sincerely prepared to walk away from the deal. The owner can usually sense whether the contractor is bluffing. Similarly, experience enables the contractor to ascertain whether the owner is sincerely in need of reducing the price or just bid shopping.

4.3. The Lump-Sum plus Units Contract

Many requests for bids require that the bid be submitted as a lump sum for a group of fixed items plus a unit price on each of a group of variable items. A fixed item might be cleaning and grubbing; a variable item might be earthwork.

Often the unit-price bid portion may contain a trap or serious problem for the bidder. The bid proposed will ask for unit prices for overruns and underruns and then supply a single space for your answer. Consider a case in which a unit price is requested for an overrun and underrun for structural concrete. An overrun in concrete in a foundation mat might cost about $80/cu yd, while an overrun in concrete in an elevated floor that required elaborate formwork and scaffolding might cost about $400/cu yd. The bid form is asking for a single price for any concrete. Which price would you bid? If you bid high, the owner may delete a floor slab. If you bid the low price, he may add one. The correct answer to fill in the blank is "see attached sheet." Then attach a price list for all types of concrete work involved in the job. You can use different prices for overruns and for underruns, since usually the effect on your costs is quite different. This procedure takes a little extra time; but it can eliminate a risk.

4.4. The Unit-Price Contract

The unit-price contract is most useful when the quantities of work to be done are indeterminate. For example, in an earth-moving project, if only an estimate of the quantity of earth to be moved is known, the contract price can be a unit price per cubic yard, to be applied to the actual amount moved. For example, it may not be determinable beforehand whether a 1:2 or a 1:3 side slope will be stable. Thus the contractor is freed from the risk of quantity variations.

With unit-price bidding, there is always the possibility of unbalancing. Despite the provision in many specifications which states that "unbalancing of unit bid prices may be cause for rejection," the practice is still widespread, and few contracts have been denied because of unbalancing.

Unbalancing can be employed by the contractor to minimize risks, to provide early cash flow, and to increase profit in the event that the final quantities differ significantly from the bid quantity. However, the practice can sometimes backfire on the contractor and lead to a serious loss of revenue.

A California contractor had a bad experience with unbalancing on a bridge job. During the cost estimating, the quantity of reinforcing steel was taken off and a 30 percent underrun was found. Therefore, the bid on this item was unbalanced low; that is, the reinforcing steel was bid at cost. When the bridge design engineer saw the low prices on the steel, he modified the design of the bridge to use much more steel and a little less concrete. While this redesign saved the highway department some money, it seriously hurt the financial outcome for the contractor.

Consider the case below where a contractor is bidding unit prices to furnish and drive piling. The engineer's estimate calls for 1000 piles, each 60 ft in length. Assume that the contractor's experience indicates piles 40 ft in length will achieve adequate bearing capacity due to the soil strata and driving conditions. This situation presents an excellent opportunity for unbalancing by the contractor. By recognizing the potential underrun in pile lengths and by unbalancing the bid accordingly, in this case bidding the "furnishing piles" item below cost, the contractor was able to almost double his profit as shown in the following analysis. Note also that the owner saved $70,000 in Case A but only $2000 in Case B.

Following is Contractor A's balanced bid (assuming 60-ft piles):

	Amount	Contractor's Unit Cost	Contractor's Cost	Bid Unit Price	Contractor's Bid Price
Furnish	60,000 ft	$3/ft	$180,000	$3.50/ft	$210,000
Drive	1,000 piles	$200/pile	200,000	$250.00/pile	250,000

Contractor's total cost	= $380,000
Contractor's total bid price	= $460,000
Indicated profit	$ 80,000

If Contractor A had been successful bidder, but piles actually only ran 40 ft in length:

	Amount	Contractor's Unit Cost	Contractor's Cost	Bid Unit Price	Total Income
Furnish	40,000 ft	$3/ft	$120,000	$3.50/ft	$140,000
Drive	1,000 piles	$200/pile	200,000	$250.00/pile	250,000

4.4. The Unit-Price Contract

Total income = $390,000
Total Cost = $320,000
Indicated Profit $ 70,000

Here is Contractor B's unbalanced bid (assuming 60-ft piles):

	Amount	Contractor's Unit Cost	Contractor's Cost	Bid Unit Price	Contractor's Bid Price
Furnish	60,000 ft	$3/ft	$180,000	$0.10/ft	$ 6,000
Drive	1,000 piles	$200/pile	200,000	$450.00/pile	450,000

Contractor's total cost = $380,000
Contractor's total bid price = $456,000 (*low bid*)
Indicated profit $ 76,000

Contractor B was awarded the job. Here is how it turned out when the piles ran only 40 ft in length:

	Amount	Contractor's Unit Price	Contractor's Cost	Bid Unit Price	Contractor's Bid Price
Furnish	40,000 ft	$3/ft	$120,000	$0.10/ft	$ 4,000
Drive	1,000 piles	$200/pile	200,000	$450.00/pile	450,000

Contractor's total cost = $320,000
Contractor's total income = $454,000
Profit $134,000

Obviously, a question of ethics is involved. Contractor B would say that the owner still got a satisfactory pile foundation, and it was the owner's own fault since he was the one who selected the 60-ft length. The owner will, of course, be upset (unless he never realizes what has happened to him).

Some public owners in the past used to always set up unit-price bids with 10 percent extra quantity in almost every bid item. Then their contracts always underran, which made them look good to the public. In this case, it was the contractor who suffered unless he could find a way for protective unbalancing.

Unbalancing can be utilized to help cash flow by earning extra money on early items in the schedule, by bidding these unit prices high. In effect, it can act as a means of collecting a mobilization payment. This practice is known as front-end loading. The Corps of Engineers has taken steps to eliminate or at least reduce this practice by providing a separate mobilization item and, in some cases, by limiting the unit price that can be bid on such early items as "developing water supply" or "clearing and grubbing."

The Bureau of Reclamation's approach to this problem has been to break the major items of work into two items. Consider a dredging job in which approximately 1,200,000 cu yd are required. The first item may read "Dredging the first 500,000 cu yd," and the second item, "Dredging, quantity over 500,000 cu yd."

However, the contractors, faced with heavy demands for early expenditures, continue to bid the early items of work high, reducing the price accordingly on late items so as to preserve the same total. In times of high interest rates, the contractor may even carry out a "present value" analysis to help him evaluate the effect of front-end loading.

As with many ethical questions, it seems to be a matter of degree that is involved; some unbalancing may be just good business judgment, whereas gross unbalancing may be manifestly improper.

There were two interesting California cases where this practice was carried to an extreme. In the first case, on a state highway contract, a major California contractor placed essentially its entire markup on the earliest pay item, clearing and grubbing. Then it bid prices that were essentially at cost for the rest of the bid items. The state accepted this contractor's bid, and the entire amount for clearing and grubbing was paid in a timely fashion. This device enabled the contractor to avoid financing cost and even make money on bank interest. Thus, it could give a low, highly competitive bid.

In another case, with a different California contractor, extreme front-end loading did not work. In this case, also, essentially all the markup was placed on the earliest pay item—clearing and grubbing. However, this time the owner did not pay the contractor the full price on the item when it was completed early in the job. Instead, he only made partial payments on this item as the rest of the work progressed. He would simply make one excuse after the other as to why clearing and grubbing was not complete. For example, an inspector could always find a few tufts of grass that had not been cleared and grubbed in accordance with the contract document. While the whole matter was petty and questionable practice for both parties, it successfully defeated the contractor's plan for early income. Perhaps if the contractor had used more finesse, such as dividing the front-end loading between several early items, he would have been more successful.

In another similar case, the public authority agreed to award an obviously unbalanced bid only after negotiations had achieved a secondary understanding that the lump sum for clearing and grubbing would be paid when 50 percent of the remainder of the contract was completed.

4.4. The Unit-Price Contract

One of the oldest San Francisco foundation contractors possessed a particular knack of knowing what a job would entail. When he was starting out in business, a contract was bid for the foundation piles for a large post office in San Francisco. The plans and specifications showed 100-ft piles; this length was also predicted by a test pile. All other bidders, except this one contractor, assumed this data to be valid. They put all their overhead and profit on lump-sum items and bid the unit price of the piles at the cost of materials only (so as not to be hurt if the piles underran in length). This contractor, however, was intimately familiar with this particular geological area by having had other piling jobs here in the past. He carefully studied the boring data that was available and conducted his own site investigation. He concluded that the test pile had been driven at the shallow corner of the site. Therefore, he bid the lump-sum items at cost and put his overhead and profit on the unit price of the piles (price per linear foot). This tactic paid off: the piles averaged 135 ft. He made sufficient profit on this job to buy much needed basic equipment to really get himself started in the construction business.

On a state highway bridge unit-price contract, a contractor who specialized in bridge piers suffered a bankruptcy because of a major quantity change and his failure to use unbalancing as a means of reducing risk. The contract called for unit-price bids for "excavation for bridge piers," this item being required to include all incidental work thereto, including the cofferdam and dewatering. Six cofferdams were required, for which the contractor estimated a cost per cofferdam, including dewatering and excavation, at $400,000 each. Therefore, he needed a total of $2,400,000 to build the cofferdams and complete the excavation. Once the cofferdam was built, the cost of actual excavation was trivial, being only about $4000 each pier. Each cofferdam required 200 cu yd of excavation, or a total of 1200 for the six. The contractor then divided $2,400,000 cost by 1200 yd and obtained the unit cost of $2000/cu yd for cofferdam excavation. Adding overhead and profit gave him a bid unit price of $2450/cy for a total of $2,940,000. That winter, shortly after award of the contract, there was a flood that eroded the site and reduced the amount of material to be excavated from 200 cu yd per pier to 70 cu yd. This reduced the contractor's income to a third of what he anticipated, or only $1,029,000. Meanwhile his cost to do the work was almost the same because the main cost determinant was building cofferdams, not excavation. Therefore, it would appear to be essential in such cases to put much of the bid money in items whose quantities cannot change. The moral is to beware of items for which the cost does not bear a reasonable relationship to the quantity that will be measured for payment.

Following this principle on a unit-price contract for the East Bay Municipal Utility District, a contractor put most of his bid price on some concrete work that he thought he could not change, in order to try and protect his total income. The price for the item became so high that the District made every effort to reduce the pay quantity, deducting for chamfers, the volume occupied by reinforcing bars in the concrete and by the volume of the pile heads in the concrete. This resulted in a claim, long drawn-out negotiations, and an unsatisfactory settlement for the contractor.

4.5. The Cost plus Fixed-Fee Contract

Under this type of contract, the contractor is paid on the basis of his actual costs to perform the work plus a preagreed fixed fee, the latter to cover his main office overhead and profit. The contract may be open-ended, in which case there is no contractual ceiling in the costs, or it may have a guaranteed maximum ceiling, with the contractor agreeing to stand any costs that exceed this ceiling.

In general, contractors like the cost plus fixed-fee contract, especially if it is open-ended, as it eliminates many risks. Owners usually try to avoid open-ended cost plus fixed-fee contracts—this type of contract has a history of overruns. Costs can get out of hand. The main advantages to the owner are the quick start, the ease of making changes, and the elimination of the adversary status between owner and contractor. One contractor—W Construction Company—always assigned its best personnel and equipment to a cost plus fixed-fee job because it wanted the job to go well for the owner. W Construction valued that client's goodwill and wanted him to give it another cost plus fixed-fee job in the future. Unfortunately, this is not the case with all contractors and all owners.

Any agreement between an owner and a contractor to perform work on a basis of cost plus a fixed fee should carefully define the scope of the work. It is useful to have drawings and specifications to define scope, but in negotiated work these may be sketchy or incomplete. The formal proposal should have an explicit definition of the scope of the work. It should, as applicable, include a laundry list of those things included and those things excluded. This scope is then incorporated in the negotiated agreement. The reason for this is to provide a basis for adjustment in the fixed fee if there is a major change in scope.

A variation on a cost plus fixed-fee contract is a cost plus percentage contract. This type of contract is rarely used because of the negative incentive for the contractor to save money. An old saying in the industry was that contractors like a cost plus percentage contract to overrun the estimate because "the more cost, the more plus." A cost plus percentage contract is still used in maintenance contracts, in repair work when the scope cannot be ascertained, and in an emergency when very quick action is essential; for example, after a fire or a railroad derailment.

One way to provide both a cost-savings incentive for a contractor and budgetary protection for the owner is to use the cost plus fixed-fee with guaranteed maximum price contract. An example of this type of contract is shown in Figure 4.1. A variation of this contract is to establish a formula for

4.5. The Cost plus Fixed-Fee Contract

sharing any savings under the guaranteed maximum between the contractor and the owner. This will give an added incentive for the contractor to reduce costs. Cost plus fixed-fee with guaranteed maximum price contracts are often used in building construction. They are more easily sold to the owner because he can use the guaranteed maximum price for budgeting and financing. Occasionally, when a private owner has called for lump-sum bids, a contractor may find that offering an alternative contract with a guaranteed maximum price "allows" the owner to select him even if he was not the low bidder.

Inexperienced contractors are often unduly afraid of the risks of a guaranteed maximum price (GMP) contract. They feel that they are taking all the risks of a lump-sum contract with none of the benefits. However, this is not necessarily always true. First, the GMP contract may help land the job. Second, you may be able to get a slightly higher price; that is, you may be able to negotiate a "contingency allowance." Third, since you are both on the same team, and both are keeping costs, the owner knows where you stand all the time. If you are doing the best job possible, he knows you have spent the money. This makes requests for change orders easier to negotiate. Occasionally (very occasionally), an owner has even voluntarily raised the guaranteed price when he sees that the contractor has had an overrun in cost that was not the contractor's fault.

The last contract of this type to be discussed is the "cost plus fee contract converted to a lump-sum contract." This contract has the advantage to the contractor of potentially yielding a healthy profit with much-reduced risk and the advantage to the owner of a quick start. It is a relatively easy contract to budget and administer and obviates many of the restrictions and constraints which are imposed on governmental agencies in their administration of straight cost plus fee contracts. It has the disadvantage to the owner of a high probable cost.

A main problem on all cost plus projects is defining the items of cost, especially those that are difficult to ascertain directly. Examples are:

1. Material costs. To whom does the discount belong if the contractor uses his own funds to pay ahead of schedule?
2. Insurance premiums. Should these be charged at manual rates, or should they reflect the lower costs due to the contractor's more favorable experience in past years? What if his premiums are higher than manual?
3. Main office (off-site) costs directly related to the job; for example, purchasing, engineering, cost control, interest cost on working funds, general insurance.
4. For work performed at contractor's yard, who pays the rent? Who pays for the use of yard facilities and supplies?
5. To whom does salvage belong at the end of the job? How should the income from the sale be credited against the cost?

STANDARD FORM OF AGREEMENT FOR CONSTRUCTION
COST PLUS FIXED FEE, GUARANTEED MAXIMUM PRICE

1. BEN C. GERWICK, INC., A California Corporation, hereinafter called Contractor, agrees to perform for ....

....,

hereinafter called Owner, the following work; ....

....

in accordance with specifications ....

and drawings ...., which are hereby made a part of this agreement. Owner shall furnish property, access thereto, and all permits needed for the construction work.

    1a. Work excluded: ....

....

    2. For this work Contractor shall be paid by the Owner an amount equal to his direct costs plus a fixed fee of $..............................
Contractor guarantees that the sum of the direct costs plus the fixed fee will not exceed the Guaranteed Maximum Price of $..................
    If the sum of the direct costs plus the fixed fee exceed the Guaranteed Maximum Price, then all such overrun shall be borne solely by the Contractor.
    If the sum of the direct costs plus the fixed fee are less than the Guaranteed Maximum Price, then the savings shall be shared on the basis of 75% to Owner and 25% to Contractor.

    3. The Guaranteed Maximum Price shall be adjusted up or down for changes ordered by the Owner. The Fixed Fee shall not be changed unless the scope of the work as shown in Section 1 is changed.

    4. Direct Costs shall be defined to include the following:

    a. Materials and construction supplies used in the work, including sales and use taxes, and including freight or delivery charges and transportation taxes in accordance with vendor's invoices.

    All discounts shall be taken for Owner's benefit, except that cash discounts shall be for his benefit only when he makes payments on time in accordance with this agreement.

    "Materials and construction supplies" shall include items furnished from Contractor's stocks for the benefit of the work. Such items shall be priced in accordance with current out-of-stock market prices for similar materials in similar condition.

    b. Labor wages, plus travel expenses and fringe benefits, in accordance with current agreements with labor unions or men involved. Taxes and insurance based on payroll shall be included.

    c. Salaries of supervisors, engineers, and other personnel assigned directly to the job at the job site, plus actual travel expenses, plus fringe benefits in accordance with current agreements with such employees, plus insurance and taxes as set forth in 4b above.

    d. Equipment rentals for major floating equipment, eg. barges, piledrivers, derricks, shall be charged in accordance with the Associated General Contractors' of America Manual of Equipment Ownership Costs, using the present replacement value as the "value in the formulae shown." Present replacement values for non-standard major floating equipment shall be in accordance with the latest appraisal of independent marine surveyors.

    For equipment which it is anticipated will be used on this project, the following rates will apply, having been computed in accordance with the method set forth above. Rental rates shall be charged from the day the equipment leaves Contractor's yard until it returns.

    e. Equipment rentals for standard items of equipment having an individual value more than $250.00 shall be charged as follows:

    (1) Where procured from a third party, at actual cost.

    (2) Where furnished by Contractor at 85% of the rates recommended in the latest edition of the Associated Equipment Dealers Manual. Rentals shall be charged from the day equipment leaves Contractor's yard until it returns.

    f. Small tools and hand tools with an individual replacement value of less than $250.00 shall be charged at 5% of direct field labor.

    g. Delivery and return of equipment and tools.

FIGURE 4.1. Cost plus fixed-fee, guaranteed maximum price contract.

h. Towing charges at actual cost of tug procured from third party, or at.................... per hour when tug furnished by Contractor. Hours to be computed from time boat leaves home port until it returns to home port.

i. The equipment rental rates in sections d, and e, above include insurance, taxes, major overhaul, painting, readiness for service, and depreciation. However, any repairs of a character normally performed on the job including labor and replacement items, including steam hose, steel lines, manila lines, welding and burning supplies and parts replacements, shall be charged as a direct cost. Any pile hammer or follower breakage on the job shall be charged as direct cost. In the event of damage to equipment, the costs of repairs borne by the owner of the equipment under the "deductible" clause of standard insurance policies shall be charged as direct cost, but this charge shall not exceed the deductible.

j. For items manufactured at Contractor's precast and prestressed concrete manufacturing plant, for incorporation or use in the work under this contract, charges shall be in accordance with the "Schedule of Prices" attached as an appendix to this agreement. Units manufactured and stored at the plant, and materials to be incorporated in the units, which are stored at the plant, shall be included in the monthly progress estimates and payments.

k. Where demolition of existing structures is involved, the costs of hauling away and disposing of materials removed shall be included as direct cost.

Special provisions relating to demolition and salvage credits, may be attached as an appendix to this agreement. If none are attached, then salvage value will be considered to be nil.

l. The cost of work performed off or away from the job site, such as loading materials to barges, etc., which is directly connected with the job shall be charged as a direct cost to the job. Where the amount of such off-site work will be a major item, special arrangements shall be made for the recording of costs. Otherwise, the charges shall be at the rate of 2 times payroll, which shall include the cost of yard equipment, insurance and taxes on labor, etc. Any material involved will be charged under Section a.

m. Any other job insurance, including fire or builder's risk insurance. Premium on any bonds required.

n. The costs of any permits, licenses, tests, or fees paid for by the contractor, and directly related to this specific job.

o. The cost of field office supplies, postage, blueprints, telephone, telegram, first aid supplies, sanitation, water and power, etc.

p. The cost of any subcontracts. The Contractor will obtain Owner's approval prior to subcontracting any substantial item of work.

q. Any other costs, miscellaneous expenses, or services, directly related to the job and approved by Owner's representative.

r. Where engineering services are required in addition to construction, the costs of such shall be charged in accordance with appendix attached.

5. The Contractor's fixed fee shall be compensation for the following:

a. Contractor's main office overhead and administration.

b. Supervisory personnel of Contractor's organization not assigned to the job at the job site, such as General Superintendent, and officers of the contractor.

c. Contractor's profit.

6. Piledriving operations inherently involve noise and vibration. Where piledriving is required by plans and specifications, Owner is to hold Contractor harmless for any damage or alleged damage due to noise or vibration.

7. A monthly tabulation of costs shall be submitted to Owner on the 15th of each month, covering the work performed during the preceding month. Since it is recognized that all costs may not be billed promptly to Contractor by vendors, Contractor shall submit all available data in his report on the 15th of each month, and shall submit costs subsequently received in the report of the following month.

a. Final statement, with a complete summary of all costs, shall be submitted within 30 days after the end of the job.

8. Payments shall be made as follows:

On the 20th of each month, Contractor shall submit his estimate of the costs to be incurred that month including materials received, and the proportion of the fee. Owner shall pay this estimate, less retention, by the 10th of the following month.

The retention shall be 10% until 5% of the Guaranteed Maximum Price is retained, and from then on the retention shall be 5% of the Guaranteed Maximum Price. The retention may be reduced near the end of the job if in the opinion of the Owner, such reduced retention will be adequate to protect his interests.

Final payment, including all retained moneys, shall be due and payable 35 days after completion.

9. Time is of the essence in the performance of this agreement, and all work shall be completed on or before........................... Should the performance of the work be delayed by strike, storm, or other causes beyond Contractor's control, an appropriate extension of time shall be granted. Should the Owner delay the work, or should materially changed latent or subsurface conditions be encountered, in addition to extension of time, an appropriate adjustment shall be made in the Guaranteed Maximum Price.

IN WITNESS WHEREOF, the parties hereto have excuted this Subcontract the day and year first hereinabove written.

....	BEN C. GERWICK INC.
By.. Owner	By.. Contractor
Title:..	Title:..
Address:...	755 Sansome Street
....	San Francisco 11, California

REV. 64

FIGURE 4.1. (*continued*)

6. Equipment rental. Should this be at published equipment rental manual figures or at a percentage thereof; or should the rates for each item of equipment be set forth in an appendix? Who pays for repairs? Who pays for downtime?
7. Who pays for tools? for lost and stolen tools?
8. How are interdivision charges handled? or charges by wholly owned subsidiaries of contractor?

All the above, as applicable, should be sorted out ahead of time to prevent later disagreement.

One reason contractors like the cost plus fixed-fee type of contract is that they feel that there is no risk to the contractor on cost. However, there can be risks. As indicated above, there is a risk to the contractor as to what may or may not be classified as a cost. Are costs from errors and accidents "costs" that may be charged to the job? The contractor has one answer and the owner often has another to this question. Consider a typical accident of a crane boom collapsing. While insurance may pay for the crane boom, who pays the additional cost of the delay in having the job shut down for a week or two? The owner may take the position that he is not paying the contractor to make mistakes, and the contractor may wind up paying all or part of the cost of this delay.

A building contractor on a cost plus fee job installed the forms and reinforcement for a retaining wall, ready to place the concrete. It was discovered on final check that the forms were spaced too closely. The forms had to be unbuttoned, the steel relocated, and the forms reinstalled. The owner refused to pay for these costs; he took the position that he was paying for proper supervision and he should not pay for errors. The contractor took the position that some errors and readjustments were common to such construction and were properly part of the job cost. He did not succeed in convincing the owner, who withheld payment on this item.

As yet another example, a nationally known foundation contractor had received a large cost plus contract for several miles of sheet pile bulkhead in one of the Southern states. The sheet piles had been driven, and the tie backs had been installed; but sections of the bulkhead kept failing. The reason for the failures turned out to be a brittle fracture of the tie backs due to corrosion; the shop-applied protective coating was not of the quality specified. The purchase agreement, between the contractor and tie-rod supplier, was typical of most steel purchase contracts. It limited responsibility of the purveyor to the replacement of the material. Therefore, all that the steel company would do was to send replacement tie rods with the right quality of steel and right protective coating. The owner refused to pay for the mistake. The delay was costing him a great deal of money. Finally, a settlement was negotiated whereby the contractor agreed to reconstruct the bulkhead at his own expense, while the owner absorbed the cost of the delay.

What can a contractor do on a cost plus contract to limit his risk? Sometimes when a specific risk is recognized as great, that risk can be insured. The contractor can ask the owner to carry the insurance as part of the cost of the job. If the owner does not want to pay the insurance, the contractor can exercise one of two options. First, he can buy the insurance himself if he feels the risks are too great. This, of course, cuts into his fees. Second, he can perhaps reach agreement with the owner that since the owner has elected not to insure, it is agreed that in the event of loss, the reconstruction will be considered as part of the cost of the work and that the guaranteed maximum price will be correspondingly adjusted.

4.6. The Day-Rate Contract

A day-rate contract is sometimes very attractive to an owner. In small-scale applications, day-rate contracts are often used by insurance companies for disaster work—for example, clean up and repair work after a fire. For a fixed price per day, the contractor furnishes everything that is needed to do the work—the labor, crew, equipment, tools, and supplies. A familiar example is the rental of a truck crane, on a fully operated and maintained basis. The cost of materials is generally billed to the client as a separate charge. For large-scale applications, day-rate contracts are sometimes used if the work cannot be clearly defined; for example, if it depends on the weather. In offshore construction, for example, for $125,000 a day, a contractor may agree to furnish a complete offshore pipeline spread (lay barge, boats, crew, and so forth), and the amount of work it completes will depend to a large extent on the weather. However, in day-rate contract work, there is a potential problem to the contractor with disallowed days; if the spread breaks down, the owner may refuse to pay for those days the spread is down.

Sometimes, with a lump-sum contract, certain specific items of work will be set forth separately to be formed on a day-rate basis. For example, work to be done in a bad weather season is performed on a day-rate basis (if the owner wants work attempted in weather when productivity can be severely affected).

4.7. The Target Estimate Contract

A target estimate contract is often attractive when negotiating for a large project. A target estimate can be difficult to describe with words alone; therefore, a graph similar to the one shown in Figure 4.2 is often included in the actual contract documents. In the example, a target estimate was established at $4 million. The contractor, on the basis of his estimate, believes the project will cost $4.1 million, while the owner feels that the project should only cost $3.9 million. Therefore, they agree to establish a "gray area" be-

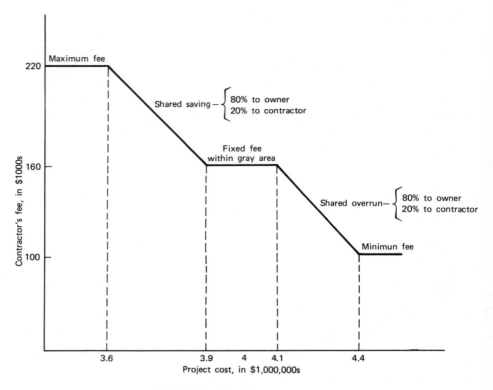

FIGURE 4.2. Target estimate.

tween $3.9 and $4.1 million. They agree that as long as the project cost ends up within this gray area, the contractor will be paid a $160,000 fee. A feature of the target estimate type of contract is a built-in incentive for the contractor to save money. If the contractor can complete the work at a final cost below the lower limit of the gray area, the savings are shared: 80 percent to the owner and 20 percent to the contractor. These percentages may be varied, depending on the contract. For example, the savings could be shared fifty-fifty. A lower limit or cutoff is established: in this example, at $3.6 million so that if final costs are less than $3.6 million, the contractor still gets only $220,000 fee. It is assumed that if the costs end up below $3.6 million, the scope of the project must have been reduced or the character of the work changed, so that the contractor is entitled to a maximum fee of $220,000. Now, if the costs overrun; for example, if the project costs exceed $4.1 million, the losses are also shared—in this case, the owner takes 80 percent of the loss, the contractor 20 percent. A minimum fee cutoff is also established. In this example, the contractor's fee can be eventually reduced to $100,000 but no lower. It is assumed that the contractor has administrative overhead costs that must be paid and that this minimum fee will cover these. In some target estimate contracts, the contractor's fee may be reduced to nothing—so that the contractor must be prepared to take the loss of home office overhead but no further loss on costs.

A target estimate can be an extremely useful marketing tool, especially on large complex projects and projects in remote areas.

4.8. Consulting Contracts

There exist several different types of contracts that are commonly used for consulting work. Table 4.3 shows different types of consulting contracts.

An hourly rate plus expenses contract is often used. The client is billed on an hourly rate for manpower—for example, $80 per hour for a principal of the company, $50 per hour for an associate engineer, $28 per hour for a draftsman, and so forth. Then job-related costs are added in—for example, a price for computer time, a price for blueprints, and so forth. Subconsultants and outside expenses are usually set forth separately and multiplied by some factor, say 1.15, to cover office overhead. This type of contract has the advantage to the consultant of no risk; however, this type of contract also limits profits. Time is the only product for which the consultant is paid. The main disadvantage to the client is there is no control on total cost.

To get around this problem for the client, an hourly rate plus expenses with a maximum price contract is sometimes used. This contract provides budget control for the client, but obviously involves risk of overrun for the consultant. However, this may be the only type of contract the consultant can sell in a buyer's market. To limit his risk, the consultant should carefully define the scope of work. Besides the maximum price clause, the other contract terms are the same as an hourly rate plus expenses contract.

Another type of contract that is commonly used in consulting is the salary times a multiplier contract. This contract is similar to the hourly rate plus expenses contract. The difference is that with a salary times a multiplier contract, the actual wage of a person working on the project is multiplied by a burden (to cover insurance, taxes, and benefits), by a general and administrative cost (G and A) factor to cover overhead, and, finally, by a third multiplier to cover profit. Frequently, these three multipliers are combined into one, for simplicity. For example, if an engineer is being paid $15 per hour, his services to the project might be charged at $15 per hour times 2.25, or $33.75 per hour. Some prominent consultants have used multipliers as high as 3.0. They market their service as higher-quality work than the competition. Higher prices to many people do imply better work, especially with professional services. This pricing policy can be an effective sales tool.

Many design engineering consultants have found that clients are readily willing to pay a reasonable fee, for technicians and moderate-level professional staff, but resist high rates for the top-level personnel. It actually works

TABLE 4.3. Consulting Contracts

Type of Contract	Contractor's Viewpoint		Owner's Viewpoint	
	Positive	Negative	Positive	Negative
Hourly rate plus expenses	No risk	Limits profit	Easy to administer Easy to make changes	No control on total fee
Hourly rate plus expenses with guaranteed maximum	Easy to sell Can negotiate	Risk Low fee	Fixed limit on fee Low fee	
Salary times multiplier	No risk Takes care of escalation	Low fee	Easy to administer Easy to make changes	No control on total fee
Percent of construction cost	Easy to sell Can negotiate	Low fee Limits options	Easy to administer	Adversary status possible Limits alternatives which consultant can consider
Lump sum	Can maximize return on work in which he has competence	Risk of overrun Easy to administer	Fixed price	Adversary status possible High price

out on many such contracts that the profit is earned largely or wholly on these lower- and mid-level personnel.

Similarly, clients appear willing to pay overhead and profit on subcontractors.

Hence a scope of work that requires a large number of technical, subprofessional, and moderate-level professional effort, and substantial subcontractor involvement, is usually much more profitable than a scope that involves only the principals and top-level professionals.

In building construction, it is typical for the owner to engage an architect, paying him on a percentage of the construction cost, say 6.5 percent. The architect then engages a structural engineer, a mechanical engineer, an electrical engineer, and perhaps a landscape architect as subcontractors. The structural engineers fee may be around 1.5 percent, out of which he may have to pay his own foundation engineering (geotechnical) subcontractor, as well as pay all his own costs of design engineering.

Because the fee schedule is normally very tight for all parties concerned, there is seldom adequate funding for any of the parties concerned to achieve more than a bare minimum profit. Thus structural engineers of high reputation and occasionally geotechnical engineers frequently try to negotiate a contract directly with the owner rather than serve as a subcontractor to the architect.

The limited fees generally available make it impracticable for an engineer to evaluate many alternatives. Contractors submitting alternative proposals should consider this: in some cases it has turned out that their alternative has been rejected not because of lack of merit but because it required extra design engineering. One contractor, recognizing this, when submitting an alternative or value-engineering proposal, used to say that he would pay the engineer and architect, through the owner, the extra engineering costs for checking its adequacy, whether or not it was approved. He found that his percentage of approvals more than doubled.

Figure 4.3 is an example of a specialty or personal consulting contract. Many of the clauses are relevant to design engineering contracts also.

4.9. The Construction Management, or Project Management, Contract

The construction management (CM), or project management (PM), contract is being increasingly utilized as a means of providing overall management for a large or complex project. This is a professional services agreement, in which a general contractor or a specialized management organization or an individual works directly for the owner, coordinating both the design and construction activities and acting as an agent of the owner in administering the contract.

This type of contract is normally negotiated directly between the prospective construction manager and the owner. Hence many large contractors find

AGREEMENT FOR CONSULTING SERVICES

THIS AGREEMENT is entered into in _____ as of the _____ day of _____ 19____, between _____
a _____ whose principal place of business is located at _____
_____, herein referred to as Client, and _____ a civil engineer licensed to practice under the laws of the State of _____ whose address is _____ herein referred to as Consultant.

    1. Consultant agrees to perform for Client the consulting services described in the annexed Schedule A. Such services shall be performed during the period prescribed in the Schedule, and at the times and locations specified therein.

    2. For his services, Consultant shall be paid by Client the compensation provided for in Schedule B. Consultant shall submit to Client, on or before the 10th day of the month, a detailed statement of time spent, services rendered, and expenditures incurred on behalf of the Client during the preceding month. Client shall pay consultant the full amount of his monthly billing within 10 days after receipt thereof, unless Client objects to any portion of the invoice, in which case the amounts not in dispute shall be paid and the disputed portion submitted to arbitration, as set forth below.

    3. In performing services under this agreement Consultant shall operate as and have the status of an independent contractor and shall not act or be an agent or employee of the Client. Consultant shall not be entitled to Workers' Compensation or any other benefits available to employees of the Client unless the parties specifically agree to the contrary. Consultant shall make his own arrangements for insurance, including hospital and medical costs and comprehensive liability.

    4. As an independent contractor, Consultant shall be solely responsible for determining the means and methods for performing the consulting services described herein. Client agrees that it has no right to control or exercise any supervision over Consultant as to how the services will be accomplished and Consultant shall determine the amount of time, the place and the manner in which he will accomplish the services. Client will receive only the results of Consultant's performance of such services.

    5. Consultant agrees that he will perform his consulting services with that standard of care, skill and diligence normally provided by a professional in the performance of such consulting services on work similar to that hereunder. Client shall be entitled to rely on the accuracy, competence and completeness of Consultant's services hereunder but Consultant shall not be regarded as a guarantor with respect to estimates, systems, plans, or any work product performed hereunder. Consultant's professional liability hereunder to the Client, to all contractors and subcontractors on the project, and to any third parties due to Consultant's negligent acts, errors, or omissions, shall not exceed in total aggregate the amount of $50,000, and shall be limited to a period ending one year after termination of this contract.

FIGURE 4.3. Personal consulting contract.

Client agrees that if Consultant is named as a defendant in any legal action arising out of the project which is the subject of this contract, Client will hold Consultant harmless, indemnify and defend him against any claims which may be asserted against Consultant, unless such claims are attributable to proven negligence on the part of Consultant.

In the event that Consultant is requested or authorized by Client to retain or associate a third party as an expert or sub-consultant or otherwise, Consultant shall not be responsible for the acts or omissions of such third party for performance of that party's duties. No negligence on the part of the third party shall be inputed to Consultant.

6. Client shall be entitled to use any concept, product or process, patentable or otherwise, developed by Consultant in the performance of this Agreement, but such right shall be nonexclusive. Consultant shall not be restricted in the use, during or after the term of this Agreement, of such engineering and construction concepts as he may develop while working for Client, provided Client has not elected, within the terms of this agreement, to obtain patents upon them, and provided further that the use of such concepts by Consultant would not divulge information of a confidential or proprietary nature insofar as Client is concerned. If requested by Client, Consultant agrees to do all things necessary, at Client's sole cost and expense, to obtain patents or copyrights on any processes, products or writings developed or produced by Consultant in the performance of this Agreement, to the extent that the same may be patented or copyrighted and further agrees to execute such documents as may be necessary to implement and carry out the provisions of this paragraph. When the scope of the patent or copyright includes concepts developed outside of or prior to this agreement as well as concepts developed under this agreement, then Consultant shall have a non-terminable royalty-free right to use the concepts developed outside or prior to this agreement on other projects and the right to make a similar grant to others for whom the Consultant may perform consulting services. All materials prepared or developed by Consultant hereunder, including documents, calculations, maps, sketches, notes, reports, data, models and samples, shall become the property of Client when prepared, whether delivered to Client or not, and shall, together with any materials furnished Consultant by Client hereunder, be delivered to Client upon request and, in any event, upon termination of this Agreement; provided, however, that Consultant shall be entitled to make copies of all such materials, at Consultant's expense, should he so desire.

8. Consultant shall have the right to publish technical papers and reports incorporating data from the work carried out hereunder, except that he shall not report any data or information which is of a confidential or proprietary nature. He shall not use the Client's name nor identify the Client without written permission. If so requested by the Client, he shall submit the proposed report to the Client for a review and approval prior to publication, but Client

FIGURE 4.3. *(continued)*

shall not unreasonably withhold publication of scientific or technical information that will advance the state-of-the-art.

9. It is understood that Consultant has entered into, and intends to enter into, consulting agreements with third parties who may be engaged in the same general fields of activity as Client. Nothing herein contained shall be construed to prevent Consultant from accepting such consulting assignments.

10. Consultant agrees that he will not divulge to third parties, without the written consent of Client, any information obtained from or through Client in connection with the performance of this Agreement unless: (a) the information is known to Consultant prior to obtaining the same from Client; (b) the information is, at the time of disclosure to Consultant, then in the public domain; or (c) the information is obtained by Consultant from a third party who did not receive the same directly or indirectly from the Client.

11. This agreement shall be effective as of the date of its execution and shall continue until the date set forth in Schedule C, or if no date is entered, then until such time as either party notifies the other, in writing, of his desire to terminate the agreement, such termination to be effective thirty days from the date of notification. If all of the services to be performed under this agreement have not been completed within one year from the date of its execution, the parties shall thereupon and annually thereafter renegotiate the payments to be made hereunder, so as to reflect inflationary and other factors bearing upon the rate of compensation. Upon termination, Consultant shall be paid in full for all amounts due him as of the effective date of termination, including any expenditures incurred on Client's behalf, whether for the employment of third parties or otherwise.

12. In the event of any dispute with regard to the provisions of this agreement, or the services rendered pursuant hereto, or the amount of the Consultant's compensation, the matter shall be submitted to arbitration at the instance of either party, under such procedures as the party may agree upon, or, if they cannot agree, then the Construction Industry Rules of the American Arbitration Association. The laws of the State of   shall apply unless the parties agree specifically to the contrary in writing.

13. This agreement represents an entire and an integrated understanding between the Client and the Consultant and as such supersedes all prior negotiations, representations or agreements, either written or oral, between the parties. It may be amended only by an instrument in writing signed by the parties and may not be assigned by either party without the prior written consent of the other. Unless otherwise agreed by the Client in writing, the Consultant shall personally perform the services specified herein.

14. In the event that a government other than the United States of America, or a political subdivision thereof, assesses any tax on the payments made to Consultant hereunder, or withholds a portion of such payments, and if such foreign tax exceeds the applicable U.S. tax credit received by Consultant, then Client shall reimburse Consultant for the amount of such excess. For the purposes of this provision and the determination of any foreign tax liability, it is understood and agreed that Consultant's overhead costs constitute 50% of the daily compensation paid to Consultant, and 10% of the reimbursed cost for outside services and consultants.

FIGURE 4.3. *(continued)*

15. This agreement includes and incorporates by reference the annexed Schedules, including any additional provisions which may be incorporated therein.

_____
Consultant

_____
Owner

SCHEDULES

Schedule A—Consulting Services

Schedule B—Compensation

        per day, including travel days (Portions of one day to be charged in direct proportion).

Reimbursement at cost for all travel expenses, including board and lodging and necessary out-of-pocket expenses, long distance telephone calls, telexes, cables, etc., and reproduction services.
All air travel shall be first class.
Reimbursement at 1.10 times actual cost for outside services and consultants.

Schedule C—Termination of Agreement—(See Clause 11)

Schedule D—Additional Provisions

August 1978

FIGURE 4.3. (*continued*)

that they can obtain such management contracts at satisfactory fees yet involving low risk. This management contractor can emphasize the firm's experience and administrative capabilities.

A typical organization responsibility chart for a project is shown schematically in Figure 4.4. Under such an organization plan, the CM may obtain competitive bids for each of the construction contracts below, thus assuring the advantages of competition in obtaining low prices. The construction manager may under some circumstances perform one or more of the construction contracts.

"Construction management" is an example of how an added capability may be utilized effectively by contractors in marketing both to obtain contracts and to increase the scope of your services. Many large construction contractors have found that the key aspect of their services is the application of management skills. As a result, some building contractors do almost no actual site work but rather assemble, manage, schedule, and control subcontractors and suppliers so as to ensure, insofar as possible, the satisfactory completion of the project, on schedule, on budget. The construction manager or project manager, as he is alternately called, can be likened to the orchestra leader, who, although he plays no instrument, is the critical figure in properly integrating the work of the individual musicians. Another metaphor could be the coach and his staff directing an athletic team.

Similarly, large architectural and engineering firms have found that their service to the owner could be greatly enhanced if they were to act for the owner in a construction management capacity. Thus a number of such firms have established separate divisions or subsidiaries to perform the CM function, as an added service, for an added fee.

For the general contractor, he may feel that this is the function which he has always been providing, along with actual construction operations as well. However, as projects have grown larger and more complex, the technique and methodology have also become more sophisticated and the requirements for management increasingly demanding.

Many general contractors find that they can obtain significantly greater contract volume and reduce their risks by performing as a CM rather than as a traditional general contractor. For example, on international work, the CM can often negotiate a fee contract and then contract out (as agent of the owner) the actual construction work to local national contractors. He then bears no risk of cost overruns due to delays, political constraints, or escalation.

Some large general contractors combine construction management and design engineering as one package.

What services does a CM offer?

The essential ones most often included in the scope of his services are:

1. Budget control
2. Schedule control

4.9. The Construction, or Project Management, Contract

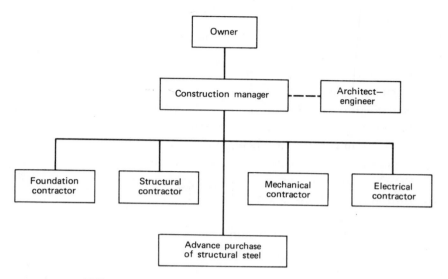

FIGURE 4.4. Organizational responsibility chart.

3. Phased contracts ("fast track")
4. Value engineering
5. Advance procurement
6. Management of construction

Budget control and schedule control apply to the entire project, including the design phase, contract phase, construction phase, and sometimes the starting phase. Schedule control, of course, includes critical path scheduling. Phased contracts (fast track) and advance procurement enable portions of the work to be placed under construction while design is being completed on other later phases. Value engineering embodies the consideration of alternative materials and solutions, preparation of detailed comparative estimates, and evaluation of functions and utility. Value engineering may also include risk analysis of different alternatives and preparation of contingency plans. The management of construction includes the organization of all contractors and subcontractors, their integration, and the solution of interface conflicts and omissions. It may include provision of general services, such as vertical transport and site preparation.

Other functions can be added as they appear appropriate to the situation. These include:

7. Feasibility studies
8. Preliminary estimates and budget estimates
9. Labor relations

10. Project financing
11. Governmental permits, licenses, and so forth
12. Site procurement
13. Environmental impact studies
14. Safety
15. Settlement of contract disputes, claims settlement
16. Processing of interim and final payments
17. Training of construction personnel
18. Management assistance to local and minority contractors
19. Warehousing and storage
20. Insurance (wrap-up) and bonding
21. Start-up of operations
22. Training of operating personnel
23. Preparation of operating manuals

Feasibility studies and preliminary estimates are usually carried out under a separate, earlier contract. However, they often lead into a subsequent CM contract.

Training of operating personnel is often attractive for complex industrial facilities and for projects in developing countries.

The profession of CM is so new that its scope, legal position, and professional responsibilities and liabilities are still being established. Hence it is very important that the contract for such services by very carefully drawn, delineating in considerable detail the services to be provided; the authority, responsibility, and liability of the various parties; the relationship with other parties; the definition of reimbursable costs; and the fee and methods of adjusting the fee.

Many firms, both large and small, local and regional, domestic and international, are finding that the redefinition of the scope of their services from that of general contractor to that of construction manager has increased their profitability while reducing their risks. These firms have arrived at the CM role by correctly asking and answering the question, "What is my business?"

One cannot complete this section without noting the abuse of the CM role in some areas, that is, high fees, inadequate capability, large cost overruns, and excessive bureaucracy.

The responsible CM will keep accurate, current cost controls so that the owner will always be up to date as to where he stands. Communication is one of his most important functions. Experience shows that CM contracts turn out most successfully for the contractor when he sets up regular monthly meetings with the principals of his and the client's company. This is in addition to the more frequent staff meetings that are commonly held. Communications must result in understanding, not just "snowing" the client with data and computer printouts.

CM and PM contracts are not without their difficulties, however. They require the commitment by the contractor of top-management personnel, men and women who often can earn much more profit for the company on standard construction contracts. This drain on scarce management resources can prevent a company from realizing its full income potential.

Liability remains an uncertain risk. To what extent is a CM liable if the final cost overruns the initial estimate? Can he be charged with negligence or improper management? Is this a potential for a professional liability suit? Can he obtain adequate insurance coverage? In the case of architects who also perform construction management, does this liability for meeting schedule and budget now extend back into their design fee? In the case of accidents, what is the liability?

From the owner's point of view, is CM worth the extra cost? Is the CM duplicating services which would otherwise be provided by the contractors directly? Is the bureaucracy increasing efficiency or needlessly impeding performance?

These problems have led in some recent cases to some contractors successfully offering a counterservice, namely, one of "complete service, with all work to be done by our own forces," a return to the traditional general contract.

4.10. "Turnkey" Contracts (Design and Construct)

Under this type of contract, the same contractor (who may adopt the name "constructor") is responsible for both design and construction. This has the advantage from the owner's point of view that it assigns total responsibility to one organization. It may expedite the work, permitting construction to start before the design is completed.

From the contractor's point of view, this type of contract is usually (but not always) negotiated; hence he can emphasize his experience and capabilities. It allows him to implement new techniques and designs, taking full advantage of new developments and innovation. It enables the construction personnel to influence the design details so as to assure practicality.

Usually, the total cost to the owner cannot be determined until he is already fully committed. Thus the contractual arrangements for payment are usually on some type of cost plus fee arrangement: target estimate, cost plus fixed fee, or cost plus fee converted to lump sum.

For any of these, the contract may contain a guaranteed maximum price.

In smaller or specialized work, where the requirements are readily determined ahead of time, a turnkey contract may be performed on a fixed-price basis, either lump sum or unit price.

One disadvantage to the contractor is in regard to professional liability. If a problem occurs, there is usually no one else to share the blame. Further, the amount of liability in such a case may extend to the total contract price. Professional liability insurance on a turnkey project may be unobtainable.

4.11. Subcontracts

The various subcontracts on a project are usually let on a fixed-price basis, that is, lump sum or unit price. There are occasions however, such as when the owner has nominated a particular subcontractor, that one of the cost plus fee type of contracts is employed.

Because, by definition, the subcontractor is responsible for only one segment of the work, it is extremely important to define the scope of his subcontract, to clearly enumerate the services that will be provided by others and to address the interfaces with other subcontractors and with the general contractor. The integration of the subcontractor's schedule of work, with that of the other participants on the project, requires special care.

Many supply contracts, such as those for furnishing of fabricated structural steel or precast panels or elevators, are purposefully written as subcontracts, making the supplier responsible for the installation. In such a case, the value of installation services may be only 10 percent of the value of the total subcontract.

The purposes of including installation are several: to ensure single responsibility for fit; to avoid sales tax on the purchase of the fabricated material; and to utilize the specialized expertise that the supplier has developed.

Conversely, many purchase orders or purchase contracts for the supply of material to the building site may require some activity of the supplier at the site, even if it be only the furnishing of a technical representative or the repair of minor defects. Many general contractors have found it prudent to write such a "supply contract" or purchase order on a subcontract form, or at least include many of the clauses of a typical subcontract form, in order to cover the matters of responsibility and liability in case of an accident on the site.

4.12. Packaging a Contract

If you were marketing, for example, perfume, its packaging would be of utmost importance. This same principle can be true for contracts. Consider two examples of this point.

Consider first a negotiated cost plus percentage contract. The contractor has determined that he wants a 10 percent fee to perform the work. The contractor could offer the terms:

   cost plus a 10 percent fee

The client may judge this to be arbitrary and exorbitant. However, the contractor could offer the same terms, with a little salesmanship and packaging, as:

3.5 percent administrative fee
Plus 5 percent markup
Plus 1.5 percent financing

The same client may then feel he is getting the work done at a fair price.

Smart contractors, architects, and engineers develop and print their own contracts. You should always use a printed contract as opposed to a typewritten contract! A printed contract can have a tremendous psychological advantage. Given a typewritten contract, the other party's lawyers have a tendency to go through the contract document line by line, making all sorts of changes that are not to your side's advantage. On the other hand, given a printed contract, the other side and even their lawyers get the subjective feeling that the contract is standard, and they are far less likely to make or suggest changes. What should you do if the contract is drafted the night before it is to be signed? Printing can be done on a rush basis by paying the extra or overtime costs.

4.13. Risk Sharing

By nature of the business, construction contains risks. One way to be more competitive is to find some way of sharing the risk with another party—for example, the client. Consider the following methods by which the risk in an undertaking can be shared:

1. Add money for a contingency. From the contractor's experience, he may know the probability of the risk and can self-insure against it. This is the traditional method, but, of course, it makes you less competitive.
2. Insure against the risk. There are many types of physical and contractual risks that can be insured. This option is often overlooked. For example, insurance for a 10- or 100-year flood is usually available for a bridge construction contract. The cost of the premium can then be included as a cost to the project. Alternatively, require the owner to pay the premium directly or to procure the insurance and name you as an additional insured.
3. Use a contractual clause to return some or all of the risk to the owner. To "sell" this idea, it is often effective to find a "standard" contractual clause used by some government agency or large corporation and sug-

---

The sign every business concern looks for most keenly is the one that appears on the dotted line.

Forbes Epigrams

TABLE 4.4. Risk Sharing

Type of Risk	Possible Counteraction
A change in a quantity of work required	Unit prices for both overruns and underruns
A change in the kind of work required	A changed conditions clause
A change in conditions under which the work may be performed	1. Unit prices for different conditions 2. A changed conditions clause 3. Weather standby payment 4. A delay clause—extension of time for causes beyond contractor's control; extension of time and standby costs for delays due to owner's actions
Low productivity	1. Planning work and setting objectives 2. Keeping crew size constant 3. Personal involvement
Strikes	1. A delay clause—extension of time for strikes 2. A no-strike job agreement
Escalation	1. Escalation clause 2. Estimate of additional cost due to escalation 3. Firm prices from material suppliers and subcontractors or limits on the percentage by which they may raise prices 4. Bid prices tied to indices 5. Critical items purchased in advance.
Wildcats and slowdown	1. Planning work and setting objectives 2. Personal involvement 3. Prejob meetings with unions
Physical damage to structures	1. Insurance 2. Contingency 3. Protection
Loss of the use of vital equipment	1. Direct insurance 2. Indirect overinsurance for the delay, about 10 to 20 percent over the equipment cost usually permitted
Subcontractor's failure to perform	1. Bond 2. Proper contract terms, permitting replacement of subcontractor in event of failure to perform 3. Clearly defined contracts
Subcontractor's failure to pay bills	1. Bond 2. Proper contract terms 3. Requiring certification that these suppliers have been paid before making subsequent payments to supplier
Delays in payment or nonpayment by owner	1. A provision for advances by client 2. Irrevocable letters of credit 3. A.I.D. insurance

TABLE 4.4. (*continued*)

Type of Risk	Possible Counteraction
	4. Contract terms for payment for all but the disputed amount in the case of disputes
	5. Arbitration clause
	6. Owner required to put money in escrow while settling
Environmental damage	1. Contract clause passing responsibility for certain events to client
	2. Insurance
War and riot	1. A.I.D. and O.P.I.C. insurance
	2. Contract clauses
	3. Payment for force majeure delay costs
Accidents to workers, public, or subcontractors	1. Insurance
	2. Contingency plan
	3. Safety program
Materials delays	1. Contract clauses
	2. Purchase key materials in advance (by owner or contractor)

gest that clause. For example, a "changed conditions" clause or a "flood damage" clause, such as those used by the U.S. Corps of Engineers, may be adopted.

4. Subcontract the risk item for a fixed price. Some specialty subcontractors know the risk and how to handle the risk better than the general contractor.

In Table 4.4, a number of potential risks are listed, together with possible methods for reducing that risk to the contractor. Obviously, not all these risks are applicable to any one specific project. There may be other possible solutions. The intent of the table is to stimulate constructive approaches to risks, as opposed to the alternatives of not submitting a bid or just adding more money. This latter is really a form of self-insurance and, as such, generally violates the rules of insurance; namely, that you obtain a large enough volume of similar risks to develop a statistical distribution of events. The contractor, bidding on one unique job after another, is never able to develop such a statistical experience except in the crudest form.

5
Meeting Competitive Pressures

5.1. Meeting Competitive Pressures

In order to meet competitive pressure, you have to fight hard. Fortunately, the pressure is on the competition too. However, do not fight so hard that you are not fighting fairly. You should play the construction game according to rules; you do not want to make enemies. Today's competitor may become tomorrow's joint-venture partner or client. For example, often a competitor will grow and become successful in another area. Or the competitor may merge with a conglomerate. You will then be offering your services as a subcontractor to this competitor. Then you will want him to respect you and your abilities, rather than treat you as an enemy. Experience shows that you can still win while fighting fairly.

One of the temptations in a competitive business is to tell of a competitor's mistakes and shortcomings, or, to use a colloquial expression, to "badmouth" him. This usually has the opposite result from that intended: the listener starts to defend the competitor and to argue with you.

A much used and, from a marketing point of view, more successful ploy is to damn him with faint praise: "ABC is a fine young company—it's been doing some excellent work on the smaller jobs"; "Out-of-Town Construction certainly does fine work in its home territory"; or "George was one of the most competent contractors around. It's a shame he's never found some good younger engineers to carry on the tradition."

Best of all is the honest answer, the finding of something sincerely good to say in a positive way. If you can afford to give deserved praise to your competitor, you must be confident of your own ability. You raise your stature in the listener's eyes.

If you have a better service, you can win. If you treat each job as your most important job and find a dynamic noncompetitive solution, you can win. In

5.1. Meeting Competitive Pressures 75

Chapter 2, many cases were presented in which a contractor successfully found an alternative solution by varying the scope of work, contractual approach, or geographical location. You can meet competitive pressures by finding your most favorable alternatives.

One way to meet competitive pressure is to accept a risk that the competition will not. A major subcontract on the Arkansas River Lock and Dam Project was obviously hindered by an access problem created by an old and inadequate highway bridge across a creek. All the subcontract bidders, but one, qualified their bids, requiring that the bridge be replaced by the general contractor. Getting the bridge replaced would cause the general contractor a lot of problems and headaches. The state would probably demand that the bridge be replaced with a permanent concrete bridge costing over $500,000. The one subcontractor determined that the creek usually ran dry eight months a year. Therefore, by constructing a low-level access road across the dry creek bed, 90 percent of the equipment and material could be brought in by heavy trucks during the dry season. He was taking a calculated risk that unusual weather would not flood out his access. The general contractor accepted his bid because it was the only one not qualified. Thus by accepting a calculated risk, he got a fine subcontract, one that turned out profitably for all concerned.

Certain contracts have an obvious potential claim built in. By recognizing this, some contractors will bid low and then use the claim to negotiate to make a profit. One example of this occurred on a project in Arizona. A bidder learned that the presence of a large number of Indian archeological relics on the site might require the project to be moved to another site. He bid low and then used the changed site situation to negotiate a more favorable price. This type of tactic can be playing with fire, for if the owner (the government in this case) determines that the costs of relocation will be too high, it may decide not to relocate.

Many years, ago, a well-known West Coast contractor made large profits on Corps of Engineer dam jobs through clever use of claims for changed subsurface conditions. It would bid low, get the job, and then wait for changed foundation conditions to be uncovered as excavation proceeded. There used to be a joke about this practice: "The first man on the job was not the project manager—it was the lawyer."

Neither of the two companies above is still in business. They could not make it when the going got tough.

The construction industry is always changing, and solutions to new marketing problems are always needed. A problem that has recently reemerged is inflation. What is needed today are creative approaches to handle this problem.

The problem of the effect of inflation on material prices, for example, is sometimes correctly handled by the early purchase of key materials. Consider a 550-day building contract. The prices for some of the finish items, such as acoustical tile or resilient flooring, can greatly increase over the job period. Therefore, many subcontractors will purchase the material at the time of award to hedge against inflation. Some contracts allow the subcontractor to be paid for the materials when properly stored on the site. With other contracts, the subcontractor or contractor must pay for the warehousing and financing of the advanced purchase. Even in this case, this practice can pay when inflation is at a high rate.

On a major bridge project, the contractor bought and stored all the prestressing tendons upon award of the contract. The primary reason for this action was to hedge against inflation. Another reason was to protect himself against shortages caused by a renewed or increased demand from the energy industry. He was developing a solution to potential problems of both inflation and material shortages.

In 1928, a contractor we will call Witt Construction received a contract from a major department store to build a large store building in downtown San Francisco. At the time the contract was let, there was a boom economy and inflation was rampant. Labor and materials were very high at the time. Therefore, the contract went for a high price. Shortly after this contract was let, the stock market crashed, and the Great Depression began. Thus Witt's construction costs were well below those which had been estimated. After the job was complete—although Witt Construction had no legal or business responsibility to do so—it returned a portion of the excess profit to the owner. Witt did not want to take advantage of a situation over which neither it nor the owner had any control.

It is not surprising that to this day, more than 50 years later, this department store (now a chain of stores) always goes to Witt Construction to perform all its construction work on a negotiated basis. An ethical, honest contractor can win in the long run.

5.2. Use of Allies

The use of allies can give you a competitive edge. Material suppliers, fabricators, and subcontractors can become your allies. Generally, the people who are or who work for material suppliers, fabricators, and subcontractors are neglected and sometimes even taken advantage of by general contractors. These suppliers should be treated with respect for the vital service they are providing. Clients are accustomed to entertainment and flattery by general contractors; suppliers and subcontractors are accustomed to having the general contractors take out their frustration on them.

As a result of this pattern, suppliers and subcontractors really respond when they are treated as important contributors. If you have a good relationship with suppliers and subcontractors, they will often come to you with suggested alternatives and their good ideas. Most important, they can often give you good leads on future work. For example, a process equipment supplier may hear about a new process plant first. Perhaps the owner needs to find out how that new centrifuge works or what delivery schedule is possible. Likewise, a soils engineer may hear about new foundation work first. Courtesy, a thank-you letter, and respect can win many important allies. Fair and equitable treatment of suppliers and subcontractors is the most effective way of doing business.

There is one bad ally—the person who wants a finder's fee. Do not pay attention to people who want finder's fees. They may walk into your office and say that they can lead you to a big job controlled by their close friend for a 5 to 10 percent fee. Usually it develops that they have just heard that there might be a job, and they are talking to all your competitors, too, and hoping to get a fee from someone. These leads are generally useless and worthless. This "ally of opportunity" is particularly prevalent in the Mideast.

The contractor can often establish a special relationship with a subcontractor or supplier—the "tied-in sub." The subcontractor, working closely with the general contractor, may be able to come up with a low subcontract bid price, which gives the general contractor a competitive edge. The contractor agrees that he will take a bid only from the tied-in sub, and the tied-in sub will give a bid only to the contractor to whom he is tied. This is normally a relationship that is based on past relationships and respect. However, both must play the game with the intention to win. If the tied-in sub ever gives the general contractor a high bid, this may be the end of a good relationship.

Another way of meeting competitive pressures is through perseverance and imaginative adaptation to a dynamic situation. A West Coast contractor was particularly interested in a large foundation and underpinning subcontract in conjunction with the MARTA project in downtown Atlanta. It negotiated an agreement whereby the West Coast contractor went as an exclusive subcontractor with a large Southeastern contractor. It spent a large amount of time and effort working closely with the Southeastern contractor to prepare an engineering and construction plan. Everything worked out fine except that when the bids were read, the Southeastern contractor was the third bidder.

Instead of giving up, the West Coast contractor asked the Southeastern contractor for permission to contact the low bidder. Since the possibility of the first two bidders dropping out was low, the Southeastern contractor gave permission. As it happened, the low bidder was not committed to a foundation subcontractor. It had planned to do the work themselves; but it was certainly interested in a proposal from the West Coast contractor. The low bidder received the West Coast contractor's proposal but felt that it was much too high. However, both parties negotiated further. By reducing the scope of work to about three-fourths of the original subcontract, the West Coast contractor finally received the foundation contract at a price that could yield an adequate profit. So it pays not to give up too soon on a job that appears to be lost.

5.3. Brain Picking

Many contractors and consultants do not like to give free advice. They feel that the potential client is just picking their brains and that they do not have the time or the money to provide this free service. The ethical potential client may even state that he is unable to give any competitive edge in the bidding in return for free service. Some clients do not like to get free advice. One estimator used to say, "Free advice is worth what you pay for it." This estimator might get some free advice now and then, but he would really scrutinize that advice to make sure that it was worthwhile.

On the positive side, the giving of free advice can be turned into a source of intelligence and a business development tool. There are several reasons that the policy of giving free advice can pay off. First, by giving free advice, the contractor or consultant gets to learn about the project before his competitors and, hence, may be able to develop a rationale as to why the client might be advised to switch to a negotiated contract. Second, if the contract is going to be competitively bid, the contractor or consultant may, by giving free advice, be able to ensure that his preferred alternatives are included. Giving free advice can be one way to get your foot in the door early, in the design process.

A San Francisco construction engineer had been giving substantial free advice to a mining company for an offshore loading terminal structure in a remote area in Tasmania. The owner was a consortium of international companies which had made a policy decision for competitive bids on an international basis. The construction engineer's boss at this company decided that they were not going to submit a fixed-price competitive bid on the project because of its many variables, risks, and contingencies, and they only wanted such a distant and difficult job on a negotiated cost plus fee basis. However, the engineer was technically interested in the project and continued to give free technical advice on construction methods to the mining company, showing it how the several phases of work could be efficiently integrated. When bid time came, the mining company invited the construction engineer's company, along with other international contractors, to submit a lump-sum bid. His company regretfully declined, stating that it felt a negotiated cost plus fee contract would be a more appropriate solution and that it felt that a lump-sum bid would inevitably be highly qualified, so that it would become only a basis for negotiation. The construction engineer presumed that the job was lost as far as his company was concerned. Two months later, the mining company called the contractor to come to Australia to negotiate the contract. The mining company found that the lump-sum bids it had received were all so qualified as not to be really "fixed price" at all. It was able to convince its partners that the only honest and valid response had been that of the construction engineer's company. So it negotiated the contract, on a cost plus fee basis, because of the goodwill that had been established earlier through the free technical advice and because the mining company recognized that this company had been sincere and honest in its evaluation of the project requirements and difficulties.

One problem that sometimes develops during the early stages of a project is that the client will obtain detailed drawings from you as part of the "free advice" and then take these drawings to your competition and sometimes even go out to bid with them. This is perhaps due to lack of ethics on the part of the owner, or it may be that the individual to whom you have been talking has been overruled by his superior. A contractor can minimize this possibility by not going into too much detail on drawings given to a client. The use of a proprietary note on the drawings may help but the authors have even seen one case where the unethical owner, in photocopying the drawings, blanked out the proprietary note! While your free engineering may sometimes be unethically exploited, nevertheless experience shows that it is best to play for a win and give the best information you can, even if occasionally you are treated unfairly.

After a proposal for a package of work has been submitted, and if this is being favorably considered, the owner is very likely to want to reassure himself about the contractor's ability to perform. He will probably again ask about the construction plan, size, location of equipment, production rates, crew sizes, and schedule: the owner is entitled to reassure himself about the contractor's ability to perform.

The smart contractor will answer the owner by such means as citing recent track records on similar jobs and inviting him to check with satisfied clients. The contractor should beware, however, of giving out too much specific technical data—particularly, if he does not know the client too well or if the client has his own engineering staff. Some clients will use the specific technical data to the contractor's disadvantage. For example, the client could give the detailed technical data to a competitive contractor with whom he is particularly friendly, in order to receive a lower-priced proposal. In a few cases, the client decides that he now knows enough about the project to do it himself.

About two years ago, a West Coast foundation contractor prequalified and, in due course, was invited to submit a proposal for a large project for a utility company in the Midwest. The project was to build a slurry wall for a nuclear power plant. The contractor prepared a detailed construction plan and a detailed estimate of cost. Within a few weeks, it was advised that its proposal was strongly being considered and the owner's engineers would like a meeting to discuss the details of the proposal. Since the bid price was eight figures and it needed the work, the contractor readily supplied all the needed information to close the deal. At the end of this meeting, however, it was unable to close the deal. The same thing happened in two subsequent meetings when the contractor supplied additional technical information.

About six months later, the contractor read in *Engineering News Record* that the slurry wall contract had been negotiated with a local contractor. One of the contractor's engineers later had the opportunity to visit the job site and learned that the local contractor had used their construction plan that had been presented to the owners, down to the last detail, from the exact equipment recommended to the same slurry mix.

Recently, this same West Coast foundation contractor received an invitation from a major engineer-constructor, acting as construction manager to bid on the piling for the foundation for a new high-rise office building in San Francisco. The soils engineers engaged by the CM conservatively recommended 16-in. square prestressed piles (to support a design load of 100 tons per pile. Since the foundation contractor had driven a number of piles for foundations in the immediate vicinity, and was familiar with the soils and the local code requirements, it concluded that 12-in. square prestressed piles would be adequate. These met the building code requirements. Therefore, it submitted a bid on an alternative pile, stating that it guaranteed its piles to meet the design load criteria as well as the building code. Almost immediately, the contractor's representatives were summoned to the CM's office and asked to explain why they were the only contractor who did not bid on the plans and specifications. They concluded from the tone of the conversation that they must have been quite a bit lower than their competitors or they would not have been asked to this meeting, which was attended by some of the CM's top executives. During a lively and heated conference, they refused to reveal the nature of the alternative pile until they received a letter of intent. Finally, they agreed that if the alternative pile failed under the specified pile-load test, the foundation contractor would install the originally specified 16-in. piles at no change in price. Incidentally, the 12-in. square piles stood up to the test with far less than the allowable settlement.

The two examples above, with opposite outcomes, are directed at the stage when the contractor has already submitted a proposal. The conclusion that can be drawn is that one has to give enough details and assurance to land the job, without giving away all the innovative ideas that have been generated. It is also important to evaluate your position carefully, so as to determine whether or not you have a substantial edge or are merely fighting to hold on against a close or even lower competitor.

When a major airline was planning its new hangar facilities at San Francisco Airport, advance word was picked up by Hammer Construction through a telephone call from an engineer in New York asking about their experience in driving long steel H piles. Hammer Construction soon realized that this was a very large job and that there was a major economic advantage to using prestressed concrete piles in lieu of steel H piles at this particular site. So personnel flew to New York, presented data on prestressed piles to the engineer, and persuaded him to change the design. The airline, following its standard bidding procedure, instead of negotiating the job with Hammer Construction, invited bids on prestressed piles from several contractors. Unfortunately, Hammer was second bidder and a competitor got the job. So Hammer's brilliant concept and trip saved the owner several hundred thousand dollars but did nothing for Hammer.

Another hangar was subsequently advertised at the same airport, and this time Hammer Construction was smarter. When the owner requested bids on steel H piling, Hammer bid a lump sum on an "alternate pile, guaranteed to meet the design criteria, and the city code." Of course, Hammer had to know the site well and be sure of its facts. This time Hammer did succeed in landing the job and made an excellent profit.

5.4. Monitoring Your Market

You need to keep in touch with what is happening in the market. You need to monitor new bidding patterns as they emerge and new relationships between subcontractors or material suppliers and general contractors. In a well-run office, these trends are not just kept in somebody's mind; there is systemized record keeping. You also need to monitor the prospective programs of potential clients and contracts to be bid. In marketing, it is essential to find out about new work as early as possible. Sources of long-lead information on new work have been discussed. You need to establish some sort of record-keeping system to take advantage of this information. Some contractors will use a simple card file system that is updated monthly. It is easy to find out about a possible contract several months away and then forget about it. You should keep records and write informal memos on your sales calls and meeting with clients. You or your associates may need records of those meetings in the future and, especially, of the names and positions of key personnel whom you met or contacted.

5.5. How to Fight Unethical Competition

There is often some cheating in the construction game, as in other games. The legitimate contractor should be aware of unethical or dishonest practices: while he may not indulge in such practices himself, he must take steps to protect himself.

Consider the following case which happened many years ago. It is bid day for a government dam job, and the bids are opened and listed below:

$14,240,000	Morrison-Knudsen
$13,980,000	Guy F. Atkinson
$13,972,000	Peter Kiewit Sons' Co.
$12,400,000	Brand X

Brand X is awarded the job; it bid at cost, with no profit. The next day Brand X goes back to the government engineers saying that it left a page out of its bid. It presents the page as proof. The page numbers agree with the claim.

The page that was left out just happened to be for $1,452,000. Brand X is still low bidder. The government "corrects" Brand X's bid, and then awards it the contract. The other three contractors do not even find out about this for several weeks. They complain to the government but to no avail: the government claims it has the legal right to waive informalities and to correct "obvious" mistakes. But Brand X was so enamored by its clever scheme to get jobs that it tried the same tactic again. The other contractors were more alert this tme; they filed a protest the day of the bid. Ultimately, they forced the government to change the rules so that now a bid with a mistake can only be withdrawn—not corrected. That was in the 1950s, but in the 1960s a large building contractor used this same "trick" on several private multimillion dollar hospital jobs in a row. In this case, its ruse was to "leave out" one subcontractor's bid. It took four jobs in succession for the competition to wake up to what was happening to them.

Unethical practices by the competition in the construction business are often obvious and are particularly disturbing to the young engineer entering the construction business. However, there are ways to fight unethical procedures by your competitors.

Consider a case where a material supplier is leaking information to your competition. You call up Brand X Readymix Concrete and ask for a bid on 9654 cu yd of concrete for a contract. This is the exact figure plus 2 percent for wastage. Brand X Readymix gives you a bid and records 9654 cu yd. A week later your competitor telephones Brand X and says it has just decided to bid on the contract. Brand X Readymix gives them your figure of 9654 cu yd. You have just performed a free takeoff service for your competition! To avoid this problem, change the number of cubic yards of concrete you give to Brand X Readymix a little. You could ask for a bid on 11,000 cu yd of concrete, but this is obviously a rounded off number. Ask for 10,642 cubic yards of concrete, so your competition may use this higher figure or, at the least, may be worried about its own takeoff. If you suspect Brand X is intentionally feeding your takeoffs to the competition and if this job is one in which your firm is not interested, ask for a bid on 8462 cu yd of concrete.

"Loose lips lose jobs." It is natural; everyone likes to talk about his job—especially when he is doing something innovative. Even hints about what you are doing can give away your job to the competitor. Remember that the competitor is just as smart as you are. You do not need to lay out the full and detailed construction plan in order for your competition to recognize what you are working on. For example, complain about working weekends, and the competition will immediately know you are working on one of two big jobs in a limited market. Then the competition goes to a material supplier and finds out that you wanted a bid on 21,000 cu yd of concrete. That is enough; the competition now knows which job you are working on. Leaks are everywhere in construction—subcontractors, suppliers, and maybe your secretary. One San Francisco construction company had a young secretary who, as it was later found out, was giving a copy of every letter to one of its competitors. The competition had an employee whom the secretary was dating. And how did they find this out? The employee recipient bragged about it at a cocktail party to one of the joint-venture partners of the construction company in question.

Estimators often talk too much. They are members of a highly specialized and demanding profession. They know their counterparts in their competitors' firms, and some may even belong to a local professional cost engineers' society. Sometimes, instead of talking about how they estimate, they let slip information about what they are estimating. Also, they can sometimes be baited into responding to statements by other estimators, such as, "This job is too tough" or "We are not going to bid on that." The novice will answer, "Well, we're not afraid of that job" or "You're right; we are not bidding on that, either. The experienced estimator says, "What job?" Loose lips lose jobs!

Most leaks are unintentional—the leaker would not intentionally hurt his company. Therefore, you need to train your staff to keep their mouths shut and act dumb. Keeping your mouth shut can be hard when you are talking to a skillful prodder.

Junior engineers, because they do not know better, and senior engineers, because they like to display their knowledge, often talk too much and too freely at prebid conferences and again at bid openings. It is important to remember not to ask questions that will help the competition unless doing so is in your best interest. A request for clarification of a minor point may tell a clever competitor which approach you are taking or reveal an alternative scheme you are developing. While you may wish to have the point clarified, you should evaluate the risk involved and find some way to disguise the purpose behind the question. One way to cover up your intent is to prepare and ask several questions, only one of which is relevant to your bidding approach.

Most contractors go to prebid conferences to listen to others' questions and the answers. They are aware of the fact that when competitors are working night and day on the same subject, a relatively small clue may be all they need in order to divine their intentions.

One day, the president of Hungry Construction was called into a client's office. The client informed him that he would like the contractor to perform some work recently bid on—a large pile-driving job. The client really liked the approach, schedule, and equipment that were proposed; however, the client sadly stated, "I do not know if we can give you this job; your price is too high." Just as the client was asking if there was anything the contractor could do to lower the price, the client's secretary came in and told him that he had a long-distance telephone call. The client excused himself by saying he would have to take the telephone call in another office. Now, sitting on the desk, upside down to Hungry's president, was a paper with the following:

$$\begin{align} \text{Hungry} &— 408{,}700 \\ \text{Raymond} &— 392{,}000 \\ \text{Healy} &— 387{,}000 \end{align}$$

The contractor looked at this paper. His bid of $408,700 was correct. He wondered how his competitors could bid so low. The client came back to his office and continued the negotiations on price. At first, the contractor held firm and then finally reduced his price to $390,000. The client responded, "You are getting close. Keep trying." Finally, the contractor agreed to do the job for $385,000 (this was now at his cost, with no profit). Later he wondered why he had agreed to take the job at such a low price.

A few months later the three competitors happened to get together for a joint labor negotiation meeting. During lunch, the subject of the particular piling job came up and Hungry's representative asked the others why they had bid so low. Healy's representative truthfully informed the table, "I did not bid that; our bid was $420,000." Raymond said, "$392,000! I was $440,000!" Hungry Construction had actually been low bidder all along! But Hungry could not complain; he had stolen a look at the paper on the other man's desk. He had done something unethical. Later, it turned out that the client had been using the same price-chiseling practice on all the subcontractors.

This type of price chiseling is not that rare. One of the authors was asked to fly to Western Australia to review a bid proposal his company had made for a major corporation. He was asked to work out the price modification for a number of small changes and was given a private office to work in. On the table in the office lay a proposal with the name of his principal competitor. Although severely tempted, he did not look at the file for three reasons. First, if he took advantage of the file, his price would most likely be too low. Second, the owner was most likely baiting him. Third, it would be unethical. So he immediately called in the owner's engineer and handed him the competitor's proposal. A year later he mentioned this to this competitor; and his competitor informed him that when he went to Western Australia, a copy of the author's proposal was sitting on the same table!

In the construction business, unethical practices may pay off on the short-term basis, but actual experience, as well as moral philosophy, shows that they never succeed on the long-term basis. The ethical contractor is the one who can stay in business over the years. As previously discussed, some general contractors are unethical in their dealings with subcontractors. For example, the general contractor will not take the low subcontractor's bid but instead will negotiate a subcontract price with a preselected "favorite" subcontractor. In this case, the general contractor will not tell the other subcontractor what he is doing. He is, in fact, just using the other subcontractors to get a check figure for his negotiations. While this may work in the short term, eventually the other subcontractors will figure out what is happening. They certainly will not want to provide a free cost estimate for the general contractor when they have no chance of getting the work. Eventually, they will blackball the general contractor and no longer bid to him or, more effectively, will submit high bids. To get around legal restrictions which may exist, requiring that bid prices be the same to all bidders, they make minor modifications in the scope or terms so as to justify the higher bid. Unethical practices just do not work in the long term.

5.6. "Opportunism"

You should look for opportunities and develop imaginative approaches to the client's unusual needs. For example, disasters tend to present good opportunities for marketing. As an example, Pacific Oil's old wooden pier caught fire and burned on a Saturday afternoon. As soon as a certain West Coast contractor heard this, it ordered a barge to be loaded with piles and to set sail, along with a derrick barge, upon San Francisco Bay. The contractor's vice president then called Pacific Oil and told them they had a rig in north San Francisco Bay and wanted to know if "we could do anything to help." (Without a pier to unload oil tankers, the whole refinery would shut down.) Pacific Oil asked them to come right over, and the contractor was driving dolphin piles before the fire was completely out. Oil tankers were tying up to the temporary pier within two days. There is a difference between "ambulance chasing" and recognizing a legitimate opportunity to help. Two months later Pacific Oil asked the contractor to negotiate a contract to construct a new multi-million dollar pier and later awarded them the contract on a cost plus fee with guaranteed maximum price basis. The contractor had gained a valuable client.

The identification of a collusive market presents an opportunity. A company we will call Bell Construction won an insurance building contract in Houston, Texas, by identifying a collusive market. Bell's costs to perform the work were the same as the competition's, but it found a competitive edge. The landscaping cost for similar buildings was running about $300,000. There were three landscaping contractors, and their prices were always about the same. All three landscaping subs seemed very prosperous. Bell had never done any landscaping, but this time it figured its own landscaping bid, which was only $55,000. To be safe, it doubled the figure and used $110,000 in its estimate. This gave it a $200,000 edge on a $25,000,000 building, and it won the contract. Its actual landscaping costs turned out to be $70,000. Bell had broken a collusive market.

In the old days, a utility district had a problem with collusion between the two main suppliers of large concrete sewer pipe. The district would receive high bids for all sewer interceptor work. Finally an enterprising contractor saw the opportunity and bid a large job, based on manufacturing his own pipe. The two colluding contractors filed a complaint that the contractor had no experience in pipe manufacture, but the enterprising contractor had anticipated this and had hired a retired expert as a consultant. This broke the collusive market and gave a competitive edge on several subsequent contracts as well.

Every once in a while there exists a special case when there is no competition. One large Mideast building contractor had, by coincidence, a number of Muslim employees, including a Muslim vice president. In fact, at the time it was one of the few large companies in the world that had both executives and employees who were Muslims. Now, Islamic law dictates that only Muslims can enter the holy cities of Mecca and Medina. Therefore, when a very large hotel project in Medina was offered to international competition, this company was the only U.S. company that could bid effectively on the work. Consequently, it succeeded in obtaining a major project with limited-price competition. As could be expected, the Muslim personnel gave devoted and dedicated effort to this project and, reportedly, the results were outstanding.

However, other major international contractors missed the boat. They could have formed joint ventures with some of the middle-sized but highly competent Muslim construction companies in the Mideast and thus been in a position to bid the project.

5.7. Reacting to Market Conditions

If price competition is becoming excessive or ruinous, do something different; change your tack—change the type of contract, change the scope. Do not just sit there. If things get too bad, change your areas of work. If jobs are going on the basis of low bid prices, try a cost plus fee alternative. Conversely, if your competition is selling their services by negotiating contracts or using other marketing tools, such as those described in this book, you can try price. Perhaps you can offer a lump-sum contract whereby you accept some risks. It is important to be dynamic.

Conversely, if you suddenly find that you have a competitive edge in a particular market, exploit it promptly. Competition will not stand idly by.

There are lots of ways to be dynamic: new ideas, product development, new packages. For example, Bechtel has been successful in offering a complete building construction service that includes not only design and construction but also financing.

5.8. Price Chiseling

Let us first examine a case of price chiseling when Brand X Subcontractor (BXS) does it to the Ethical Subcontractor Company (ESC), assuming all the general contractors are completely ethical. ESC prepares a subcontract bid of $394,000 for the steel erection of the Tower Highrise contract and submits his bid of $394,000 at 11:00 A.M. to all seven general contractors that are going to bid on the contract. BXS calls general contractor A at 1:15 P.M. and submits a bid of $410,000 for the steel erection. BXS asks general contractor A how its bid looks. General Contractor A tells BXS it is high. After talking some more to General Contractor A, BXS guesses its bid may be about 5 percent high. Now BXS calls general contractor B at 1:25 P.M. and submits a bid of $398,000 and asks how does this bid look. General contractor B tells BXS that it "is close, but no cigar." Now BXS calls general contractor C at 1:30 P.M. and submits a bid of $390,000 and asks how does this bid look. General contractor C tells him that it looks good. Now BXS calls the rest of the general contractors and gives them also a bid of $390,000. Also, BXS calls back general contractor A and general contractor B and tells them it made a mistake. It wants to bid $390,000. Obviously, one way by which ESC can minimize this risk is to call its bids in much closer to the 2:00 P.M. bid deadline.

Now let us examine the case of price chiseling where Brand X Subcontractor meets an unethical general contractor—general contractor Z. The bidding situation is the same as the last example; ESC has called its bid of $394,000 at 11:00 A.M. as before. BXS now contacts general contractor Z first. At 1:15 BXS submits a bid of $410,000 to general contractor Z. Then, BXS asks general contractor Z how does its bid look. General contractor Z says to himself, "I have found a pigeon." So general contractor Z tells BXS it is too high. General contractor Z then lets BXS talk some more. Finally, general contractor Z tells BXS he will give it a special deal; he will give BXS the subcontract for $380,000. This low sub-bid will give general contractor Z a competitive edge; and if general contractor Z gets the job, BXS gets the job. However, BXS has to agree to give bids at $420,000 to all the other contractors. Both parties hold to their agreements—there is usually honor among thieves.

Now let us examine a final case of price chiseling. Brand X Subcontractor went bankrupt on the last job it agreed to do for $380,000 and is out of the picture. However, the unethical general contractor Z is still in business and he wants lower bids as in the first example—so far $394,000 is the low bid. However, general contractor Z, using the $394,000 figure as a guide, uses an arbitrary figure of $380,000. Z gets the job. When the subcontractors contact him, he tells them he used his own estimate and plans to go into the steel erection business. General contractor Z even makes a few telephone calls about renting a heavy crane; he really acts like he is going into the steel erection business. The waiting period is just too much for one of the subcontractors who finally agrees to the job at $380,000. General contractor Z says, "I really hate to give it to you because I have always wanted to get into the steel erection business, but this time, I guess we will make a deal."

There are ways to fight price chiseling. One of the best ways is to establish personal contact with the highest person you can in the client's office. Many people do not seem to mind chiseling when they have received a multiple-copy letter or an impersonal voice on the telephone. However, people do not like to be unfair to someone they know personally. Therefore, it is important for the subcontractor to meet the client personally. This takes time, but it will pay off. A second way is not to give out bid prices too early. G Construction received a key subcontract bid (that was low and got it the general contract) at 1:58 P.M. for a 2:00 P.M. bid opening. Smart subcontractors often give their bids to the general contractors at the last opportunity; however, 1:58 P.M. is cutting it a little close.

This practice brings up the logistical problem of how a general contractor can turn in a bid at the last minute. Sometimes, two people are sent to where the bids will be submitted. This will provide redundancy in case of a flat tire or no parking place; then they find out where the room is in which the bids are to be submitted. Often the room where the bids are opened is not the same room as that in which the bids are to be submitted. Find out where the right room is. A few years ago, on the Broadway Tunnel contract in San Francisco, one contractor tried to submit a bid at the last minute, but could not find the proper room. The engineer delivering the bid asked one of his company's competitors whom he recognized where the bid room was: this "friend" obligingly sent him to the wrong floor! This cost his company the job, for they would have been the low bidder! The second bidder was awarded the job. The job unfortunately turned out to be a bad job, and the second bidder had a multimillion dollar loss. So two years later this fact helped ease the heat on the man who had failed to submit the bid properly. It was jokingly reported that his company gave him a bonus and fired him.

Once you determine the proper room in which to submit the bid, you grab the nearest pay telephone, with a pocketful of quarters, and establish a communications link to your office. This ties up the phone, so your competitor will have to use the phone down the hall. On your bid form, leave the bottom line open so you can make a last minute change. At the last minute, your office will tell you to deduct, say, $110,000, from your bid. Some contractors do not want to state bid figures over the telephone; there can be leaks. After deducting $110,000, for example, you can add up the sum of the digits and give that number over the telephone as a check on the figure and the total.

One way that a subcontractor can fight price chiseling is to offer something extra to the general contractor. Perhaps the subcontractor can offer to provide his own access or job layout. Discussions on such modifications should be done by telephone or even in person before the bid day. The last minute is too hectic to offer extra services and to be understood.

A general contractor should ask for written proposals before the bid date from each subcontractor. (These proposals do not include prices, just a detailed description of the work to be done.) This is done because not all subcontractors are covering exactly the same scope of work and services. For example, one steel fabricator might include painting and another steel fabricator might not. The general contractor has to put subcontractor bids on an equivalent money basis.

Sometimes a general contractor may try to gain an "edge" on his competition by providing key information on another subcontractor's bid to the low subcontractor. For example, consider the case when Drywall Company A bids $260,000; Drywall Company B bids $250,000; and Drywall Company C bids $220,000. A general contractor who has a close working relationship with C could let C know it is the low bidder and that if it has not already quoted the field, it can raise its price $20,000 to the other general contractors and still be the low subcontractor with them. C is usually happy to do this because if another general contractor gets the job, C makes an extra $20,000.

And, of course, the general contractor believes he has a $20,000 "edge" over the competition.

The above, of questionable ethics, has been practiced many times, with some success. However, since others are playing games at the same time, last minute price cuts by other subcontractors could cost both companies, the general contractor and Drywall C, the job.

An amusing case arose when two general contractors and two subcontractors were playing the same game on the same job. All four lost!

6
Brochures and Other Marketing Aids

Brochures are used to establish the standing, prestige, and strength of a contractor or purveyor of a professional service. They are invaluable in marketing for purposes of sales and also for prequalification.

6.1. Three Categories of Brochures

There is often a misunderstanding about the purposes of a brochure. A corporation will spend a great deal of money and a great deal of time on producing a brochure without fully understanding what that brochure is intended to accomplish.

In construction, brochures can be divided into three categories:

1. Corporate brochures, directed to a major segment of the general public
2. Brochures to describe special types or classes of construction
3. Brochures for a specific project (probably in competition with other submittals)

The brochure of category 1 may be utilized to establish corporate identity. It is useful for banks, bonding companies, and stockholders but has limited effectiveness in direct marketing to prospective clients. It can also be very useful in recruiting and in the orientation of professional staff.

It is interesting to note how much time and effort are put by many companies into preparing and printing their corporate brochures. Such brochures are usually replete with charts showing the growth of sales volume and profit (usually ignoring the component of inflation), with pictures of the president and key personnel and a few general pictures of major facilities constructed. Often the company may have played only a secondary or subordinate role in the overall facility pictured. One very large construction company makes a habit of showing pictures of every large project with which it had the slightest connection, with nothing to indicate that it did not design and construct the entire project, even though, in fact, it may have been only a minority member of a joint venture or, perhaps, one of their subsidiary companies acted as a subcontractor.

These brochures of category 1 are also increasingly used in an effort (often in vain) to convey to the general public the corporation's advances in complying with social pressures and regulations, such as increasing participation by minorities, women in executive ranks, emphasis on safety. Not that such social advances are not commendable, but society eventually judges a company by more tangible evidence than a self-laudatory brochure.

Obviously, these corporate brochures are directed at generating prestige. Unfortunately their greatest success is often with their own executives: it flatters their egos but does little to sell a project to a modern, sophisticated client in this highly competitive world.

It is the brochures in categories 2 and 3 that will directly generate sales or business. A category 2 or 3 brochure is valuable because it can demonstrate that your firm has experience with the client's problem. This point should be one of the first, if not the first thing, that the prospective client sees in the brochure. Brochures that make vague, over-encompassing statements about the firm's business, such as "engineering" or "general construction," tend to be ineffective. Be specific, let him know that you have what he needs or that you can solve his type of problem.

Although a brochure can interest a client in your services and can satisfy him as to your past performance and competency, a brochure should be thought of primarily as a door-opening device. Your brochure is most effective for the establishment of new client relationships, especially when the client is unfamiliar with the construction industry or engineering profession.

6.2. Directing the Brochure to Reach the Right Audience

The brochure must be tailored to fit the audience or the client. Think about the market you are trying to reach and the person who will make the primary decision. Then, also, think about others who participate in the decision process. What are their interests and their motives?

The Hidden Persuaders contains an interesting example of a brochure directed to the wrong client.*

A maker of steam shovels found that sales were lagging. It had been showing in its ads magnificent photos of its mammoth machines lifting great loads of rock and dirt. A motivation study of prospective customers was made to find what was wrong. The

---

* Vance Packard, *The Hidden Persuaders* (New York: D. Mckay, 1957), p. 64.

first fact uncovered was that purchasing agents, in buying such machines, were strongly influenced by the comments and recommendations of their steam-shovel operators, and the operators showed considerable hostility to this company's brand. Probing the operators, the investigators quickly found the reason. The operators resented pictures in the ad that put all the glory on the huge machine and showed the operator as a barely visible figure inside the distant cab. The shovel maker, armed with this insight, changed its ad approach and began taking its photographs from over the operator's shoulder. He was shown as the complete master of the mammoth machine. This approach is 'easing the operators' hostility.'

Make sure the brochure is aimed at the audience or the client and not your ego.

6.3. Designing the Brochure

The design of the brochure requires preparation by the marketer or sales manager of the company involved, since only he knows what is needed and desired. In many companies, the entire brochure is first produced in-house, in draft form. Once this draft brochure has been produced, the professional advertising specialist can be called in to check and redesign (if necessary), so as to enhance the psychological impact through style, color, and approach.

Here are some guidelines which may make the brochure effective as a sales tool:

1. Brochures are normally looked at and not read in detail. Therefore, the brochure should present the message as a series of visual and verbal images, quickly, clearly, and concisely.
2. Stress one point: clearly tell what that point is, develop it from three or four approaches, and then tell the point again. The objective is to leave one clear impression—repetition is effective.
3. For category 2 brochures, include some technical content or data: the brochures then stay in engineers' and architects' files instead of being filed in the wastebasket. To reach architects, give design details and dimensions; to reach engineers, give design tables, weights, and properties. When including technical content, do not forget to include a disclaimer statement to relieve you from legal liability. A typical disclaimer statement is as follows: "The technical information presented herein is given solely to indicate how our product has been actually or conceptually applied in the past and/or is believed suitable for application in the future." Now to follow one's own rules: do not use the above statement without verification by your attorney!
4. Use pictures and sketches. The pictures and sketches must support the text: they should stress the point you want to make. Unimportant details should be airbrushed out. It is effective to use small pictures of finished

work and larger pictures of projects under construction, especially illustrating a key point. Pictures and sketches mixed together are very effective. Pictures showing work in action, with workers, attract more attention than those of completed or partially completed structures. It is imperative that every picture has a caption. This caption should describe what your firm did and not just what the picture is about. And finally, there is an old marketing adage, "A picture is worth a thousand words, but a bad picture is worse than no picture at all." One of the authors once engaged a professional photographer to take a picture of a pre-stressed girder. The picture was of exceptional clarity and it was used for a company brochure. After the brochure had been circulated, people noticed a large manure pile that was clearly visible in the background. Some people in the industry began to refer to the manure pile in the brochure, and it became a much-repeated joke, not complimentary to the company. This was not the company image that it had intended to project. A good photographer would have had the subject in focus and the background out of focus. Double-check very carefully all the pictures that will be used in a brochure.

5. Use brevity, one item at a time, and simplicity. The brochure should not be too big or involved—it should be easy to read. The paragraphs of text should not be too long, and the emphasis should be put on key words or phrases. The page should not be cluttered or look crowded. The use of headlines, color, and large print is advisable. Empty space between photos or key phrases can serve to emphasize them.

6. Check the technical aspects to make sure that the company name and addresses, as well as the telephone, telex, and cable numbers, are indicated on the brochure. Be sure the brochure is dated and that pages are numbered, so a prospective client can refer to a particular section.

7. Introductions in brochures are usually a weak way to start. You can hold your reader only a minute or two, so do not waste time: get right to the most effective point and stress it.

8. In designing your brochure, it is important to emphasize that you have the ability to solve your client's problem. This is what the client really cares about. There is a natural tendency for engineers and contractors to want to describe in too much detail how they will solve the problem. This is normally not an effective selling point. For example, a sick person is more interested in learning if the doctor can make him well than in knowing the physiological aspects of his illness.

A packaged approach, or format, is often used. A packaged approach where category 1 and category 2 brochures are combined might save some printing costs, but it is much less effective as a selling tool. This type of brochure does not sufficiently address a client's particular needs. Category 2 brochures often take a package approach where several categories of work are indicated

FIGURE 6.1. Turner City 1979 brochure.

	Projects	Architects/Engineers
1	Tufts University, Barnum Hall, classroom and laboratory building, Medford, Mass.	Kubitz & Pepi Architects, Inc.
2	Y.M.C.A. addition, for the James B. Chambers Memorial, Wheeling, W. Va. (Joint venture with John W. Galbreath & Co.)	Van Buren & Firestone Architects, Inc.
3	Radnor Corporation (a subsidiary of Sun Company, Inc.) Radnor Four, office building, Radnor, Pa.	Geddes Brecher Qualls Cunningham
4	Gallaudet College, dormitory, Washington, D.C.	The Kling Partnership
5	Douglas Plaza, office building, a development of McDonnell Douglas Corporation, Irvine, Calif.	William L. Pereira Associates
6	Hacienda Hotel, addition, for Del Ray Capital Corporation (Frank A. Klaus), El Segundo, Calif.	Albert C. Martin & Associates
7	Exeter Towers, 96–unit apartment building, Boston, Mass. (Joint venture with Exeter Towers Associates)	Steffian · Bradley Associates, Incorporated
8	Warner-Lambert Company, corporate headquarters addition, Morris Plains, N. J.	Welton Becket Associates
9	Harleysville Insurance Company, office and warehouse facilities, Harleysville, Pa.	Geddes Brecher Qualls Cunningham
10	John Fitzgerald Kennedy Library, presidential archives, public exhibit areas and theaters, Dorchester, Mass.	I. M. Pei & Partners
11	Washington Plaza, apartment building for the elderly, for The First Washington Plaza Co. (New Frontier Developments), Springfield, Ill.	Miller Melby Anderson Architects
12	U.S. National Aeronautics and Space Administration, NASA Lewis Research Center, Cleveland, Ohio	Madison-Madison International
13	Park Place, casino, hotel and parking garage, for Bally of New Jersey, Inc., Atlantic City, N.J.	Skidmore, Owings & Merrill
14	U.S. Department of Energy, combustion research facility, Sandia Laboratories, Livermore, Calif.	Garretson-Elmendorf-Zinov-Reibin
15	MCA, Inc., 70 Universal City Plaza, office building and garage, Universal City, Calif.	Skidmore, Owings & Merrill
16	Alco Standard Corporation, corporate headquarters, Valley Forge, Pa.	The Kling Partnership
17	Ball Corporation, Ball Metal Container Group, manufacturing plant addition, Williamsburg, Va.	McOG Architects
18	Houston Independent School District, Barbara Jordan Technical Institute, Houston, Texas	Caudill Rowlett Scott
19	City of Detroit, renovation of Tiger Stadium, Detroit, Mich. (Joint venture with The Emanuel Company)	Rossetti Associates/Architects Planners
20	Springfield Metropolitan Exposition and Auditorium Authority, Prairie Capital Convention Center, Springfield, Ill.	Harry Weese & Associates
21	Contra Costa County, detention facility, Martinez, Calif.	Kaplan & McLaughlin
22	Gutrich office park, for Gutrich Development Company, Inc., Greenwood Village, Colo.	Seracuse Lawler and Partners
23	Springfield Metropolitan Exposition and Auditorium Authority, Prairie Capital Convention Center parking garage, Springfield, Ill.	C. F. Murphy Associates
24	General Public Utilities Service Corporation, corporate headquarters, Parsippany-Troy Hills, N.J.	The Grad Partnership
25	Florida Junior College at Jacksonville, Kent Campus, Jacksonville, Fla.	Kemp, Bunch and Jackson, Architects, Inc.
26	911 Wilshire Building, office building, for Cabot, Cabot & Forbes Co., Los Angeles, Calif.	Skidmore, Owings & Merrill
27	Turner Development Corporation, 1100 Lake Shore Drive, condominium apartment building, Chicago, Ill.	Harry Weese & Associates
28	Centre Hotel, for H. H. Sheikh Hamdan bin Mohammad al Nahayyan and Pakistan International Airlines, Abu Dhabi, United Arab Emirates	L. B. Cassia and Associates; Abbey and Hanson Rowe & Partners
29	The Park, office complex, for Exchange Square Associates and Prudential Insurance Company of America, Los Angeles, Calif.	Daniel L. Dworsky, FAIA Architect and Associates
30	Yamaha Motor Corporation, USA, corporate headquarters, Cypress, Calif.	William L. Pereira Associates
31	Toyota Motor Sales, U.S.A., Inc., regional office and parts depot, Mansfield, Mass.	Ballinger
32	Milwaukee County, General Mitchell Field parking structure, Milwaukee, Wisc.	Mochon Schutte Hackworthy Juerisson Inc.
33	Polaroid Corporation, manufacturing and office facility, Norwood, Mass.	Ganteaume & McMullen, Inc.
34	United States Postal Service Air Mail Facility, John F. Kennedy International Airport, Jamaica, N.Y.	Kahn & Jacobs/HOK
35	City of Attleboro, advanced wastewater treatment facilities, Attleboro, Mass.	Whitman & Howard, Inc.
36	Passaic Valley Sewerage Commissioners, Newark Bay Pumping Station, Influent Facilities (joint venture with Cayuga-Caycon) and Sludge Thickening Facilities (joint venture with Cayuga-Caycon and Wolff & Munier), Newark, N.J.	Charles A. Manganaro
37	Southern California Edison Company, Engineering and Services building, Rosemead, Calif.	Owner
38	South Central Bell Telephone Company, data processing center, New Orleans, La.	Curtis and Davis
39	Montgomery Ward Development Corporation, retail store and auto service center, West Mifflin, Pa.	Loeffler-Johnson & Associates
40	Kulicke & Soffa Industries, Inc., manufacturing plant addition, Horsham, Pa.	Bass & Elias
41	Coeur d'Alene Development Company, Inc. (a subsidiary of Honeywell Inc.), office building and parking garage, Bellevue, Wash.	John Graham Company
42	Honeywell Inc., electro-optics center addition, Lexington, Mass.	Sullivan Design Group
43	U.S. Department of Health, Education and Welfare, Ada Comprehensive Health Facility, Ada, Okla.	McCaleb, Nusbaum, Thomas Architects Engineers, Inc.; Architects Plus, Inc.
44	Medical Center of Beaver County, Rochester, Pa.	Perkins & Will Partnership

FIGURE 6.1. (*continued*)

45	Duke Hospital North, Duke University, Durham, N.C.	Hellmuth, Obata & Kassabaum, Inc.
46	King Faisal Specialist Hospital, power plant expansion, Riyadh, Saudi Arabia	Gresham, Lindsey, Ried
47	The Moses Taylor Hospital, hospital addition and parking garage, Scranton, Pa.	Perkins & Will Partnership
48	Vanderbilt University Medical Center Hospital, Nashville, Tenn.	Schmidt, Garden & Erikson
49	City of Holland, Mich., Holland City Hospital addition, Holland, Mich.	Caudill Rowlett Scott
50	Regional Medical Center of Hopkins County, Madisonville, Ky. (Joint venture with South East Construction Services, Inc.)	Nolan & Nolan, Inc.
51	Hillcrest Hospital, additions, alterations and education wing, Mayfield Heights, Ohio	Dalton-Dalton-Newport
52	Parking Authority of the City of Perth Amboy, Perth Amboy General Hospital New Brunswick Avenue Parking Garage, Perth Amboy, N.J.	The Office of Karel B. Philipp, AIA
53	Memorial Hospital, patient tower addition, Logansport, Ind.	The McGuire & Shook Corporation; Henningson, Durham & Richardson, Inc.
54	The Children's Hospital Medical Center, multidisciplinary intensive care unit addition, Boston, Mass.	The Architects Collaborative Inc.
55	The Portsmouth Hospital, additions and alterations, Portsmouth, N.H.	The Ritchie Organization
56	Mercy Hospital, hospital addition and parking garage, Scranton, Pa.	Burns & Loewe

FIGURE 6.1. (*continued*)

Turner City 1979

Turner City 1979: a rendering of the 95 buildings completed by Turner Construction Company during the single year 1979. More than 15-million square feet of construction in 53 different cities in 23 states and the District of Columbia, as well as two foreign countries. Projects representing the best efforts of the 1800 Turner staff members around the world to build facilities of all types and sizes and to complete them on schedule and within budget.

57	Carson-Tahoe Hospital, hospital additions, Carson City, Nev.	Casazza, Peetz & Associates
58	Wills Eye Hospital, hospital and research facilities, Philadelphia, Pa.	Ballinger
59	Ohio Valley Medical Center, patient tower addition, Wheeling, W. Va. (Joint venture with John W. Galbreath & Co.)	Dalton, van Dijk, Johnson & Partners

FIGURE 6.1. (*continued*)

In addition to the projects pictured in Turner City 1979, Turner Construction Company completed major renovation, alteration and consulting projects not suitable for visual portrayal for:

Owners	Architects/Engineers
The Aerospace Corporation, El Segundo, Calif.	Owner
His Highness The Amir of Qatar	Various architects
Atlantic Richfield Company, Los Angeles, Calif.	Kaneko-Laff Associates
Bank Hapoalim B.M., Philadelphia, Pa.	Swanke, Hayden, Connell and Partners
Blue Cross of Southern California, Woodland Hills, Calif.	Albert C. Martin & Associates
Brattle-Palmer-Church Realty, Ltd., Cambridge, Mass.	Sert, Jackson & Associates, Inc.
British Consulate, Cleveland, Ohio	Owner
The Brooklyn Hospital, Brooklyn, N.Y.	Rogers, Burgun, Shahine & Deschler
The Winifred Masterson Burke Relief Foundation, White Plains, N.Y.	Lothrop Associates
Chicago Dock & Canal Trust, Chicago, Ill.	Shaw & Associates
City of Cincinnati, Ohio	KZF Incorporated
Cleveland Cliffs Iron Company, Cleveland, Ohio	—
Colonial Penn Group, Philadelphia, Pa.	H2L2
Dunfey Hotels, Berkshire Hotel, New York, N.Y.	Peter Gisolfi
Dunfey Hotels, Dunfey Atlanta Hotel, Atlanta, Ga.	John Mixon Architect
Eaton Corporation, Cleveland, Ohio	James F. Hawver & Associates
Equitable Life Assurance Society, Cleveland, Ohio	Owner
FMC Corporation, Chicago, Ill.	Hague-Richards Associates Ltd.
Harvard University, Boston, Mass.	The Architects Collaborative Inc.
Heidrick & Struggles, Inc., Chicago, Ill.	Salom K. Shaheen
Hospital of the University of Pennsylvania, Philadelphia, Pa.	Mirick Pearson Batcheler Henry
Hughes Aircraft Company, El Segundo, Calif.	Owner
Howard Hughes Medical Institute, Nashville, Tenn.	Schmidt, Garden & Erikson
Kaufmann's Department Store, St. Clairsville, Ohio	Loeffler-Johnson & Associates
Massachusetts Mutual Life Insurance Company, New York, N.Y.	Various architects
Robert Morris College, Chicago, Ill.	Space Design Group, Inc.
New York News, Inc., New York, N.Y.	Technical Services Corporation
Ohio Bell Telephone Company, Cleveland, Ohio	Barnes-Neiswander Associates; Blunden-Barclay
Pacific Telephone Company, Pasadena, Calif.	Ronald T. Aday, Inc.
Pasadena Redevelopment Agency, Pasadena, Calif.	Charles Kober Associates
Peterson Ross Schloerb & Seidel, Chicago, Ill.	Hague-Richards Associates Ltd.
Provident National Bank, Philadelphia, Pa.	Adcock & Matz and Associates; Edward Zimmerman Associates; The Kling Partnership
St. Ignatius College Prep School, Chicago, Ill.	Owner
St. Vincent's Hospital and Medical Center of New York, New York, N.Y.	Ferrenz and Taylor, Inc.
Sonnenschein, Carlin, Nath & Rosenthal, Chicago, Ill.	Skidmore, Owings & Merrill
SRI International, Menlo Park, Calif.	William L. Pereira Associates
Lula Belle Stewart Center, Inc., Detroit, Mich.	Sims-Varner and Associates
TRW Systems Group, Inc., Redondo Beach, Calif.	Owner
Underwriters Bank (Overseas) Ltd., Hong Kong	Innerspace Design Limited
United States Gypsum Company, Chicago, Ill.	Environmental Systems Design, Inc.
U.S. Navy Regional Medical Center, San Diego, Calif.	Welton Becket Associates
University of Pennsylvania, Philadelphia, Pa.	Geddes Brecher Qualls Cunningham
Vanderbilt University, Nashville, Tenn.	Schmidt, Garden & Erikson
V.S.I. Corporation, Pasadena, Calif.	Neptune & Thomas Associates
Warner-Lambert Company, Elk Grove Village, Ill.	David G. Dearlove
Washington Squash Associates, Washington, D.C.	Skidmore, Owings & Merrill
Howard S. Wright Development Co., Los Angeles, Calif.	Cannell & Chaffin Commercial Interiors

Cover Photo:
John Fitzgerald Kennedy Library,
Dorchester, Mass.

FIGURE 6.1. (*continued*)

Turner Construction Company

Corporate Headquarters
150 East 42nd Street
New York, New York 10017   (212) 573-0400

Offices:

Atlanta 30303
229 Peachtree Street, N.E.   (404) 522-7120

Boston 02116
38 Newbury Street   (617) 421-5700

Chicago 60601
180 North La Salle Street   (312) 558-7600

Cincinnati 45202
511 Walnut Street   (513) 721-4224

Cleveland 44114
100 Erieview Plaza   (216) 522-1180

Columbus 43215
180 East Broad Street   (614) 228-3251

Denver 80203
700 Broadway, Suite 1130   (303) 832-2390

Detroit 48202
932 Fisher Building   (313) 871-7070

Houston 77098
3336 Richmond Avenue   (713) 529-3511

Los Angeles 90071
445 South Figueroa Street   (213) 683-1430

New York 10017
150 East 42nd Street   (212) 573-0400

Philadelphia 19102
1528 Walnut Street   (215) 545-2838

Pittsburgh 15219
601 Grant Street   (412) 765-2114

San Francisco 94104
44 Montgomery Street   (415) 391-1310

Seattle 98121
2033 Sixth Avenue   (206) 624-7101

Tampa 33609
5401 West Kennedy Blvd.   (813) 879-6632

Washington 20006
1725 K Street, N.W.   (202) 861-4200

Turner International Industries, Inc.
405 Lexington Avenue
New York, New York 10017
Telex: 424619 TCCO UI   (212) 883-0300

Dubai, U.A.E.
Post Office Box 4425
Telex: 46988 TURNR EM   283246

Hong Kong
Turner (East Asia) Limited
33 Queen's Road Central
Telex: 73948 TEASQ HX   H-252091

Singapore 1
Turner (East Asia) Pte. Limited
150 Cecil Street, Rm. 703
Telex: RS21863 TEA NCP   2200311

An Equal Opportunity Employer   Printed in USA

FIGURE 6.1. (*continued*)

in different sections of the brochures. While this type of system works, individual brochures are even more effective. Category 3 brochures often consist of a loose-leaf brochure system. The appropriate standard pages, along with certain specialized information, are assembled for each particular project or need. This system can achieve the same effectiveness as a typed brochure with original pictures.

Brochures are put together in many forms, from a typewritten brochure with photographs attached to a highly elaborate, colored brochure produced by a professional advertising agency. In considering whether to use a printed brochure or a brochure that is typed with original photographs, there is always a break-even point. For many moderate-sized firms that produce, for example, six brochures a year, a handmade brochure, although it costs $500 per copy, can be more economical than a printed brochure if the thousand copies printed are not needed.

There are four major elements which make up the cost of a printed brochure. The following are given by Coxe.*

35% — concept development, data gathering, writing and supervision
30% — illustrations (photography and/or drawings)
15% — graphic design
20% — production (typesetting, printing and binding)

Some engineering firms have a tendency to make their own printed brochures completely in-house. Experience demonstrates that while the cost is about the same as for a brochure designed with professional help, the production time is longer, the quality lower.

For both a beginning contractor and an established contractor who wishes to produce a brochure for a specific project, an inexpensively prepared typewritten brochure, along with appropriate pictures on carefully designed stationery and enclosed in an attractive cover, can be particularly effective. This type of brochure conveys a subtle message that the company who prepared it is very interested in a personal way in providing its services for the client and that the client will receive attention by top management who is particularly interested in that specific job.

6.4. Other Marketing Aids

Besides the brochures, there are other marketing aids—models, movies, slides, photographs, and technical charts. Of these, photographs and technical charts are easiest to use. Slides have been much used in the past but incur many practical difficulties in an office environment. Movies are even harder to project effectively except when elaborate arrangements are made for screen,

* Coxe, *Marketing Architectural and Engineering Services* (New York: Van Nostrand-Reinhold, 1971).

FIGURE 6.2. Turner—Serving America's Growing Health Care Needs brochure.

Because only the highest standards of construction are acceptable. A health care facility, whether a hospital, research laboratory, or nursing home, demands a special "feel" for the building's complexities with consideration for the protection of public and existing property, cleanliness, safety, and noise control.

An environment planned to meet growing community needs and the rapidly changing technology of medical science must, in addition to meeting design specifications, be built on time and on budget—it must be the best!

Turner's record of building excellence has been proven during its 70 years of experience. In the last decade alone, our company has managed the construction of almost 10-million square feet of health care facilities.

Turner's approach to the construction process is professional. We do more than put up buildings. Our career specialists have the technical credentials and field training to provide superior planning, organizing, estimating, purchasing, and on-site supervisory services.

When a health care sponsor selects Turner, he hires a complete construction department prepared to evaluate costs on a continuing basis, fix schedules, and offer objective

/Bronx, N.Y.; Henry L. Moses Research Institute.
Architects: Philip Johnson Associates

The Children's Hospital Medical Center/Boston, Mass.; Basic Pediatric Sciences Building.
Architects: The Architects Collaborative Inc.

FIGURE 6.2. *(continued)*

advice from the formative design stage through project completion.

He has, in fact, hired the best.

Today, recognizing the specialization that has taken place in the construction industry, more and more sponsors of health care facilities call for Turner's participation in the drawing board stage. With planning for health care facilities starting long before foundation work actually begins, such application of Turner's resources can prove invaluable in terms of time and money saved.

Under the negotiated fee or construction management type of contract, as opposed to the lump-sum bid award, Turner works as a "partner" on the building team—owner, architect, engineer, and contractor— to provide thoroughly practical advice relating to budget, scheduling, and material selection and availability.

Working within this framework Turner, as general contractor or construction manager, complements the efforts of the architects and designers by counseling on the cost and availability of labor and materials, project phasing, and by proposing alternate schemes to reduce cost and time. Turner's unique nationwide position enables us to leverage a tremendous purchasing power to the client's advantage early in the design phase. Savings so generated can be used to provide other benefits in the original design.

As a result of this preliminary participation, the owner may also derive important benefits from savings in time and money afforded by the "fast-track" concept which telescopes the design and construction phases. By detailed construction pre-planning, existing facilities can remain fully operative while work proceeds even as the design is being finalized.

Our widespread involvement in institutional projects gives us a keen insight and appreciation of successful financing techniques which may be applicable to your particular project. Fund-raising targets demand accurate

Cleveland Clinic Foundation/Cleveland, Ohio; new hospital, research facilities, and doctors' offices.
Architects: Flynn, Dalton, vanDijk & Partners

The Jewish Hospital/Hamilton County Hospital Commission/Cincinnati, Ohio; hospital addition.
Architects: George F. Roth & Partners

Jeanes Hospital/Philadelphia, Pa.; hospital addition.
Architects: Lee & Thaete Associates

FIGURE 6.2. (*continued*)

budget estimates which will reflect the most economical solution while assuring the latest innovations in the health care field. As a member of the planning group, Turner constantly reviews the budget as the design is developed and follows this up with up-to-date cost reports as the actual work is accomplished. With fundraising so crucial to successful health care projects, the construction management approach takes on even greater value.

Turner offers a professional solution to the medical profession's construction problems.

Additional Turner Clients in the Health Care Field:

American College of Surgeons/*Chicago*
American Dental Association/*Chicago*
Peter Bent Brigham Hospital/*Boston*
The Winifred Masterson Burke Relief Foundation/*White Plains, N.Y.*
Children's Hospital/*Cincinnati*
Christ Hospital/*Cincinnati*
Cardinal Cushing General Hospital/*Brockton, Mass.*
Delaware Hospital/*Wilmington*
First Community Senior Citizens Village/*Upper Arlington, Ohio*
Holzer Medical Center/*Gallipolis, Ohio*
University of Illinois Medical Center Library of Health Sciences/*Chicago*
Jewish Orthodox Home for the Aged/*Beachwood, Ohio*
Long Island College Hospital/*Brooklyn, N.Y.*
The Long Island Jewish Hospital/*Glen Oaks, N.Y.*
Massachusetts General Hospital/*Boston*
Massachusetts Memorial Hospitals/*Boston*
Memorial Hospital (Sloan-Kettering Institute)/*New York*
Mercy Hospital/*Pittsburgh*
New England Deaconess Hospital/*Boston*
State of New Jersey
  Ancora Mental Hospital
New Rochelle Hospital/*N.Y.*
New York Hospital, Laurence G. Payson House
State of New York
  Central Islip State Hospital
  Creedmoor State Hospital/*Queens Village*
  Kings Park State Hospital
  Pilgrim State Hospital/*Brentwood*
  Rockland State Hospital/*Orangeburg*
Ohio Presbyterian Home/*Columbus*
The General State Authority (Pennsylvania)
  Eastern Pennsylvania Psychiatric Institute/*Philadelphia*
  Embreeville State Hospital
Princeton Hospital/*N.J.*
The Rockefeller Institute for Medical Research/*New York*
St. Francis Hospital/*Poughkeepsie, N.Y.*
Sinai Hospital of Baltimore
United States Army, Cushing General Hospital/*Framingham, Mass.*
University Hospital/*Boston*
University Hospitals of Cleveland

Case-Western Reserve University/Cleveland, Ohio; medical school complex including Schools of Dentistry, Medicine and Nursing.
Architects: Barnes, Neiswander & Associates and John Williams & Associates

State of New York, Health and Mental Hygiene Facilities Improvement Corporation; Lincoln Medical and Mental Health Center/Bronx, N.Y.
Architects: Max O. Urbahn Associates, Inc.

Los Angeles Orthopaedic Foundation/ Los Angeles, Calif.; Diagnostic treatment and hospital facility.
Architects: Albert C. Martin & Associates

FIGURE 6.2. (*continued*)

New construction is always a serious business. The building product, be it a new corporate headquarters, manufacturing plant, civic center complex, or institutional facility, testifies to the foresight and confidence of its owner and calls for the allocation of large sums of money. The prime prerequisite of new construction is control—cost control, schedule control, and quality control.

As perhaps the nation's largest general contracting and construction management firm, Turner Construction is building controlled projects for a roster of clients including many of the nation's leading banks and corporations, hotel chains, retail stores, educational and health care institutions, and federal, state, and city governments. Most of these projects are being built under the Turner management contract, which gives both the client and the architect the maximum benefit of Turner's total resources throughout the planning and construction phases to assure the most effective balance of cost, quality, and time.

The fact that 60 percent of our construction contracts come from former clients and more than 80 percent are awarded to us on a selected basis speaks for our reputation for quality, service, and performance. With fully-staffed territory offices in nine major cities across the country, Turner offers a truly nationwide capability unique in the construction industry. This regional identity provides Turner with great sensitivity to the particular problem of getting the job done under local conditions while offering the centralized support services available only from a national organization.

For 70 years we have grown together with the very America we are helping to build.

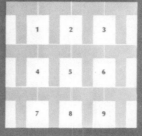

Cover photos

1. Methodist Home for the Aged/*Cincinnati, Ohio*
   Nursing wing
   Architects: A. M. Kinney Associates
2. The Children's Hospital Medical Center/*Boston, Mass.*
   Diagnostic and treatment center
   Architects: The Architects Collaborative Inc.
3. Oak Pavilion, Inc./*Cincinnati, Ohio*
   Nursing home
   Architects: Charles James Koulbanis Associates
4. Atomedic Research Center, Inc./*Flushing, N.Y.*
   Hospital prototype
   Architect: J. Paul Gilmore
5. Nazareth Hospital/*Philadelphia, Pa.*
   Hospital addition
   Architects: Henry D. Dagit & Sons
6. St. Luke Hospital/*Fort Thomas, Ky.*
   Hospital addition
   Architects: Potter, Tyler, Martin & Roth
7. Pavilion Nursing Home, Inc./*Cleveland, Ohio*
   Architects: Charles James Koulbanis Associates
8. San Marcos Medical Properties/*Santa Barbara, Calif.*
   Santa Barbara Medical Center
   Architects: Ellerbe Architects (William R. Shannon, Jr.)
   Associate Architects: Cooke, Frost, Greer & Schmandt
9. Riverside Congregational Homes/*Riverside, Calif.*
   Housing and medical facilities for the elderly
   Architects: William J. Fleming & Associates

FIGURE 6.2. (*continued*)

Turner Construction Company (FOUNDED 1902)

Corporate Offices
150 East 42nd Street
New York, New York 10017

Territory Offices

BOSTON 02116
38 Newbury Street
(617) 266-2600

CHICAGO 60603
105 West Adams Street
(312) 346-0921

CINCINNATI 45202
511 Walnut Street
(513) 721-4224

CLEVELAND 44114
100 Erieview Plaza
(216) 522-1180

COLUMBUS 43215
100 East Broad Street
(614) 228-3251

LOS ANGELES 90017
445 South Figueroa Street
(213) 683-1430

NEW YORK 10017
150 East 42nd Street
(212) 573-0400

PHILADELPHIA 19102
1528 Walnut Street
(215) 545-2838

SAN FRANCISCO 94104
44 Montgomery Street
(415) 391-1310

AN EQUAL OPPORTUNITY EMPLOYER

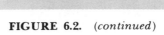

FIGURE 6.2. (*continued*)

FIGURE 6.3. Ben C. Gerwick, Inc. brochure.

FIGURE 6.3. *(continued)*

BEN C. GERWICK, INC.

PRECAST PRESTRESSED CONCRETE DIVISION

755 SANSOME STREET · SAN FRANCISCO 11

YUKON 2-9779

Many improvements in the manufacture, design and use of pretensioned prestressed concrete piles have been developed since the publication of our first pile brochure several years ago.

Considerable credit for these improvements must be given to those progressive engineers whose use of pretensioned concrete piles, under many different conditions, has led to their growing application in all types of jobs.

For foundations on land, they have been used as high capacity bearing and friction piles and as combination pile-columns with excellent economic and structural results. For marine projects, their great durability, column strength, and ease of handling have contributed to their worldwide acceptance as a superior type of concrete piling.

We are confident that whenever tested design procedures and practices for pretensioned concrete piles are followed, the result will be improved performance, reserve capacity and reduced costs.

Please call or write if we may be of service to you on projects which require piling, and on any other application of precast prestressed concrete units.

FIGURE 6.3. (*continued*)

ADVANTAGES OF PRETENSIONED CONCRETE PILES

High Load Carrying Capacity

Because of their superior strength, high axial loads, well within the allowable limits of regulating codes, can be carried by prestressed piles to the extent allowed by soil conditions. Higher loads per pile mean fewer piles, smaller footings and, consequently, in most instances, lower costs per ton carried.

Durability

The dense, high-quality concrete, maintained permanently under stress, is relatively free from shrinkage and other cracks, and is relatively impervious to water. Experience plus accelerated tests have proven pretensioned concrete piles to be exceptionally durable under even the most severe conditions of exposure.

Ease of Handling

Fewer picking points and the great strength of pretensioned piles greatly facilitate their transportation and handling and contribute to lower driving costs.

Ability to Take Hard Driving

The ability to withstand hard driving allows the use of heavier hammers for driving through denser soils, thus permitting the development of adequate bearing in the soil.

Greater Column Strength

The long-column behavior of prestressed concrete piles is excellent, and thus permits high axial loads even when piles are also subject to eccentricities or lateral forces and moments. When bending resistance is critical the pile capacity can be increased by raising the effective prestress up to the recommended maximum.

Resistance to Uplift

Pretensioned piles can be utilized very effectively in tension where uplift must be resisted, and the transfer of tension to the pile can be easily accomplished.

Quality Control

Close supervision, by experienced personnel, of materials and workmanship in a centrally controlled plant insures a high quality product with a minimum 6,000 psi concrete. Guaranteed strengths to 8,000 psi are obtainable at slight additional cost.

Economy

Prestressed concrete piles, when properly used, have shown substantial economic advantages for pile foundations.

USES OF PRETENSIONED CONCRETE PILES

Among the many types of projects on which pretensioned concrete piling may be efficiently used are:

Piers and Wharves	Mooring Structures	Railroad Bridges and Trestles
Industrial Structures	Building Foundations	Aqueducts
Intake Structures and Outfalls	Bulkheads	Off-Shore Drilling Platforms
Drydocks, Locks and Dams		Highway Bridges and Causeways

APPLICATIONS OF PRETENSIONED CONCRETE PILES INCLUDE:

Bearing Piles	Sheet Piles	Struts and Ties
Fender Piles	Friction Piles	Structural Columns
Dolphin Piles	Uplift Piles	Combined Pile/Columns
Batter Piles		Soldier Beams

Ben C. Gerwick, Inc.

FIGURE 6.3. (*continued*)

DESIGN PROCEDURES

FOR

PRETENSIONED PRESTRESSED CONCRETE PILES

Introduction

This pile design brochure is issued as a general guide only. The matter presented herein is in general accord with existing codes and regulations; however, the code regulations differ slightly among themselves and are constantly being revised.

The design of pile foundations for any specific project must, like any other structural element, be designed by professional engineers to meet the particular conditions and applicable codes.

The pile design loads considered here are based upon structural capacity alone and the ability of the soil to carry these loads must be established by actual load tests or evaluated by competent soils engineers.

The structural strength and behavior of pretensioned prestressed concrete piles have been confirmed by test data as well as by actual experience records available from many installations. It is possible to analyze their structural behavior so that pretensioned piles can be designed on a rational and reliable basis.

The following design procedures for pretensioned concrete piles are based essentially on their ultimate strength, both under direct load, and in combination with bending. For working loads, the limiting stresses for both tension and compression are determined by the elastic theory. For long column piles, criteria for buckling is developed and used as an additional requirement for design. Stresses during transportation, handling and driving are considered and proper factors of safety are applied to all ultimate load calculations. Experience has proven that the stresses imposed on a pile during driving will seldom limit the design load on the pile. The controlling factor is generally the frictional and bearing resistance of the soil.

While these design procedures closely parallel those originally published several years ago, several changes have been made. Following the trend towards higher concrete strengths, the values used in sample calculations have been increased to 6,000 psi. The sections on long columns, and combined bending and direct load have been revised. Allowable bending moments are evaluated on the basis of tensile stresses at two different levels. Sections on bending and connections at the head have been added. A wider range of pile sizes and sections is considered.

The allowable loads tabulated herein are maximum loads for 6,000 psi concrete under average conditions. For installations where the pile may be exposed to sea water, in the region of maximum bending, it will be desirable to decrease the allowable tensile stresses and the tabulated bending moments to provide a greater factor of safety against cracking. In cases where long unsupported lengths are used with combined bending and direct loads, it may be necessary to reduce the allowable direct loads and moments to maintain the required safety factors, particularly if the pile deflection is appreciable.

For ideal conditions, such as for foundation piles fully protected or embedded, or with short unsupported lengths, a smaller safety factor may be permissible and the allowable loads can be increased.

The designer should modify the tabulated values to fit the actual service conditions and to maintain reasonable safety factors consistent with the character of the applied loads and whether the piles are for marine or land installations.

BEARING PILES

I. STRENGTHS UNDER DIRECT LOADS

1.01 Concentrically Loaded Short Columns

(a) Ultimate load—If the cylinder strength of concrete is f'_c, it can be safely assumed that the concrete stress at failure of a concrete pile will be $0.85 f'_c$. The amount of prestress remaining in the tendons must be deducted. Assuming an ultimate concrete strain of 0.003 it can be shown that only about 60% of the effective prestress, f_e, is left in the member when it reaches its ultimate load. Thus, if a 6,000 psi concrete pile is prestressed to an effective prestress of 700 psi, the ultimate strength can be computed as:

$$N_u = (0.85 f'_c - 0.60 f_e) A_c \quad \ldots\ldots\ldots\ldots\ldots\ldots\ldots (1\text{-}1)$$
$$N_u = (0.85 \times 6,000 - .60 \times 700) A_c = 4,680 A_c \ldots (1\text{-}1a)$$

Where A_c = gross sectional area of the concrete in square inches, and N_u is the ultimate concentric short column load in pounds.

FIGURE 6.3. (*continued*)

(b) Design load—For a concentric load on a short column pile, a factor of safety of between 3.0 and 4.0 is usually applied.

Current building codes and AASHO-PCI standards allow design loads on concentrically loaded, short column prestressed piles as follows:

$$N = 0.2 f'_c A_c \quad \quad (1\text{-}2)$$

N is the allowable design load in pounds. Thus, for 6,000 psi concrete

$$N = 1,200 A_c \quad \quad (1\text{-}2a)$$

If the design load is $1,200 A_c$, the ultimate safety factor for the pile as given in section (a) will be $\frac{4,680}{1200} = 3.9$, which is considered quite adequate.

For a pile with $f'_c = 6,000$ psi and an effective prestress of 1,200 psi, the corresponding ultimate safety factor will be approximately 3.6. The value of $0.2 f'_c$ is considered to be about the desirable upper limit for the prestressing force, ($0.25 f'_c$ for fender piles), and a value of 700 psi is recommended as the desirable lower limit for all piles over about 40 ft. in length. The ultimate safety factors for this range of prestress fall within the acceptable limits as indicated above.

1.02 Concentrically Loaded Long Columns

(a) Ultimate load—For long columns where buckling may occur, the pile capacity computed by Euler's Formula is:

$$N_{cr} = \frac{\pi^2 EI}{L^2} \quad \quad (1\text{-}3)$$

Where N_{cr} = the critical buckling load, in lbs. E = modulus of elasticity = 5,000,000 psi for 6,000 psi concrete. For sustained loading, a creep coefficient of 2.5 is used giving E = 2,000,000 psi. I = moment of inertia of the gross section of concrete in in.[4] L = length of pile in inches, assuming pin-connected at both ends (to be modified for other end conditions).

Equation (1-3) may also be written in the following form:

$$N_{cr} = \frac{\pi^2 E}{(L/r)^2} A_c \quad \quad (1\text{-}4)$$

Where r is the radius of gyration of the section and the other terms are as defined above.

(b) Design load—A factor of safety of 2 is generally considered sufficient for the buckling load. This factor should be placed on the load, and not on the length of the pile. Analysis of equation (1-4) for E = 2,000,000 psi, indicates that with an L/r value of approximately 90, the critical buckling load is $2,400 A_c$. With a safety factor of 2, the design load is $1200 A_c$. Thus, it is apparent that for 6,000 psi concrete the maximum L/r value for which the short column formula (1-2) can be used is 90.

In order to avoid a sudden change in safety factor, it is suggested that design loads be based on the short column value of $.2f'_c A_c$ at $L/r = 60$ and that a straight line function be used between $L/r = 60$, and $L/r = 120$. At the latter value, $L/r = 120$, based on a safety factor of 2 for equation (1-4) and for E = 2,000,000 psi, the allowable load reduces to approximately $690. A_c$.

Thus, for a 6,000 psi concrete, this relationship may be expressed as follows:

for L/r 0 to 60, $N = 1,200 A_c$ . . . . (1-2a)

for L/r 61 to 120, $N = (1,710 - 8.5 L/r) A_c$ . . . . (1-5)

Where N is the allowable concentric load on the pile in lbs., and L and r are as previously defined.

For L/r values greater than 120, it is recommended that the pile be investigated for elastic stability using recognized methods of analysis, taking into account the effect of creep and deflection.

The L/r value is used rather than the conventional L/d ratio in order to take into consideration the variable "r" values for square, circular, and octagonal solid or hollow core piles.

For piles considered fully fixed at one end and hinged at the other end, it is suggested that L be taken as 0.7 of the length between hinge and assumed point of fixity. For piles fully fixed at both ends, L may be taken as 0.5 of the length between the assumed points of fixity.

II. MOMENT RESISTING CAPACITIES

2.01 Elastic Theory

(a) Allowing no tension. Under this criterion, if the effective prestress is $f = F/A_c$ the moment capacity is, $M = fI/c$. This criterion of "no tension" is much too conservative for prestressed piles under normal conditions. Zero tension will result in an ultimate factor of safety of about 3 or more, which is greater than required for bending and greater than the safety factors used in steel or reinforced concrete design.

(b) The modulus of rupture for 6,000 psi concrete is generally taken to be $7.5 \sqrt{f'_c}$ or approximately 600 psi. Allowing tension up to about 50% of the modulus of rupture, the allowable tension for normal bending can be taken as $4 \sqrt{f'_c}$ or 300 psi. If the prestress is 700 psi, the total stress available for bending is 1,000 psi and the moment capacity is, $M = 1000 I/c$.

This usually gives a factor of safety of 2.5 or more. For earthquake and other transient loads this is rather conservative and the allowable tension may be increased to 600 psi. This gives a moment capacity of, $M = 1300 I/c$.

FIGURE 6.3. (*continued*)

2.02 Ultimate Theory

(a) The ultimate moment capacity of a pretensioned pile can be computed by the ultimate strength of the tendons multiplied by a proper lever arm. As an approximation, the total ultimate strength in all the tendons can be used. The lever arm is approximately 0.37d for solid square piles and 0.32 for solid circular and octagonal piles. For hollow piles, the lever arm will be a little longer, approximately 0.38d for square piles and 0.34 for circular and octagonal piles.

Thus, the approximate ultimate moment, is given by:

$M_u = 0.37 d A_s f'_s$ for solid square piles.................(2-1)

$M_u = 0.32 d A_s f'_s$ for solid circular and octagonal piles..(2-2)

$M_u = 0.38 d A_s f'_s$ for hollow square piles...............(2-3)

$M_u = 0.34 d A_s f'_s$ for hollow circular and octagonal piles (2-4)

Where A_s = total steel area of all the tendons in square inches, f'_s = ultimate strength of the tendons in psi, and d = diameter of the pile in inches.

2.03 Design Moment

In general, the design moment should be based on an allowable tensile stress based on the modulus of rupture as in 2.01 and checked by the ultimate moment as in 2.02 to insure a factor of safety of 2 for normal loading. For wind, earthquake, or other short-time loads a safety factor of 1.5 is considered adequate. In corrosive conditions the Engineer should make an evaluation to determine whether to reduce or eliminate the allowable tension.

III. COMBINED MOMENT AND DIRECT LOAD

3.01 Elastic Theory

By the elastic theory, the existence of direct load delays the cracking of the concrete piles and thereby increases the moment carrying capacity. For example, if the prestress in the concrete is 700 psi, an external load induces 400 psi, and the allowable tensile stress is 300 psi, the total fiber stress available for bending moment is 1400 psi, and the moment capacity is:

M = 1400 I/c

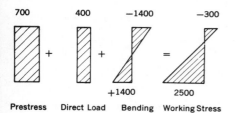

Prestress   Direct Load   Bending   Working Stress

$$\frac{F}{A} + \frac{P}{A} - \frac{M}{I/c} = \begin{cases} f_c \leq .45 f'_c \\ f_t \leq 4\sqrt{f'_c} \end{cases} \quad (3\text{-}1)$$

Piles controlled by compression should also be checked against the Interaction formula:

$$\frac{fa}{Fa} + \frac{fb}{Fb} \leq 1 \quad \dots\dots\dots\dots\dots\dots\dots\dots\dots\dots\dots\dots(3\text{-}2)$$

Where Fa is the allowable direct stress and Fb is the allowable compressive stress in bending.

3.02 Ultimate Theory

The ultimate moment capacity of a prestressed pile is reduced by the presence of external direct load, because the area of the compression zone is increased and the available lever arm for the resisting steel is correspondingly reduced.

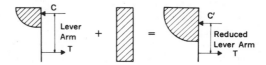

Prestress + Bending + Direct Load = Reduced Ult. Mom.

It can be roughly estimated that, for an external load producing 700 psi in the concrete, the moment capacity is given by:

$M_u = 0.29 d A_s f'_s$ for solid square piles...............(3-3)

$\quad\quad = 0.25 d A_s f'_s$ for solid round and octagonal piles..(3-4)

For hollow piles the lever arm is a little longer, 0.30d and 0.26d respectively.

At the pile head, where combined moment and direct load may be critical, the transfer in which the prestress increases from zero to full value (about 20″) must be considered.

3.03 Design Moment

With the presence of external direct load, the moment capacity of the piles should be first computed by the elastic theory as in 3.01 and then checked by the ultimate theory as in 3.02 to insure a factor of safety of 2. For wind, earthquake, or other short-time loads a safety factor of 1.5 is considered adequate. In corrosive conditions the Engineer should make an evaluation to determine whether to reduce or eliminate the allowable tension.

FIGURE 6.3. (*continued*)

IV. DESIGN EXAMPLES

4.01 Design Example A

For an 18" square solid pile, compute the allowable load and moments.

Given:
$A_c = 322$ sq. in.
$I = 8,597$ in^4
$I/c = 955$ in^3
Prestress with 14-7/16" ⌀ 270 K strands.
Effective prestress = 17,600 lbs./strand = 765 psi
$A_s = 1.63$ sq. in.
$f'_s = 270,000$ psi or 31,000 lb. per strand
$f'_c = 6000$ psi

(a) Direct Load

Using the formula $N = 0.2 f'_c A_c$ ....(1-2)
Allowable $N = 0.2 \times 6,000 \times 322 = 386$ kips or 193 tons.
Ultimate load is given by:
$N_u = (0.85f'_c - 0.60f_e) A_c$ ....(1-1)
$= (0.85 \times 6,000 - 0.60 \times 765) 322 = 1495$ kips.

Thus, factor of safety $\dfrac{1495}{386} = 3.9$

This is higher than required and under suitable soil conditions, the allowable load could be increased.

(b) Moment Capacity

For an allowable tension of 300 psi:

$M = f \dfrac{I}{c} = (300 + 765) 955 = 1017$ kip in.

Ultimate moment capacity:
$M_u = 0.37 d\, A_s f'_s = 0.37 \times 18 \times 1.63 \times 270,000 = 2930$ kip in.
Thus, factor of safety $\dfrac{2930}{1017} = 2.88$, which is higher than necessary. Thus, for transient loads, the allowable tension could be increased beyond 300 psi.

(c) Allowable Unsupported Length

This will vary with the load. Following our rule, the maximum L/r for which the short column formula can be used is 60.

$r = \sqrt{\dfrac{I}{A}} = \sqrt{\dfrac{8597}{322}} = 5.2''$

Thus, for a sustained load of 386 kips and assuming pile fully fixed at both ends, the allowable unsupported length

$L = \dfrac{60r}{0.5} = 52$ ft.

(d) Combined Direct Load and Bending for Long Column

Assuming an external direct load of 140,000 lbs., the compressive stress induced in the pile is

$\dfrac{N}{A} = \dfrac{140,000}{322} = 435$ psi.

By the elastic theory, the resisting moment as governed by tension is increased to

$(300 + 765 + 435) \times 955 = 1432$ kip in.

Note that this is higher than the value computed in (b). The compressive stress is then $\dfrac{1432}{955} + 765 + 435 = 2700$ psi

This is at the limit of the permissible compressive stress, equal to $0.45 f'_c = 0.45 \times 6,000 = 2700$ psi.

Check, using the Interaction Formula.

$F_a = 0.2 f'_c = 0.2 \times 6000 = 1200$ (l/r < 60)

$F_b = 0.45 f'_c = 0.45 \times 6000 = 2700$

$\dfrac{f_a}{F_a} + \dfrac{f_b}{F_b} = \dfrac{435}{1200} + \dfrac{1500}{2700} = .362 + .566 = .928$

which is less than unity.

By the ultimate theory, extrapolating between coefficients 0.37 for zero direct compression and 0.29 for 700 psi direct compression to get a coefficient for 870 psi and a factor of safety of 2.

$= .37 - .08 \left(\dfrac{870}{700}\right) = 0.27$

Thus, the ultimate resisting moment is reduced to approximately $M_u = 0.27\, d\, A_s\, f'_s$.

$= 0.27 \times 18 \times 1.63 \times 270,000 = 2150$ kip in.

FIGURE 6.3. (*continued*)

With a factor of safety of 2, the allowable moment should be limited to

$$\frac{2150}{2} = 1075 \text{ kip in.}$$

If a factor of safety of 1.5 is applied to both the direct load and the moment for earthquake loads, we have direct compressive stress of $435 \times 1.5 = 652$ psi., which reduces the coefficient for ultimate resisting moment to

$$0.37 - .08 \cdot \left(\frac{652}{700}\right) = 0.295 \text{ and}$$
$$M_u = 0.295 \times 18 \times 1.63 \times 270{,}000 = 2337 \text{ kip in.}$$

For a factor of safety of 1.5, the allowable moment is

$$\frac{2337}{1.5} = 1558 \text{ kip in.}$$

The cracking moment given by the elastic theory is high and does not control the design in either case. Thus, for an external direct load of 140,000 lb. and a safety factor of 1.5 the allowable external moment can be set at 1558 kip in. For a safety factor of 2 the allowable external moment is 1075 kip in.

4.02 Design Example B—Sheet Pile

For a 12" x 36" sheet pile, compute allowable moment. Given:

$A_c = 432$ sq. in.
$I = 5184$ in.[4]
$I/c = 864$ in.[3] (per pile)
Prestress with 20-1/2" 270 K strands
Effective prestress $= 23{,}550$ lbs./strand $= 1090$ psi
$A_s = 3.10$ sq. in.
$f'_s = 270{,}000$ psi or 41,300 lb. per strand
$f'_c = 6000$ psi

Moment Capacity

(a) **Elastic Theory**—For allowable tension of 300 psi.

Allowable $M = f\, I/c = (300+1090) \times \dfrac{5184}{6} = 1200$ kip in.

or $\dfrac{1200}{3} = 400$ kip inches per foot of wall.

(b) **Ultimate Strength**

Ultimate moment, $M' = 0.37 d\, A_s f'_s = (0.37 \times 12 \times 3.10 \times 270{,}000) = 3715$ kip inches, indicating a factor of safety of

$\dfrac{3715}{1200} = 3.1$, which is more than sufficient.

Thus the above allowable tension of 300 psi may be increased for transient loads, except in corrosive conditions.

FIGURE 6.3. (*continued*)

PROPERTIES OF PRETENSIONED PRESTRESSED CONCRETE PILES

Pile Size Diameter	Shape	Solid or Hollow	A_c	Weight plf	Number of Strands Per Pile	Effective Prestress (to nearest 5 psi) PSI	I	I/c	Perimeter	ALLOWABLE MOMENT 300 psi Tension	ALLOWABLE MOMENT 600 psi Tension	ALLOWABLE LOADS Based on f'_c 6000 psi	ALLOWABLE LOADS Based on f'_c 7000 psi	Pile Size Diameter
Inches (1)	(2)	(3)	Sq. In. (4)	lbs. (5)	(6)	(7)	In.^{4}	In.^{3}	Inches	Kip Inches (8)	Kip Inches (8)	Tons (9)	Tons (9)	Inches (1)
10"	Square	Solid	98	105	4—7/16"	720	790	158	38	161	209	59	69	10"
12"	Square	Solid	142	152	6—7/16"	745	1,664	277	46	290	373	85	100	12"
14"	Square	Solid	194	209	8—7/16"	725	3,112	445	54	456	589	116	135	14"
16"	Square	Solid	254	273	11—7/16"	765	5,344	668	62	711	912	152	178	16"
18"	Octagonal	Solid	268	288	11—7/16"	720	5,705	634	60	647	837	161	188	18"
18"	Square	Solid	322	346	14—7/16"	765	8,597	955	70	1,017	1,303	193	225	18"
20"	Square	Solid	398	428	13—1/2"	770	13,146	1,315	78	1,407	1,801	239	279	20"
20"	Square	11" H. C.	303	326	10—1/2"	775	12,427	1,243	78	1,336	1,709	182	212	20"
24"	Square	14" H. C.	418	450	13—1/2"	730	25,490	2,124	94	2,188	2,825	250	292	24"
36"	Round	26" H. C.	487	524	17—1/2"	820	60,016	3,334	113	3,734	4,735	292	341	36"
48"	Round	38" H. C.	675	726	24—1/2"	835	158,222	6,593	151	7,483	9,460	405	472	48"
54"	Round	44" H. C.	770	829	28—1/2"	855	233,409	8,645	170	9,985	12,578	462	539	54"

Inches (1)	Shape	Solid or Hollow (3)	A Sq. In.	Weight plf lbs. (5)	Number of Strands Per Pile (6)	Effective Prestress (to nearest 5 psi) PSI (7)	I In.⁴	S Per Pile In.³	S Per Ft. of Wall In.³	Inches	300 psi Tension Kip Inches (8)	600 psi Tension Kip Inches (8)	Inches (1)
BG–9 9" x 36"	Rectangular	Solid	324	348	14–½	1,020	2,187	486	162	90	214	262	BG–9 9" x 36"
BG–12 12" x 36"	Rectangular	Solid	432	465	20–½	1,090	5,184	864	288	96	400	487	BG–12 12" x 36"
BG–18 18" x 36"	Rectangular	W/2–10" H.C.	491	528	24–½	1,150	16,515	1,835	612	108	887	1,071	BG–18 18" x 36"

FENDER PILES

| 14" | Square | Solid | 194 | 209 | 12–⁷⁄₁₆" | 1,090 | 3,112 | 445 | N/A | 54 | 618 | 752 | 14" |
| 16" | Square | Solid | 254 | 273 | 12–½" | 1,115 | 5,344 | 668 | N/A | 62 | 945 | 1,145 | 16" |

NOTES (Applicable to All Tables)

(1) Nominal pile-size is measured through the center of the pile, except for TRICON piles, which are measured along the side of the triangle.
(2) Circular piles with comparable properties may be used in lieu of octagonal piles shown.
(3) Holes for hollow core piles are circular.
(4) Reduction in area for chamfers on square piles has been taken into account.
(5) Tables are based on regular concrete at 155 lb./cu. ft. density. The use of high strength lightweight concrete in piles for certain specific applications, such as fender piles, should be considered, when available, because its lower E value gives greater deflection and energy absorption characteristics. With lightweight concrete, an f'c of 5000 psi should be used, and the values in the table should be adjusted accordingly.

Ben C. Gerwick, Inc.

(6) Based on 1/2", 7/16" and 3/8" diameter high strength strands with ultimate strengths of 41,300 lbs., 31,000 lbs. and 23,000 lbs., respectively. If different diameter or regular strength strand is used, the number of strands per pile should be increased or decreased, in accordance with strand manufacturers' tables, to provide approximately the same minimum effective prestress shown in the table.
(7) Effective prestress assumes a uniform distribution of strands resulting in a uniform prestress. For special applications of sheet piles, eccentric prestress may be desirable and economical. Experience has shown that such eccentricity may be safely used provided the effective compression on the face with minimum prestress is above 400 psi.

(8) Allowable bending moments are listed for a permissible tensile stress of 300 psi with an effective prestress as given in the table. f'c = 6000 psi, and assuming a modulus of rupture of 600 psi. Allowable moments for earthquake or similar transient loads are based on a tension of 600 psi. Where bending resistance is critical, the allowable moment may be increased by using more strands to raise the effective prestress to a maximum of 0.2 f'c psi.
(9) Allowable design loads are based on the accepted formula of N = 0.2 f'c × Ac and are computed for f'c = 6000 psi and 7000 psi. Concrete strengths in excess of this may be used to increase allowable design loads whenever driving and soil conditions are favorable.

Ben C. Gerwick, Inc.

FIGURE 6.3. (*continued*)

TYPICAL DETAILS FOR PRETENSIONED PRESTRESSED CONCRETE PILES

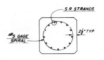

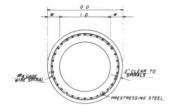

ALTERNATE PILE HEADS

Reinforcement may be specified to project from the pile into the cap or footing. If so required, attachment of the pile to the cap or footing may be made by any one of the following methods unless otherwise specified:

ALTERNATE PILE TIPS

When driving into rock or hard strata, either Type I or Type II alternate tips may be used in lieu of the standard flat tip. Size and length of steel section used shall be as determined by Engineer for adequate penetration. Type I or Type II tips may be used for either square or octagonal piles.

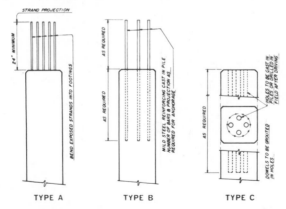

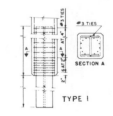

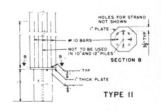

If mild reinforcing steel is used for projection into the cap or footing, the minimum area of steel required shall be 1½% of the gross cross-section of concrete pile, with not less than four bars being used for piles up to 24" in diameter, and not less than eight bars being used for piles greater than 24" in diameter. Arrangement of bars shall be in a symmetrical pattern with bars as close as practical to the sides of the pile. Anchorage of bars shall be sufficient to develop strength of bar, but not less than 20 bar diameters.

FIGURE 6.3. *(continued)*

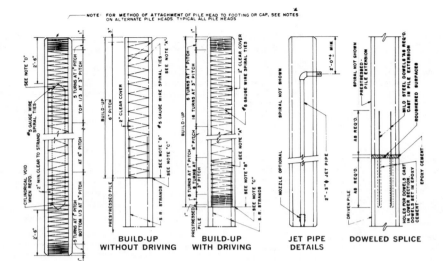

NOTES:

Note A: The minimum area of reinforcing steel shall be 1½% of the gross cross-section of concrete. Placement of bars shall be in a symmetrical pattern of not less than four bars.
Note B: Method of attachment of pile to build-up may be any of the methods given in the notes on Alternate Pile Heads. If mild reinforcing steel is used for attachment, the area shall be no less than that used in the build-up.
Note C: Concrete around top half of pile shall be bush-hammered to prevent feather edges.
Note D: Conical end fitting or form may be rounded, flat or tapered with proper taping to prevent leakage.

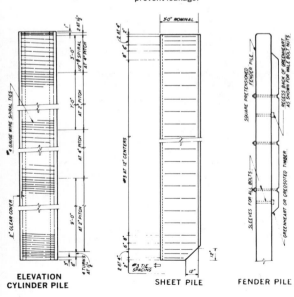

PICKING POINTS

Maximum lengths for pick-up are determined by using the following stress assumptions: Loading = 1.5 x full dead load (to allow for impact). Allowable tensile stress = $6.0\sqrt{f'c}$. These stress and loading criteria are based on normal care in handling of the pile.

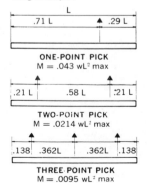

FIGURE 6.3. (*continued*)

GENERAL NOTES

HIGH CAPACITY PRESTRESSED CONCRETE PILES

Concentrated loadings and deep foundations give a strong economic and engineering advantage to high capacity piles. Prestressed concrete piles are proving particularly well adapted to this use because of their high structural capacity in direct load, bending, and uplift; their relative economy, their durability, and their relative ease of installation. The cross-section can be selected to give the most favorable transmission of vertical and lateral loads to the soil.

In most metropolitan areas, the trend in building construction is to ever higher buildings, with their increasing concentration of loads. Frequently, these new buildings are being erected in areas underlain by poor or medium-strength soils which were formerly considered adequate only for low and medium height buildings. Thus it now becomes necessary to carry the foundations to firmer soils at much greater depths.

Industrial and chemical plant construction is similarly faced with much heavier and more concentrated loadings, and their locations are more often than not in areas requiring deep foundations.

The combination of concentrated loads with the necessity for deep foundations is best satisfied by using high capacity piles. This is particularly true where the overlying soil is very weak or unstable.

In the past such demands have usually been met by steel piles, concrete-filled pipe piles, and by caissons. Recently a number of important buildings in San Francisco, Honolulu, and Melbourne have successfully employed high-capacity prestressed concrete piles.

PROPERTIES OF PRESTRESSED CONCRETE PILES

Building codes generally permit a direct compressive load equal to a $.2f'_c A_c$. Prestressed concrete piles are now generally available from established plants with a guaranteed f'_c of 6000, 7000 and, in some cases, 8000 psi. Thus an 18″ square pile with 7000 psi concrete is capable of supporting a design load of 224 tons. Under seismic and wind loads, an increase of 33% in stress is usually permitted.

Stacks and refinery columns are often subject to uplift forces from wind and earthquake. For these conditions, the effective prestress in the pile can be utilized in tension. For example, an 18″ square pile with 800 psi effective prestress can be designed for $800\, A_c = 128$ tons in uplift under wind or seismic conditions.

Prestressed concrete piles have generally high bending strength and good strength as a long column. This latter property is not usually needed for building foundations except for those that are elevated or built over water.

Prestressed concrete piles also have excellent durability which can be important for installations in corrosive or aggressive soils.

TRANSFER OF LOAD TO SOIL

High capacity prestressed piles have been employed both as end-bearing piles and as skin-friction piles.

In the case of end-bearing, the large and constant cross-section can effectively transmit the load to rock or firm material. With large hammers, these piles can be driven well into firm material; for example, two to four feet into soft rock. Steel shoes and steel H stubs are sometimes employed to ensure good seating into the rock. Certain recent experience confirms that the prestressed pile can transmit full load to the rock with much less penetration than that required for a steel H pile.

In the case of skin-friction, the large and constant perimeter is very effective for the transmission of load.

The configuration of the pile can be readily varied to give the most efficient properties for load transfer. Square, triangular, octagonal, and circular cross-sections are available.

INSTALLATION TECHNIQUES

High-capacity requires that final seating be done with a large hammer. Hammers delivering an energy of 30,000 to 60,000 ft. lbs. per blow are generally employed for developing design load capacities of 150 to 250 tons. These piles must be seated to or into a satisfactory bearing stratum as determined by the soils engineer. To ensure penetration and to prevent the excessive absorption of driving energy from upper strata, special methods must sometimes be employed, depending on the characteristics of the overlying soil. Jetting, including pilot jetting, internal jetting, and external jetting during driving, are often effective and practical. Pre-drilling with a wet drill, with or without the aid of a bentonite slurry to hold the hole, is a recent development with great promise.

NOISE ABATEMENT

Increasing attention is being paid to reduction of noise in cities. At the same time, an adequate, settlement-free foundation must be constructed and, in most cases, nothing has proven as sure as final seating with a hammer. Pre-drilling and jetting reduce the total noise by reducing the driving time. Pre-drilling and pilot jetting enable the pile to be set so its head is comparatively low during driving. Thus the hammer noise can be baffled and deflected upwards by the barricades.

The lower modulus of elasticity of concrete and freedom from resonant vibrations reduces the noise of driving. For example, acoustic tests recently performed in Los Angeles showed a 35% reduction in noise when driving a prestressed concrete pile as compared to a steel H pile.

FIGURE 6.3. (*continued*)

ECONOMY AND PRACTICABILITY

On a straight material cost evaluation, prestressed concrete usually offers a substantial advantage. For example, an 18″ square prestressed pile, delivered but not installed, may cost about 3¢ per foot per ton capacity, as compared with over 4¢ per foot per ton for steel piles on a similar basis. While installation costs may be somewhat greater than for steel H piles, the final cost in-place frequently favors the prestressed concrete pile.

With the rather general availability of larger construction equipment, prestressed concrete piles are readily transported, handled and driven under most conditions, even in the midst of crowded cities.

When compared with open-ended pipe piles, prestressed piles are generally equal or lower in material cost and usually substantially lower in the time and cost of installation.

CONNECTION OF PILE TO CAP/FOOTING

The design varies in individual cases. The connection may be designed as a hinge to eliminate moment or a moment connection may be required.

Frequently a satisfactory connection is obtained by extending the prestressing strands into the footing or capping beam. The length of strand required may be exposed during cut-off of the pile to grade, or by allowing for the extra length during manufacture. An effective moment connection can be made by embedment of the pile head into the cap by about two feet. At this depth the full moment resistance of the pile can be developed.

Where the thickness of the cap will not permit embedment, the connection is made by mild steel dowels. Where the length of piles can be predetermined within one or two feet the dowels can be either: (a) fully cast into the head of the pile and exposed during cut-off to grade; or (b) left projecting at the head and a specially notched head or follower used during driving; or (c) inserted after driving and grouted into holes formed in the head during manufacture or drilled in after driving.

Often, where the pile length is variable it is more economical and practical to drill the holes after cut-off. However, during manufacture ties must be cast into the pile along a sufficient length to be effective on the dowels in their eventual position.

SPLICES

A splice must be equal to the piles it connects in load carrying capacity, moment resistance, shear and uplift if applicable. It must be durable, economical and quickly effective so as to allow driving to be continued as soon as possible.

Several methods have been developed, of which the one favored by Ben C. Gerwick, Inc., is the epoxy-dowelled splice. The epoxy compound is placed in the drilled or formed holes in the head of the lower pile and as the top section is lowered into place the dowels, which are cast into this section, displace the epoxy into the joint, where it is retained by a form clamped around the perimeter of the piles.

Several variations of the welded plate splice have been used. Important points to watch are: (a) a central bearing plate is essential; and (b) heat dissipation during welding to prevent spalling of the concrete.

DRIVING STRESSES

These are very complex due to the number of variables involved (18 or more), and can only be solved by a computer analysis of the wave equation.

The following general remarks are based on experience, strain gauge measurements and computer analyses.

1. Heavier rams and shorter strokes reduce driving stresses.

2. Maximum tensile stresses occur in soft and irregular driving and can reach critical values. This is resisted by the effective prestress and the modulus of rupture of concrete. It is increased by any torsion or bending imparted to the pile during driving.

3. Use of soft wood cushion blocks of 6″ to 12″ in thickness will reduce the driving stresses by half or more. These compress during driving and thus do not adversely affect the transmission of energy during hard driving. New blocks should be used for each pile.

4. In soft or irregular driving it is desirable to reduce the velocity of the blow by adjusting the stroke of a single acting hammer or by reducing steam pressure with a differential hammer.

5. For piles less than about 40 feet in length, maximum tensile stresses do not occur and effective prestress values of about 400 psi may be used as a minimum value. For piles longer than 40 feet (approx.), tensile stresses will occur during soft or irregular driving and a minimum effective prestress of 700 psi is recommended. (Various authorities throughout the world use minimum values ranging from 600 psi to 900 psi.)

6. The experience of Ben C. Gerwick, Inc., on many thousands of pretensioned piles results in our recommendation of 700 psi as a minimum and emphasis on the use of soft wood cushion blocks (Item 3).

7. In certain soils, jetting ahead of the pile tip may wash out the material so as to artificially produce a "soft driving" condition. In such cases, adjust jetting practice and carefully observe procedures in paragraphs 3 and 4 above.

FIGURE 6.3. (*continued*)

SPECIFICATIONS FOR PRETENSIONED

PRESTRESSED CONCRETE PILES

SECTION I   MANUFACTURE AND MATERIALS

A. General

1. Piles, which may be either solid or hollow, shall be cast as monolithic units of homogenous high-strength concrete from head to tip and stressed with high tensile cold drawn stress relieved steel strands. Splicing of two sections of pile may be permitted when specifically noted.

2. Pretensioned piles shall be manufactured only by companies who have engaged in the regular and successful manufacture of such products for at least three years.

3. Piles may be cast in continuous lines with dividers placed at proper intervals on the pretensioning bed. Sufficient space shall be left between ends of units to permit access for cutting strand after concrete has attained the required strength and prestress transfer has been accomplished.

4. Piling shall be cured in controlled steam chambers. Steam chambers shall be equipped with sufficient recording thermometers and automatic temperature regulators to insure even temperatures throughout the curing operation. Maximum curing temperature shall not exceed 165° Fahrenheit.

5. For forming the interior of piles with hollow cores, forms shall be constructed of an approved material which will not deform or break during prestressing operations. If a moving mandrel is used for forming the inner void, special precautions shall be taken to prevent fallout of inner surface, tensile cracks, and separation of concrete from strand. Voids in hollow pretensioned piles shall be located within ¼" of the position shown on the plans.

6. For forming the exterior of piles, the use of steel forms on concrete casting beds is required. Side forms for square and octagonal piles may have a maximum draft on each side not exceeding ¼" per foot. Metal forms shall be constructed so as to permit movement of the pile without damage during release of the prestressing force.

7. Pile ends shall be plane surfaces and perpendicular to axis of pile with a maximum tolerance of ⅛" per foot transversely.

B. Concrete

1. Types I, II, or III cements, conforming to the latest ASTM specifications, may be used. (For sea-water or sea-air exposure, the alkali content shall conform to Type II requirements.)

2. Grading and composition of the fine and coarse aggregate in the mix shall conform to the latest ASTM specifications and shall be proportioned by the manufacturer so as to produce a smooth, dense, workable mixture of high strength and low shrinkage. Aggregate size shall normally be ¾" maximum.

3. Concrete shall not be placed in the forms until the placing of the strands and reinforcement has been inspected and checked.

4. Admixtures of an approved type may be used provided that they do not contain calcium chloride. Calcium chloride shall not be used.

5. Concrete shall be compacted by means of high frequency internal and/or external vibrators of a type, size and number suitable to secure adequate vibration and placement and a smooth dense surface. Minor water and air bubbles will be acceptable provided they are not more than ¼" deep. Any deeper holes shall be patched with epoxy mortar.

6. Concrete strength, as determined by cylinders cast and cured under the same conditions as the piles, shall not be less than 3500 psi at transfer of prestressing force.

7. Concrete shall have a minimum strength of 6000 psi at 28 days, and shall contain not less than 6 sacks of cement per cubic yard. Higher concrete strengths may be used and advantage may be taken of such greater strength for handling and driving stresses and column loading.

FIGURE 6.3. (*continued*)

C. **Prestressing Steel**

1. Pretensioning steel shall be seven-wire stress-relieved strand conforming to the general requirements of ASTM A416 and may be either regular or high-strength, at manufacturer's option, in accordance with strand manufacturer's published tables.

2. All strands to be bonded to the concrete shall be free of dirt, loose rust, oil, grease, or other deleterious substances and shall be accurately placed in position before concrete is poured. Minor surface corrosion of the strands due to exposure to the weather will be permitted as long as there is no pitting.

3. Strands shall be accurately held in position and stressed uniformly by a hydraulic jack equipped with an accurate reading calibrated pressure gage to permit the stress in the strand to be computed at any time.

4. Elongation of strand shall be measured at the completion of the stressing operation, and shall be checked against the hydraulic gauge reading to obtain agreement within 5%. The theoretical elongation shall be computed from tables furnished by the strand manufacturer.

5. Transfer of prestress shall be accomplished by the simultaneous gradual release of all strands by a hydraulic jack. Strands shall not be cut until prestress has been transferred.

6. Broken wires within individual strands will be permitted up to 2% of the total number of wires in each pile, providing there is not more than one broken wire per strand.

7. Certified test reports of ultimate strength and typical stress strain curves shall be furnished by the steel manufacturer.

D. **Reinforcing Steel**

1. All other reinforcing steel, including wire spiral, shall meet the standards of the latest ASTM specifications and shall be accurately placed as shown on the plans.

SECTION II DESIGN FACTORS

1. Initial tension in the strands, prior to pour of the concrete, shall not exceed 75% of minimum ultimate strength.

2. Average working stress in the strand shall not exceed 60% of the ultimate tensile strength of the strand after all losses. Working force and working stress will be considered as the force and stress remaining in the strand after losses due to creep of concrete and steel, shrinkage of concrete and other linear changes in the concrete have taken place. These losses may normally be assumed to be 20% of the initial tensioning force.

3. The minimum spacing, both vertically and horizontally, should be three times the diameter of the strand. In no case, however, shall the clear spacing between strands be less than 1-1/3 times the maximum size of the coarse aggregate. Minimum cover of strand shall be 2 inches.

4. Subject to the approval of the Engineer, prestressing may be increased as required for handling, driving, or design purposes by increasing the number or size of strand. If the unit prestress after losses is greater than .2 f'c, appropriate adjustment should be made in the allowable structural capacity of the pile.

SECTION III HANDLING AND DRIVING

1. Piles may be removed from prestressing bed for storage or transportation after concrete has reached a strength of at least 3500 psi but are not to be driven until strength has attained a minimum of 5000 psi.

2. Piles shall be lifted or supported only at the points shown in the diagram on the plans and shall be handled, driven and secured in such a manner as to avoid excessive bending stresses, cracking, spalling or other injurious results.

FIGURE 6.3. (*continued*)

3. Pile heads shall be protected from direct impact of the hammer by cushion blocks of soft compressible wood or other approved material so arranged that any strands or reinforcing bars projecting above the piles will not be displaced or deformed in driving. Under normal conditions a minimum of 6 layers of 1″ laminated douglas fir or southern yellow pine should be used for cushion blocks.

4. Jetting will be permitted and/or required when necessary to obtain the required penetration. Internal jets may be installed provided they are securely anchored to the pile and are embedded in the concrete. In hollow piles, internal jets shall be of sufficient strength and so secured as to prevent rupture during driving.

5. Piles shall be cut-off to grade where necessary. Piles driven below grade may be built up as shown on the plans or in a manner approved by the Engineer.

6. In order to minimize spalling, cut off to grade shall be made by first making a peripheral cut immediately above a clamp affixed to the pile at the required elevation. The peripheral cut may be a saw-cut or made by notching with a small pneumatic tool.

7. The driving head (helmet) shall be sufficiently large and shallow so as not to bind the head of the pile if it twists slightly during driving. Driving techniques shall be such as to minimize impact torsion under the hammer blows.

8. Hollow piles shall have vents to relieve internal hydraulic pressure.

9. For the normal range of piles the steam or air hammer used shall develop an energy per blow at each full stroke of the piston of not less than one foot pound for each pound of weight driven. Normally, the total energy developed shall be not less than 15,000 foot pounds per blow. For large and long piles the steam or air hammer used shall develop an energy per blow at each full stroke of the piston of not less than 32,500 foot pounds. Ram weights shall be as heavy as practicable, since heavy rams generally reduce driving stresses.

SECTION IV BUILD-UPS AND SPLICES

1. Build-ups, precast or cast-in-place, may be used If specified or authorized by the Engineer.

2. Two prestressed pile sections may be spliced by the use of dowels extending from the tip of the upper prestressed section, into cored or drilled holes in the lower prestressed section. For hollow-core or cylinder piles, this splice connection may be made in the walls of the pile or in a solid plug at the head and tip of the spliced sections. The dowels shall have an area equal to 1½% of the gross cross-section of pile and shall be adequately bonded into both sections.

3. The dowel holes and space between spliced sections shall be fllied with a material having properties fully equal to that of the concrete and adhesive strength equal to the shear and tensile strength of the concrete. Such properties shall be obtained within a time limit consistent with the driving requirements of the pile.

4. Concrete in build-ups and splices with driving shall have a minimum compressive cylinder strength of 5000 psi at 28-days. Without driving, minimum compressive cylinder strength shall be 3000 psi at 28-days.

NOTES

FIGURE 6.3. (*continued*)

FIGURE 6.3. (continued)

FIGURE 6.3. (continued)

6.4. Other Marketing Aids

projection, lights, and so forth. Photographs and technical charts are very effective: remember that since it is psychologically bad to point with a finger, use a pointer or a pencil when directing attention to a particular item. Note: Technical charts are not flip charts; the use of flip charts tends to "talk down" to the client. View graphs may have a similar derogatory effect.

Models can be a very useful marketing aid—particularly for the architect. They can also be of use to the engineer when describing a construction operation, sequence, or system, especially if it is sectional.

In one case, a large oil company had decided to build a very large marine terminal. It had even ordered the timber piles and timbers and had tentatively engaged contractor A to install them. Contractor Y built a model of a concrete facility, which could be assembled entirely from precast elements. The model was taken to the directors' meeting, where several directors took it apart and reassembled it. It became "the directors' concept," and a contract for the complete facility was negotiated with Y. Contractor Y was smart enough never to refer to it as "our concept!"

It is often desirable to establish and keep a positive image as a leader in technical advances. Companies are finding that a monthly news release sent to potential clients is useful. Advertising in trade journals and other appropriate publications often helps create a positive image but is only useful if it forms an integral part of an overall marketing effort. Marketing aids and brochures must never be allowed to become ends in themselves; they are aids to an effective sales approach.

Use of a company logo or symbol is an effective way of informing the public that you are on the job, building or designing. The job or product on which the logo is placed tells the story visually: the logo identifies the company behind it.

Such logos or symbols should be simple, definite, and identifiable. "Letterhead colors are more persuasive than you might guess. High-achievers, according to the psychologist David McClelland, are attracted to dark blues and regal reds. And, psychiatrist Ainslie Meares says that a corporate logo should resemble a phallus."*

These are most effective when your market is a diverse one, with multiple-private clients. An example is high-rise office building construction. Conversely, logos and symbols do little, insofar as marketing is concerned, for a contractor engaged in state highway road construction.

Logos are extremely effective for building products; for example, structural steel and precast concrete, where the market is made up of general contractors. They and the architects and engineers of the community are drawn to each new project under construction and are vitally interested as to who is furnishing important elements of the structure.

---

* John Wareham, *Secrets of a Corporate Headhunter* (New York: Atheneum, 1981).

128   *Brochures and Other Marketing Aids*

6.5. Examples of the Well-Designed Brochure

Three examples of well-designed brochures, reprinted with the companies' permission, are shown in Figures 6.1, 6.2, and 6.3. The first, "Turner City—1979," is an example of a category 1 brochure which established Turner's corporate identity as a major building contractor. The second, "Turner—Serving America's Health Care Needs," is a category 2 brochure describing a special type (hospital) of construction. (A note to the reader unfamiliar with the building construction industry: Hospital construction varies considerably from general building construction.) Third is a Ben C. Gerwick, Inc., brochure on prestressed concrete piles. This brochure was often used as a category 3 brochure aimed at a specific project.

7
Written Communication— Technical and Business Letters

7.1. Basic Principles

A technical or business letter always has a motive; that is, it intends to result in action. In a persuasive letter, the writer could be asking for more shop drawings, to have a scheduled event expedited, or for more money. In a factual letter, the writer could be documenting facts or changed conditions and asking the receiver to simply file the letter; the writer still has a motive and intends that it will eventually result in a favorable action.

The writer of the letter should consider who will act on the letter—it is not always the person who initially receives the letter. For example, protocol dictates that a letter concerning a technical change be addressed to the resident engineer on the project; however, it may be that the engineering design staff will act on the requested technical change. The writer must consider who will read the letter and be affected by it: the purchasing agent, the president of the company, or the attorney. In many contractual matters, he may also have to consider how the letter will be interpreted by an outside third party, for example, an arbiter.

The writer of the letter should also consider for whom he is writing the letter. Perhaps the letter is being written for one of his many corporate identities, joint ventures, or subsidiaries.

When a marketing letter is directed at a certain job, you should make sure that the receiver is impressed by the fact that the job is important to you and your company. There are very few times when you sit back and play hard to get; there is usually an oversupply of contractors and consultants and an undersupply of work. If the potential client feels you are strongly interested in his project, he will assume that you will do a good job.

Seven C's of a Technical or Business Letter

Correctness	Clarity	Courtesy
Conciseness	Completeness	Consideration
Concreteness		

Many letters are written in carrying out the marketing function. The client to whom you write may receive 100 letters concerning that new project; 95 of these letters may go into the wastebasket. Your letter needs to stand out; it needs to be unique. Some people with the most experience write the most uninteresting letters. Your letters need to be imaginative and interesting.

7.2. Manner of Expression

Part of marketing is presenting the truth attractively. Consider the case of selling a target estimate contract over a cost plus fixed-fee contract. You may personally believe that there is a greater incentive for the contractor to save money in a target estimate contract than in a cost plus fee contract. If you write in a letter that the target estimate contract "gives the contractor the incentive to do his best," this implies that you will do your best only on a target estimate contract. It suggests that on an open-end cost plus fee contract, you may be unprofessional and not really care about the cost to the client. A true professional will do his best on any contract. He wants the client to feel he has received the best possible job. Now, examine a way to avoid this problem. You could say that a target estimate contract "proves a mechanism by which the contractor is compensated for exerting maximum efforts to control costs." In letter writing, you need to watch your manner of expression.

The opening sets the tone of the letter; a good opening puts the recipient in the proper frame of mind. When answering a letter, for example, it is desirable to start the sentence with a positive phrase containing such words as "thank you" or "pleased" and then identify the subject. Do not go into excessive detail concerning the letter you are answering—it can detract from the opening. An example of a satisfactory opening is, "We were very pleased to receive your letter of . . . concerning. . . ." Then start writing; you have made a sufficient reference to the letter you are answering.

Some engineers and architects have a tendency to be too elaborate in their letter writing. For example, they use the phrase "in light of the fact that" instead of "since," when "since" conveys the meaning just as effectively. In today's business, conciseness is the preferred and most effective form.

There exist a series of clichés that find their way into all too many business letters. Some typical phrases are "feel free," "do not hesitate," "for your information," "don't hesitate to contact me." People seem to write these automatically without thinking. They make the letter sound like it was written without thinking. The word "hope" is also overused in business letters and often triggers a negative psychological reaction.

Consider the following expression: "our amazing accuracy in estimating." Many clients would feel this was in bad taste. Or consider this next expression: "our vast experience." Not too many contractors have had vast experience, and those who have do not need to restate it.

Word choice can project a positive or negative image. The following lists give examples of words to use and words not to use.

Positive Words		Negative Words	
exciting	share	bad	unclear
interesting	unique	claim	disagree
maintain schedule	proven	delay	damage
mutual	fee	payoff	uncooperative
new	please	profit	lawyer
sophistication	thank you	uncertain	must
innovative	consider	impossible	adversary
opportunity	facilitate	late	unfair
flexibility	endeavor	difficult	fraudulent
safe		overrun	

For example, one word that the contractor does not want to use in the letter is "claim"; he should use "change order." "Claim" is a poor word to use for both legal and sales reasons. Most owners have a policy of turning all claims over to their legal advisors; change orders, on the other hand, can be negotiated between the owner and the contractor.

In marketing, a letter of persuasion in an endeavor to get a client to change his opinion is often required. For example, the owner has asked for a lump-sum bid, and you are submitting an alternate bid. In this type of letter, the use of subjunctive and the passive can be effective. "You will have to reconsider" may be hard for the owner to swallow. "This may be worthy of your reconsideration" is more tactful and effective. In your letter writing, you may wish to try "may" and "might" instead of "will" or "would."

When trying to change someone's mind, let him save face. Do not make the potential client admit he was wrong. If you ask him to reconsider, give him a new reason to reconsider. In your letter mention "some new development," "a new technique," "it may turn out to be more economical to. . . ." Give him a reason to change his mind. You can also place the blame of "not presenting my ideas clearly" on yourself. You can make it your fault that you did not sell him the first time.

Finally, in writing a letter, take time to be polite—be sure to begin and end the letter with politeness. After all, phrases like "please," "thank you," and "we appreciate" do not add much length to the letter and greatly decrease receiver resistance. Most lawyers will agree that the words "please" or "appreciate" used in a claim letter or a letter involved in potential legal action do not detract from the legal effectiveness of the letter and do increase the chance of a favorable response from the receiver. They may, however, decrease the needs for the lawyer's services.

7.3. Business Aspects

When a letter is being written, it can be very important to use the proper letterhead. One of the most celebrated cases of using the wrong letterhead occurred just prior to World War II. The president of a major oil company was of German background and sympathetic to Germany. The board of directors asked him to resign, and he countered that he would resign if the board of directors made it worthwhile. Finally, it offered him $1 million to resign and confirmed using the letterhead of a subsidiary company. He accepted the offer and cashed the check. When he later showed up for work at the parent oil company, the board of directors was surprised—he had accepted its offer. He agreed. He had accepted the offer to resign as president of the subsidiary. For another $1 million (which was eventually paid), he would also be glad to resign from the parent company.

A contractor also needs to be careful about the letterheads he receives. A contract from Energy Oil written on "Energy Oil (U.S.A.)" stationery is presumably backed by extensive assets. A guarantee on the stationery of "Energy Oil Liberia, Ltd." may be backed by limited assets only. Be careful about letterheads.

Basically, the writer should have only one subject per letter. The British often carry this principle to extreme. The British engineer may send four letters with related subjects to the same person in one day. This is done, of course, to facilitate filing. However, a group of subjects may well be related; so in U.S. practice, one letter may address several subjects on the same project or topic. However, given the subject of a question about a concrete admixture and subject of a question of contract interpretation concerning schedule (both questions directed to the resident engineer), two different letters, each with only one subject, would definitely be in order.

There is another exception to the one subject per letter rule: if the subject will be extremely objectionable to the receiver (such as a claim), then another more favorable matter may also be included, so as to establish a more favorable attitude and reception. The inclusion of a positive subject puts the receiver in the proper frame of mind to receive the more controversial matter. You do not hit him with the bad news right away.

As is stated in other parts of this text, never bad-mouth the competition. The principle is particularly true in a letter. You should never mention the competition by name in a letter; in fact, it is usually desirable not to mention the competition at all. Rather, project the positive image that your company or firm is the right one to provide just the right service.

When you do give any technical advice in a letter, you may be held professionally liable for that advice. Therefore, be sure to qualify it in some manner, such as "Based on the information received" or "The applicability of the above to your particular project needs to be verified by the design engineer."

In a marketing letter there is a tendency to end with a statement such as "please feel free to call us if you have any questions." In most cases you will not be called, and so in several weeks you will have to call and ask the other party if they have any questions. A better ending statement would be something such as this: "I expect to be in Cincinnati on Thursday, July 17, and will call for an appointment." This way you can smoothly ask to meet with the other party to discuss and amplify your proposal.

When you sign a letter for a corporation that may have legal significance—for example, a contract or proposal—use the following form:

Very truly yours,
EB Engineering
by George P. Burdell, Executive Vice President

The form below may incur some personal legal liability for Mr. Burdell.

Very truly yours,
George P. Burdell

Be careful when you send copies. When you use "cc: Frank C. Sludge, WQCB," every recipient knows a copy has gone to the Water Quality Control Board. You can send a blind copy; for example, "bcc: Taft, Stettinius, and Hollister." This phrase does not appear on the other letters. Therefore, these other receivers do not know that the letter was sent to a legal firm. A trained business person always notes the distribution of copies.

Finally, a letter cannot be recalled (or cannot be easily recalled): it is a permanent document. Letters should not be written in anger. If you are really angry and cannot help yourself, write the letter and then throw it into the wastebasket or put it into your desk drawer. If you still want to send it a week later, then send the letter. Remember, also, to never threaten in a letter. First, a threat in a letter can create a serious legal problem for the writer. Second, if the object of the letter is to persuade, a threat most likely will not work. A person's normal reaction to a threat is to stand up to the threat or come back with another threat. The opportunity to persuade has been frustrated.

7.4. Technical Aspects

The conventional textbook salutations for the business letter are "Gentlemen" or "Dear Sirs." However, if the person's name is known, "Dear Mr. . . ."

is often used. The use of the person's name may increase the receptivity of the letter. Also, when writing to a woman executive, who represents a corporation, "Dear Ms. . . ." is smoother than "Dear Madam and Sir." If you know the person to whom you are writing on a first-name basis, and if the letter is not a formal contractual letter or proposal, then the person's first name is often used—for example, "Dear Jack." This creates a friendly tone for the letter. To cover both the formal and informal aspects, some type the address "Dear Mr. Rogers" and then write the first name in handwriting over the top, "Jack." In case of doubt, use the formal approach.

In the body of the letter the writer should use "you" or "your company" to refer to the person who represents the corporation. The writer for the corporation usually uses "we" to refer to his own company. When he wants to include a personal statement, such as "I appreciate having had the opportunity to meet with you," it is proper for him to refer to himself as "I." One good point about using "you" and "we" is that these words do not have gender, so this facilitates a letter to a businesswoman.

Obviously, a professional letter requires proper spelling and grammar. Only standard English should be used. Special language is a fad and can offend some readers; standard English offends no one. It is important to proofread your letter before signing. Typographical errors and lack of neatness detract from a professional letter. Sentences or paragraphs left out by your secretary when answering the telephone while typing the letter makes you look like an idiot—after all, you did sign the letter.

Some letter writers have a tendency to put slang in quotations. It is best not to use slang at all; however, if it is absolutely necessary to use slang, just use it and do not put it in quotations. Underlining words in a letter can offend some readers. When you underline a keyword, you give the implied message that you think the receiver is not intelligent enough to understand the import of that which you have written.

In general, in the body of a formal letter and in the addresses, abbreviations and contractions should usually not be used. For example, you should write out "Corporation" and not use the abbreviation "Corp.," and you should write out reinforcing steel, rather than "bars." Spell out "construction management," rather than using "cons't mgm't" or "CM," unless you are using CM as a specific type of contractual relationship, as is done in this book. Of course, there are exceptions to this rule, for example, titles like Mr., Mrs., or Dr. When letters are used to abbreviate a word such as "northwest," either "NW" or "N.W." is acceptable. In a business letter it is better not to hyphenate at the end of the line; hyphenation makes the word too hard to read. The exception might be with very long or compound words.

The dash is an effective means of emphasizing a key phrase; however, many engineers are unfamiliar with its use. Therefore, a review of the dash as punctuation is given. There exist three types of enclosure punctuation: parentheses, commas, and dashes. Consider the following three sentences:

7.4. Technical Aspects

George P. Burdell (a leader in the construction industry) was awarded an honorary degree from Berkeley.

George P. Burdell—a leader in the construction industry—was awarded an honorary degree from Berkeley.

George P. Burdell, a leader in the construction industry, was awarded an honorary degree from Berkeley.

All the sentences are examples of the proper use of enclosure punctuation. The parentheses are used if you wish to play down the fact that Burdell is a leader in the construction industry. The dashes are used if you wish to play up the fact that Burdell is a leader in the construction industry. The commas are used if you are neutral about Burdell's status in the construction industry. When you wish to emphasize a word, phrase, or sentence that is included in a statement, use dashes.

SMITH CONSTRUCTION COMPANY

November 5, 1981

Jones Engineers, Ltd.
777777 Dough Street
Sawmill, California 88888
Attn. Mr. J. Jones

Dear Mr. Jones:

    We are submitting four sets of drawings and calculations for the sludge pumps for the Walnut Creek Project. Our supplier, ABC Pump Manufacturing Company, says that these generally conform with the specifications, except that it has made a few minor modifications to fit its present manufacturing procedures and to correct some obvious obsolete requirements. ABC further tells us that if it has to make the pumps exactly as specified, it will cost more; therefore, we will have to receive a change order.

    Since ABC has to start manufacture now in order not to be late in delivery, we will need your immediate approval or else we will have to ask for a time extension.

Very truly yours,

SMITH CONSTRUCTION COMPANY

By _____
S. Smith

Enclosures: 4

FIGURE 7.1. Letter A—expect negative response.

SMITH CONSTRUCTION COMPANY

November 5, 1981

Jones Engineers, Ltd.
777777 Dough Street
Sawmill, California 88888

Attn: Mr. J. Jones

Subject: Walnut Creek Sewage Treatment Plant
            Contract 1001, SLUDGE PUMPS

Reference: Specifications Sections 4.01.01 and 4.02

Dear Mr. Jones:

We are submitting herewith four sets of drawings and calculations for the sludge pumps for subject project. These pumps will be furnished by the ABC Pump Manufacturing Company of Stockton, California.

We have been informed by the pump supplier that he believes that these pumps meet or exceed the intent of the requirements of the specifications but calls attention to three minor deviations for which he and we request your approval. These three items are:

1. Substitution of neoprene gaskets of Durometer 40, in lieu of the rubber gaskets specified. Neoprene is found to be more durable and is now approved by the American Pump Institute's Manual, 1972 edition, page 301, section 30.6.
2. Instrument panel frame will be made of extruded PVC instead of cadmium-plated cast steel. These frames are now furnished on all of ABC's pumps, and he believes they give equal performance, superior electrical resistivity, and longer life.
3. He proposes to use "Dilubricate Grease Fittings" as made by General Electric Company, item 16-32.03, which are currently standard on equipment of this type. The lubricating fittings shown are no longer available as standard manufacture.

No change in contract price or time is involved.

We will greatly appreciate it if you will expedite obtaining review and approval of these drawings since they are critical items and are on the critical path. If any other information is desired, please call us and we will see that it is promptly furnished.

Sincerely,

SMITH CONSTRUCTION COMPANY

S. Smith

Enclosures: 4

FIGURE 7.2.   Letter B—expect positive response.

The American letter usually has a complimentary closing. For example, the corporation letter usually ends with "Very truly yours." The proper closing represents your sentiments. For example, if you feel sincere about what you have written, particularly if the letter tends to be more on a personal theme, end with "Sincerely yours" or "Sincerely." If you are inviting a client to a social affair, you might end with "Cordially yours." If you have performed a service for a client and you are on very good terms, you might use "With best regards."

A handwritten personal note at the end, such as, "Thank you again for your help and consideration during my recent visit," can be used to create a friendly tone to your letter.

Also, if you are enclosing something, for example, three reports, in your letter, you should put at the bottom the words "Enclosures: 3." This lets the other party and his secretary know that something came with the letter—things do get lost or misfiled from time to time.

The above suggestions are made for the American business letter. Other countries have a different etiquette for letters. For example, the European may end a letter with a stereotyped statement, such as, "I assure you, kind sir, of our highest regards and our best wishes for your continued well-being and prosperity. I remain very truly yours."

Finally, there exist several letter formats: standard, letter of the future, block, and complete block. Two formats are shown in Figure 7.1 and Figure 7.2. Letter A in Figure 7.1 uses a standard format. In this letter a subject line would be optional. The salutation and complimentary closing are required. Letter B in Figure 7.2 uses a complete block format. In this format the salutation and complimentary closing are optional. The subject line, however, is required.

7.5. Examples

Consider the following two letters, in Figures 7.1 and 7.2, asking for a change—Smith Construction Company, letter A, and Smith Construction Company, letter B. Letter A is generally a negative letter. The expected response to letter A from Jones Engineers, Ltd., is that the request for a change of specification is denied and that no time extension will be granted. The writer should expect two "nos" on this type of letter. The same letter could be rewritten as letter B. Letter B will probably get a positive response, assuming the deviations from the specification are, in fact, acceptable. At worst, letter B will receive a positive and timely review.

8
Verbal Communication— The Call and the Conference

Making an effective presentation through verbal communication is an important skill required in marketing. The two most frequent types of verbal communications are the sales call and the conference. Before discussing the call and the conference separately, we will view several principles of verbal communication that are inherent in both.

8.1. Basic Principles

Receptivity and Credibility

A basic objective of the speaker is to have the audience pay attention and receive what he is saying and, most important, believe what he is saying. Receptivity and credibility can be established by pleasantness, openness, sincerity, and knowing what you are talking about. As the speaker, you are in control. Therefore, view the audience as friendly and interested in what you are saying; the audience then can feel the same way about you. Do not view the audience as hostile; they may return this emotion. The audience needs to have confidence in you to believe what you are saying—unwanted emotional signals lower the level of credibility. The audience should be treated with respect. Never talk down to an audience. On the other hand, do not try to "snow them" by displaying your own "superior" technical sophistication and use of highly specialized jargon. In business, the audience is interested, involved, and anxious to understand.

A deadpan or an overly serious delivery emits the emotional signal of nervousness. In marketing, replace this type of delivery with one of enthusiasm and confidence. Let the people know you really believe in your idea.

You should ask someone about your mannerisms (a main cause of unwanted emotional signals) and then eliminate them. Smoking can be a mannerism that detracts from your presentation. During the talk, you are trying to sell your idea and not trying to remind the other person how much he dislikes smoking. Nervous mannerisms (such as pacing, fiddling with keys, or darting and shifting eye movement) should be eliminated.

All body movements should be deliberate—not lackadaisical. For example, approach and stand at a chalkboard with authority. Also, deliberately face and look at the audience. Keep your hands still, except for purposeful gestures. If necessary, hold onto the podium (but, certainly, do not clutch it).

Although the person who has invited the caller into his office may be sitting in a very relaxed position—legs crossed and chair pushed away from his desk—the caller should still watch his own body movements. It is most effective for the caller if he sits with both feet on the floor. This is particularly true for large people; a size 11½ shoe waving in the air is distracting. What should you do with your hands? You do not want to show nervousness or create distractions. You can clasp both hands together, hold onto your attaché case, or onto a report. Be aware of your hand movements; they should be used only sparingly, but when used, should be purposeful.

In the United States, during the call, you should sit about 6 feet away from the other person. You do not want to sit closer or farther—it is unnatural for Americans and can make the other person feel uncomfortable. In other countries this distance varies. For example, in Japan you sit at one end of the room; the other person sits at the other end; and the translator sits in the middle. Also, when talking to an American, there is the 18 inches rule. That is, you should stand no closer than 18-24 inches to the other person. Some visitors from India, for example, unaware of this custom, wonder why the American keeps backing away.

To help receptivity and credibility, dress and groom suitably. Correct dress and grooming not only helps your image with the audience but also helps your confidence. You want to fit in. What is correct dress and grooming varies with the business, the community, and the level on which you are interacting.

You should try to suppress your ego. You should try to emphasize "you" and avoid "I." You can use the royal "we" when necessary, especially if you are talking about your organization and associates and their achievements.

And above all, you should be courteous; arrogance stops receptivity. Be sure that everyone can hear and that everyone can see. Ask if you can use the chalkboard or if you may approach the client's desk. Politeness increases receptivity. When you respond to a question, be polite. Sarcasm or belligerence in answering a naive or unfriendly question is a mistake. Patience and courtesy always increase the audience's receptivity and respect.

Comprehension

The audience has to understand what you are saying. It is your job to see that they do. The key is presenting your ideas in a clear, organized, and logical manner.

To be understood, speak in a clear, natural voice at your natural speed—do not speak rapidly. Also, speak in a loud voice so everyone can comfortably hear; but neither speak too loudly nor shout. You want to establish authority and confidence, not be obnoxious.

You should use words the audience understands. (Specialized jargon only favorably impresses yourself.) If you need to use technical phrases or acronyms, define them for the audience.

Finally, you should stay on the subject. Do not go off on tangents. If you can transmit one idea in the call or conference, you have been successful.

Listening

Recently, one of the authors sat in a meeting at which the chief engineer of a leading engineering firm was making a presentation of their capabilities for a very large, technically complex, and urgent project. He had several members of his staff with him; he had thoroughly prepared a complete technical presentation.

The potential client opened the meeting by saying that he had invited this engineering firm there because he and his associates were thoroughly familiar with the outstanding capabilities of the firm. He then stated that what he would like to hear from them was their planned organization and approach to the job and that his only concern was how this engineering firm was going to integrate his urgent project with their other current commitments.

The engineer and his staff then proceeded to spend the next two hours presenting in great detail their technical qualifications, just as they had rehearsed. They never answered the question of organization. They had not even listened to the question because they were so intent on carrying out the preplanned presentation. Needless to say, they did not get the job.

An engineer, who recently retired as head of a major engineering firm, said rather bluntly, "Young engineers are usually ineffective on sales calls because they do not listen to what the client is asking."

Listening often pays better than talking.

Forbes Epigrams

In another recent similar event, the engineer took the client to an expensive lunch, followed by a return to his office, where he proceeded to explain a new concept he had for reducing costs of repetitive projects in the South. The client several times interrupted to say that he was not interested in prospective work there, but he did have a project in northern Canada to which he thought this scheme could be adapted. What he needed to know was how this scheme could be modified to meet the different environment. He was willing to pay for an engineering study.

However the engineer was so wound up in his preplanned presentation that he never really heard the client and never addressed his questions but continued to talk about his original warm-weather solution.

A frustrating experience for all concerned.

Retention

At the end of the presentation, enumerate the principal point or points you have made. In order to have good audience retention, do not present too many points. The main object is to leave one or two principal ideas with the audience.

A skilled preacher said that he always modeled his sermons around a single theme, which he developed with three points. He told them early in the sermon what the theme was and named the three points. In the middle of the sermon, he developed the three points. Then at the end, he summed up the three points and repeated the theme. This has been expressed in more colloquial language as, "Tell them what you're going to say, say it, then tell them what you've said."

A famous trial lawyer was even more restrictive. She was widely esteemed for her ability to convince a jury. She loved complicated cases best; she would let her opponent present all the interactive aspects. She contained her own presentation to one key point. She never tried to answer the many sallies of her opponent. Her objective was to make sure that every juror understood and believed her one point and was convinced that this truly was the critical point of the entire case.

A brochure or special report can be left with the audience to increase retention. This should be something the audience will want to retain. For example, a brochure that contains some technical data has a good chance of being retained by an engineer.

A follow-up letter or memo on the conference that contains your key idea is the best way to assure retention. The letter should be brief, thanking the other person for the opportunity of meeting with him (or them) and then containing a sentence or paragraph which restates your principal, key point.

Motivation

Hopefully, the presentation will be motivation for some action by the listener. Perhaps the conference has led to joint agreement for both parties to investigate a negotiated contract as a team. Perhaps the client agrees to receive an alternative proposal from you. All verbal communication has some motive or purpose. It may be just a long-range presentation of your company's capabilities to meet the client's needs when they arise, or it may be a highly specific attempt to conclude a contract or settle a dispute. The trouble arises when the person making the presentation confuses the issue by trying to cover a whole range of ideas and loses sight of the principal motive for this particular call.

8.2. The Call—How to Make an Appointment

The question can be raised of how to set up the call. Today the most commonly used method is to make an appointment by telephone. The use of the telephone gives you flexibility and instant feedback. The use of the letter (to set up a call) is often inconvenient to the potential client. For example, he has an implied obligation to write back and give you an appointment and offer you a firm date. The date ties up his schedule and may later present a problem for him. Therefore, his reaction might be not to give you an appointment in order not to tie up his schedule or to set it too far in the future. If you make an appointment by telephone, he has the opportunity to be more flexible.

In making the appointment, you first need to say who you are. Then you need to suggest a time and date. Typically convenient times that you might suggest are 9:30 A.M. or 2:30 P.M. If these times are not convenient, he can offer an alternative. Also, by making an appointment by telephone, the client can gracefully change the appointment to another date at the last minute if the time becomes very inconvenient. It is always appropriate and polite to clearly state the nature of your business with the client. He may wish to refer you to someone else or to arrange for others to be present. If you are going to discuss a specific project, you should make this clear, so that he has the chance to review the file. In other cases you may state, "I am planning to be in the area and am hoping you may have some time to discuss 'your plans for your next year's capital-improvement program' or 'our new capabilities for dynamic structural analysis.' "

A "cold" call is a marketing call on someone you do not know, when you do not even know if they have a potential job. This type of call tends to be unproductive—the percentage of time you find work is too small. The one exception to this rule would, of course, be a large client who has an ongoing program.

8.3. The Call—The Initial Contact

The call, which involves a small group of about two to three people by definition, has an opening ceremony which varies according to the customs of the particular country. (In all countries it is best to set up an appointment before making the call.) The Japanese have, perhaps, the most pleasant traditional opening ceremony. A Japanese professional will have a business desk at one end of his office and a small table at the other end. When the caller arrives, he will be taken to the small table for tea first. At the small

8.3. The Call—The Initial Contact

table you can talk about the weather, your trip to Japan, and even international politics (but never local politics); but you do not talk about business. After an appropriate period of time, the host will invite the caller to sit at the business desk. Now you will talk only about business. Unfortunately, in recent years, not all Japanese have time for this opening ceremony anymore. It is an extremely pleasant way to begin a call.

In the United States, the call is usually opened with two to three minutes (but not more) of strictly general conversation before getting down to business. In many cases, particularly if the caller is from out of town, the host will offer the caller a cup of coffee. One should observe the social amenities of the call. This puts the call on a friendly basis.

As a note to the opening ceremony, be sure you get there on time—neither too late nor too early. If need be, arrive early and then walk around the block or do some other work so you will be punctual. You do not want to arrive early, as it will often present an inconvenience to the other party. For example, he may have another meeting or want to get his desk cleared of routine work. It is even worse to arrive late.

There exist in this world people who have trouble remembering names. It is an embarrassment for either party to forget the other's name. Therefore, if it is possible to do smoothly, exchange business cards early in the call. This will help the potential client remember your name and position and feel more at ease with you. Another way is to introduce yourself as you shake hands: "Mr. Jones, I am Bill Smith from Arctic Marine Services."

Early in the call, tell the other person why you are there. Sometimes a salesperson will come into an office and talk nonstop for 15 minutes without letting the other person know why he is there. Then he leaves, and the potential client wonders what that was all about.

During the call, give the person a chance to absorb what you are saying and a chance to ask questions. You should answer questions fully, with thought and tact. "Yes, that method has been very successful on a number of past cases, but recent developments . . ." works better than "That method is old fashioned!" Learn to spar when the prospective client raises objections. Learn to answer negative questions with tact; for example, "I understand perfectly well why you feel that way, but . . ." or "Your point is certainly well taken, but. . . ."

Also, if the prospective client seems to want to talk, let him do so; he may tell you how he can be sold. Some prospective clients will show far greater interest in specific points or ideas then you would expect. These points or ideas may be very important to the prospective client. While the owner is talking about his proposed project, hopefully you will discover:

1. Timing for engineering, permits, construction.
2. General description and scope of work.
3. How urgently he needs the new structure or installation.

4. Whether or not the owner has selected the architect-engineer, soils engineer, or plans to do some engineering in-house. (If he has selected any of the above, ask his permission to contact them.)
5. How much money he plans to spend. Normally, beyond a certain estimated cost, he has concluded that the project will no longer be viable, and he must abandon it completely or try for a less expensive compromise. Here is another area where you can offer help in the way of a budget estimate or alternative solutions, particularly if the project is your kind of work.
6. Whether he, or his company's policy, is receptive to a negotiated job, or whether he is automatically locked in on a policy of competitive bidding. What special points in this case might lead him to an exception—schedule, risk, potential changes.
7. Whether he has selected a site, and whether he has made any soils investigation and analysis. Has an environmental impact statement been prepared?
8. Whether he has applied for permits. Here again is an opportunity to help him in an area with which you are familiar and where he may not know the ropes.
9. Names of key people, their titles, phone numbers, addresses.

Caution: Do not press for too many items or too much detail at the first call. Listen and observe.

On the initial call, you should try to make two or three points for your company before taking leave. Do not wear out your welcome with a long recitation of all your attributes; save some of them for your next effort. Points you might make are:

1. You know the local site conditions, the local suppliers, and subcontractors.
2. Your company has skilled professionals who are very interested in this project and whose expertise can be put to work to come up with sound, practical answers to this problem.
3. You are in a special position to be helpful on this particular project; for example, you are just completing a similar project nearby.
4. Your company has 20 years of experience in this specific type of construction.

When you leave, if the prospective client has asked you to do anything, restate that which you were asked to do. Let him know you are going to take his suggestion. Tell him, "I am going to stop in to see Mr. Smith" or "I will send you the drawings next week."

After you leave, you should record the details of the call. An example of a form used by Morrison-Knudsen Company is shown in Figure 8.1.

8.3. The Call—The Initial Contact

```
┌─────────────────────────────────┬─────────────────────────────────────────────────────┐
│                                 │ NAME                              │ DATE            │
│  (MK) MORRISON                  │    J. A. McHugh                   │ 12-11-79        │
│       KNUDSEN                   ├─────────────────────────────────────────────────────┤
│                                 │ M K DIVISION / SUBSIDIARY                           │
│                                 │    Heavy & Marine                                   │
│                                 ├──────────┬──────────────┬────────────────┬─────────┤
│                                 │DATE OF   │TYPE  ☐ PHONE │ ASSIGNED       │ ☒ YES   │
│                                 │CALL      │OF            │ PRIORITY       │ ☐ NO    │
│                                 │          │CALL  ☒ VISIT │ CLIENT         │         │
│                                 ├──────────┴──────────────┴────────────────┴─────────┤
│                                 │ ASSIGNED CLIENT COORDINATOR / DIVISION              │
│                                 │              J. A. McHugh                           │
│                                 ├─────────────────────────────┬───────────────────────┤
│                                 │ DISTRIBUTION                │ E. B. Nicks           │
│                                 │ CORP MKTG & PLAN BUS DEV.   │ G. Miller             │
│  SALES CONTACT REPORT           │ CLIENT COORDINATOR          │                       │
│                                 │ Heavy & Marine Sales        │                       │
│                                 │ B. C. Burnell               │                       │
├─────────────────────────────────┼─────────────────────────────────────────────────────┤
│ COMPANY                         │ PERSON CONTACTED-TITLE                              │
│  Atlantic Richfield Hanford Co. │  Mr. V. S. Schrag, Project Engr. Manager            │
├─────────────────────────────────┼─────────────────────────────────────────────────────┤
│ DIVISION                        │ PERSON CONTACTED-TITLE                              │
│                                 │  Mr. Milt Szulinski, Principal Engineer             │
├─────────────────────────────────┼─────────────────────────────────────────────────────┤
│ STREET ADDRESS                  │ PERSON CONTACTED-TITLE                              │
│  Hanford Reservation Federal Bldg│                                                    │
├─────────────────────────────────┼─────────────────────────────────────────────────────┤
│ CITY, STATE, ZIP                │ CLIENTS' LINE OF BUSINESS                           │
│  Richland, Washington 99352     │  Nuclear fuel processing & waste management         │
├─────────────────────────────────┼──────────────────────────┬──────────────────────────┤
│ TELEPHONE                       │ PROPOSAL                 │ PREQUALIFICATIONS        │
│  (509) 946-4174                 │ ☐ REQUESTED ☐ SUBMITTED  │ ☐ REQUESTED ☐ SUBMITTED  │
├─────────────────────────────────┴──────────────────────────┴──────────────────────────┤
│ PROJECT DISCUSSED (INCLUDING MAGNITUDE, DESCRIPTION, LOCATION & SCOPE OF SERVICES)    │
│                                                                                       │
│  Hanford's deep geologic waste storage and Z-plant modifications.                     │
├───────────────────────────────────────────────────────────────────────────────────────┤
│ REMARKS                                                                               │
│                                                                                       │
│  Talked to Mr. Schrag on the evening of 12/10/79 and discussed recent efforts to      │
│  modify Z-plant (plutonium recovery facility). Mr. Schrag stated that ARHCO would     │
│  make policy statement to run Z-plant long enough to produce metal oxides for fast    │
│  flux test facilities (part of National Reactor Program). After this production run   │
│  there is a good chance the plant will be decommissioned instead of upgraded.         │
│                                                                                       │
│  Mr. Schrag stated that he was in meetings in which ARHCO has now reconciled to the   │
│  fact that funds for deep geologic waste storage at Hanford will have to come from    │
│  programs other than NWTS. This fact should add credibility to our earlier proposal   │
│  to Hanford.                                                                          │
│                                                                                       │
│  In discussion with Milt Szulinski on 12/11/79, Milt confirmed that Z-plant would     │
│  probably be kept on line for a 900 kilogram plutonium oxide production for fast      │
│  flux test facilities. He stated that the trend was now toward building a new         │
│  plutonium oxide processing facility which would probably cost in the area of         │
│  $250,000,000.                                                                        │
│                                                                                       │
├───────────────────────────────────────────────────────────────────────────────────────┤
│ FOLLOW UP ACTION (INDICATE)                                                           │
│  Routine client contact                                                               │
└───────────────────────────────────────────────────────────────────────────────────────┘
```

FIGURE 8.1. Sales contract report.

Some calls will go badly. The owner will tell you he has chosen another contractor, and he has no work for you. Or, for example, the owner will not agree to even discuss that change order and will practically throw you out of his office. No matter what happens, leave on a friendly basis. In the first case, you might thank the owner for taking his time to talk to you and then tell him you hope you can be of service to him on a future project. In the second case, you might take the blame—you did not present the facts on the change order clearly, but you will try to clarify and supplement your position in a letter. Leave smoothly and leave the door open for another call or at least a follow-up letter.

8.4. The Call—Follow-Up

There are ways to turn a negative response into positive results. For example, a negative response gives you an opportunity to send a follow-up letter with "new" information. In the case of a refused change order for subsurface soil conditions, the follow-up letter might contain "new" soil boring logs and a request for a new meeting. A negative response gives you an opportunity to come back with new information.

When you do get a negative response, do not give up completely. One time, one of the authors was turned down three times by a large public utility on a change order request. Each time he was turned down, he would come back with new data to document his request. Finally, at the fourth time the request was accepted. One reason it was accepted was that the data was far more clear and more understandable this time. An idea can be clear to the contractor in the field but not to the owner's engineer in his office. A second reason was that there was now an ample file to back up the decision to issue a change order: the owner, as a public utility, could now adequately defend itself before the state's regulatory agency.

Sometimes the prospective client will ask you for some technical information that you may or may not have. This presents an opportunity to make a second call to bring back the information for which he asked.

8.5. The Call—The Business Lunch

The business lunch has often been used in the past to increase the receptivity and pleasantness of the call. Today, however, due to the press of business activity and lack of time, it is somewhat disappearing. For the caller there are two problems: first, a person can have only one lunch a day; second, it takes too much time. For the person being called, he may have other, more important work to get finished. However, the business lunch is still popular when a long conversation (one to two hours) is required or when the caller is from another city or country.

There is no set rule whether it is better to talk about business before or after lunch. If the subject is pleasant (for example, you are explaining your company's abilities), talking about business after lunch can be good. If the subject is unpleasant (for example, a claim), talking about business before lunch is a good idea—then lunch can present the opportunity to leave on a pleasant note.

The set rule on a business lunch is that you do not talk about your business while eating. General talk about general business is acceptable. When the meal is over and it is time to talk about business, speak up and make it clear that you are now going to talk about business.

The type of person who is worth taking to lunch for marketing reasons does not have the time to waste on lunch if you are not going to talk about business. Be sure you have business to talk about, or he very likely could be irritated. He does not want to waste time on social calls. He will save any such spare time for his own good friends.

8.6. The Call—Preparation

A call involves a significant investment on your part: perhaps a trip, a long wait, a commitment of several hours or sometimes even days. And then, like an actor in a show, you have only 15 to 20 minutes available to get your message across. Therefore, you must devote time to preparation. Redefine your objective; study and think about the prospective client's needs. Assemble pertinent technical literature, but be selective: one relevant item is much better than several general pieces. Figure 8.2 shows an example of a sales call plan of Morrison-Knudsen Company.

Decide what the one point is that you wish him to retain. How will you bring the conversation to that point, how will you emphasize it?

Do your homework—it pays.

Obviously, you need to know your subject thoroughly. Knowledge of your subject and a clearly formulated objective as to why you are there and what you hope to achieve will give you confidence and poise. Just in case you do get tongue-tied, you can prepare an opening sentence and a closing sentence. This allows you to come in and leave smoothly.

The following are useful tools to prepare for the call:

1. Tell about your quality—strict quality control and adherence to specifications.
2. You can appeal to pride. All humans enjoy an appeal to their pride if skillfully stated. The compliments should normally be directed to the client's organization, although, if truly sincere, they can be directed to the man himself.
3. Use phrases like "I am sure from your experience you have found . . ." or "The design is excellent . . ." or "Your engineers have certainly come up with an innovative solution to. . . ."

Every once in a while, a potential client will use "one-upmanship" against the caller. For example, the client sits in front of a window, so that the caller is always looking toward a glare (or halo). Perhaps he sits behind a massive

148   *Verbal Communication—The Call and the Conference*

MORRISON KNUDSEN — SALES CALL PLAN		
EXPECTED DATE OF CALL: March 18, 1979		
CLIENT NAME: Washington Public Power Supply System		
DIVISION:		
ADDRESS: 3000 George Washington Way, Richland, Washington		**TELEPHONE:** (509) 946-1611

KEY PEOPLE TO BE CONTACTED

NAME	TITLE
Larry L. Grumme	Mgr.-Tech. Engr.
J. Robert Worden	Supr.-Fuel Supply
Gordon T. Austin	Program Admin.

HISTORY OF PRIOR CONTACT
Date of last previous call: R. J. Weiler - 1/14/79; Darryl Moffett - 2/4/79; Sam Barton - 2/15/79.

ANTICIPATED PROJECTS
Uranium bearing lands acquisition program. Services of general services contractor.

OBJECTIVE OF CALL
To present suggested format for proposal which is being developed by WPPSS.

OTHER AREAS OF M-K EXPERTISE WHICH CLIENT MIGHT REQUIRE
WPPSS PROJECTS: 1) TEPA concept for fossil fueled power plant
                 2) Various civil and mechanical work packages for nuclear plants
                 3) Supply of stand-by diesel generators

SALE AIDS NOTES

SKILLS
Geotechnical department; modified AEC system for mapping

TECHNIQUES
Computer mapping

EXPERIENCE
Phillips Underground Mine; Chevron-Panna Maria; Getty-slope stability studies

PERSONNEL
Hammel-PHD; Puckett-Computer; McDonald-Manager; Bauer-Mining Proj. Engr; Wilson-Underground.

ADVANTAGES
General services contractor can do total program, no overheads from other consultants.

SALES MATERIAL REQUIRED
General write-up

SPECIAL EQUIPMENT REQUIRED

NEXT ANTICIPATED CALL
May 8, 1979

NAME: R. J. Weiler
SIGNATURE: *[signature]*

M K MAR 9/77

FIGURE 8.2. Sales call plan.

desk, while the caller sits in a solitary chair in mid-room. If the caller finds this too distracting, he can move his chair to the side so he is not looking into the light. The main point is, the caller should be prepared for possible "one-upmanship" and react smoothly and politely if it occurs.

8.7. The Conference—How to Set It Up

The conference is related to the call, but is often more difficult. It is called for a specific reason—for example, the marketing and discussion of a proposal. If the chairperson is doing his job properly, the request or invitation to attend the conference will be accompanied by an agenda. This defines the subject and gives the participants the necessary time for their preparation. A conference often contains 10 to 15 people; therefore, you need to be on time. One late person can waste the time of 10 others; this can be quite irritating. It is courteous for the chairperson to get to the conference early to arrange things and to greet people. Some chairpersons, however, take the opposite tack, using their own arrival to signify that business is about to begin.

The arrangement of the principal presenter and audience around the conference table affects the presentation. In general, the presenter is asked to sit next to the chairperson. The chairperson usually sits at the head of the table. A typical arrangement is shown in Figure 8.3. (The other variation is when the chairperson sits in the middle of one side. The principles to be discussed remain the same.) Note that the "presenter" is in position W, or X, so the chairperson can make him as important or unimportant as he wishes.

A good chairperson will get a troublemaker to sit next to him (position W). A troublemaker is defined as someone who keeps interrupting and will not let the group get down to business. (Psychologically, the W seat shown in Figure 8.3 is the least effective, most controllable seat.) With the troublemaker next to the chairperson, the chairperson can ignore him by looking in another direction and tactfully taking questions from the other end of the table. If the conference is not too relevant for you, or you are a junior person, you might wish to take a position around the end of the table opposite the chairperson—out of the line of fire.

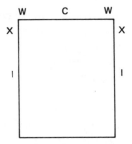

FIGURE 8.3. Conference seating arrangement. (C) chairperson's seat, (W) weakest seat, (I) most effective seat, (X) alternate seat for presenter (more effective than W).

8.8. The Conference—Robert's Rules

Most conferences are not conducted in accordance with *Robert's Rules of Order*. Voting is usually only performed on perfunctory matters, like accepting the last meeting's minutes or, occasionally, the budget. Conferences normally try to reach a consensus. For example, a conference on a joint venture would require a consensus or the dissenting party would very likely leave the joint venture. This very case happened on a joint venture that was being assembled for a large Bay Area Rapid Transit District (BARTD) contract. One of the joint-venture members did not agree with the majority decision, and therefore the embryo joint venture was never born, and no bid on the proposed contract was submitted. In conferences, even where the participants are from the same company, a consensus opinion is still sought. The company wants all its employees to enthusiastically support the decision and work as a team. A skillful chairperson can use a series of summations that can help the group reach a consensus. To do this, he needs to consider everyone's input in the summation.

Recent studies have shown that even in international politics, a consensus is far easier to reach than a majority vote—and far more productive.

8.9. The Conference—Participation

There is a certain gamesmanship required in conference participation. First, the chairperson should let everyone have his say. You should not interrupt other people. Particularly, do not interrupt the other person with small points and questions. Let the other person speak when it is his turn. Short statements by you are lost; save your comment until you have a real point to make. When you do get a chance to talk, make your point in its entirety. You can tell the chairperson ahead of time if you have a special matter on which you wish to talk.

In group theory, there are two principles that can be applied to the conferences. First, you should become part of the group: this can be accomplished by participating early. Second, groups usually consider only the first two or three alternatives; therefore, present your alternative sufficiently early so that it will be considered. An exception: If the conference is about to bog down from too much disagreement or too many diverse ideas, sometimes a succinct and clear proposal that is directed to the objective can be successfully introduced at the end.

When participating, you can use an outline, if necessary; but you should do so openly. You might wish to look at your outline before summing up. When you are preparing, study the agenda. It is perfectly proper to ask the chairperson to add an item to the agenda. Finally, in your preparation, anticipate what your opponents on the other side will come up with.

As a note to the young, highly skilled engineer with detailed knowledge of the technical subject being discussed, do not sit back and be silent—participate. Your ideas are needed to make the conference a success, and participating presents a good opportunity to establish yourself. Failure to participate, although motivated by shyness, may be interpreted as lack of competence or even arrogance.

Some conferences have a tendency to break down because of the discussion of many general points not related to the subject at hand. Pareto's law states that 80 percent of the time is spent discussing the 20 percent of the items that are least important. A good chairperson will keep the discussion heading in the right direction. If there is not an assigned chairperson, the principal participant can direct the group in the right direction. About two-thirds of the way through the conference, he can sum up all the points of view and then tactfully state the key issue(s)—from his view of his company's point of view. This approach may have to be volunteered by a participant when you do not have a good chairperson.

There exist different techniques that can be used to sell your idea to the conference. If someone else has come close to your idea, you can support that person's idea—you gain an instant ally. For example, instead of "I have a solution," try "I want to support Mr. Smith's idea. Have you thought about. . . ." Then give your modification. This technique is very effective; you reduce your ego, give authorship to the other person, and let him sell your idea. Another technique is to let the people you want to sell do the final calculations or drafting. You give them the rough calculations (draft) and let the other person finish the calculations. Then he will think he has done the work. Finally, when you disagree, ask a question. This allows the other person to answer the question and maybe change his mind—many lawyers effectively use this questioning tactic to express disagreement.

8.10. The Conference—Minutes

Finally, someone should sum up the conference with formal or informal minutes. In the minutes, the conclusions should be put first. A follow-up letter on the conference is a good way to reinforce your point, especially if you are an outsider.

The conference secretary has a unique opportunity to direct emphasis and to expand on his point, while deemphasizing the opponent's points. To an opponent's point, he may say the matter was discussed but no agreement was reached. On his side, he may expand the reasons proposed and the advantages. Ethical? Beware: A dishonest set of meeting minutes could forever destroy your credibility.

As secretary, you achieve a certain importance and standing. You become the focal point of knowledge on that particular matter. This is a good way for a junior engineer to achieve recognition.

Ninety percent of the time no one reads the minutes. The smart person will read them and write back if his interests have not been adequately recorded. He might write, "In general the minutes were complete and accurate; however, on one important item there appears to be a misunderstanding. . . ." Do not let people get away with tricks in the minutes.

Sam Smith of the Patented Pile Company used the secretarial role very effectively as a marketing tool. He would always volunteer to be secretary at the ASCE and Structural Engineer's meetings. Then the minutes would be published on Patented Pile stationery. Is it any wonder that when an engineer receiving the minutes had a piling job, he would think of Sam Smith and the Patented Pile Company?

8.11. The Conference—A Case History

The following discussion concerns an actual conference and illustrates some of the positive and negative aspects of a presentation. As background, this was a marketing conference with high-ranking members of a major oil company (the prospective client) and an engineer and several consultants who represented a construction management firm. The CM engineer began by asking his host for coffee. This caused a 10-minute interruption—the host did not have coffee in that particular conference room, and it took him 10 minutes to find some. Second, the CM engineer stated that his presentation would have to be rapid because he had an airplane to catch. This was a very important conference, and the oil company had its top people present. One of the client's staff had just flown 5000 miles to be present. Such an apparently arrogant opening had an extremely negative effect on the conference.

Fortunately, the CM engineer had a well-prepared agenda and used his notes effectively. He stuck to his well-prepared topic and put the conference on a positive track. The CM engineer had brought several consultants with him; this had the positive effect of demonstrating that his firm really did have the technological expertise. Unfortunately, he did not have any of his own people there which would have demonstrated that his own company had the right people and depth in personnel. The CM engineer did emphasize his company's experience and emphasized how its personnel and consultants could work together as a team. Now the conference was going in a positive direction.

The CM engineer then became excited and enthusiastic about the project, which was contagious. In his enthusiasm, he spoiled it all by starting to talk in detail about another similar project they had been promised by a competitor of the oil company. Now he began giving out confidential information from another client. This caused the oil company executives present, the prospective client, to quit talking altogether about their project or operations. They were justifiably concerned that if the CM engineer would talk about their competitor's confidential information, he might equally talk about their business to the competitor. Then, the CM engineer created his worst sin, by bad-mouthing other contractors. The oil company began defending these other contractors, many of whom had worked for the oil company in the recent past. Finally, the CM engineer referred to a brochure, but he could not find the proper brochure, and when he did find it, he could not find the right page. He did not know his own company's documents.

It is no surprise that the CM firm did not get the contract.

9

Prequalification

Filling out a prequalification form is often required in the marketing of construction services. This task is not just the accountant's job and should not be treated perfunctorily. Without prequalification when required, the best marketing approach in the world will only yield the satisfaction and agony of knowing you might have been considered for the job if you had prequalified. Prequalification is telling about your company, and this can take time. Therefore, prequalification should not be put off until the last minute. For the small contractor and the new contractor, understanding how to use prequalification to his advantage is essential to obtaining the right to do business. For the larger, established contractor, understanding how to use prequalification to his advantage is essential to obtaining additional bonding and bidding capacity.

9.1. Case Histories

Large companies, such as oil companies, keep an up-to-date file on the prequalification of contractors and engineers.

A marketing engineer personally made many calls on Shell, The Hague (Netherlands), where he discussed with Shell engineers his company's capabilities in installation of large-diameter submarine pipelines. He particularly mentioned his company's recent installation of a submarine line for Esso in Singapore and showed them photos. Yet when Shell took proposals on a similar line in Singapore, his company was not invited. He was terribly disappointed. Presumably his company did not think much of his efforts either. He flew to The Hague to find out why. The marketing engineer finally found out that his company's prequalification was not on file in the purchasing department and that even though the engineers had recommended him for consideration, when the purchasing agent put together the bid list, he could not find the prequalification, so he left the company off.

Why was his prequalification statement not on file? He had sent one in but had addressed it to the engineers, who thought it was probably an extra copy and had nicely filed it in the engineering office. Carelessness cost an important job!

Some 45 international contractors submitted prequalifications data to bid on a billion dollar European bridge project. Out of the 45 contractors, only 6 were to be selected for bidding. The brochures and data submitted were extensive and expensive. Each company submitted 5 to 10 volumes presenting its capabilities, financial reserves, and expertise. Just to submit this data obviously involved a significant cost, measured in the hundreds of thousands of dollars.

However, much of this effort appeared to have been wasted, through failure to properly consider the specific needs of the particular project. It was interesting to see the successful techniques used and to examine the deficiencies of the unsuccessful submittals.

For example, of the six that were chosen to bid, all used the host country's language for the enclosure letter. Many of the rejected enclosure letters were in English, and one was in German. Of the six that were chosen to bid, all had a host-country partner, whereas many of those that were rejected did not. In the selection process, size was of less importance than demonstrated ability on similar projects. Some joint ventures were not seriously considered, even though one partner was a very large international steel corporation.

Almost all the prequalifications were basically unimaginative and showed little evidence of discussion or investigation of what was required on this particular project. The international contractors tended to submit huge brochures with very little specific content. One Asian contractor even stated he planned to bring 3000 Asian workers to Europe. This was very insensitive to the local situation since one publicly stated purpose for building this particular bridge was to alleviate the high unemployment rate in the country.

9.2. Factors Involved in Prequalification

A contractor needs to evaluate the factors involved in the prequalification of his company. Just because a contractor has money in the bank and can make bond does not assure him that he will be prequalified by an owner. The following factors are used by a large university's physical plant improvement department in evaluating a contractor's prequalification request:

1. *Financial resources.* The contractor's detailed financial statement, not over six months old, should be carefully examined. The statement should indicate whether or not it is an audited statement. The "quick" assets should be ratioed to the "quick" liabilities and rated in accord-

ance with some predetermined guidelines. The contractor's bank should be identified, with the individual's name acquainted with the bidder's financial standing with the bank. An open letter of credit from the bank should be presented indicating the amount of credit the bank is willing to lend the contractor and stating the bank's lending experience with the contractor. If there is some question as to the overall financial picture, the proposed bidding should be discussed with the bank.

2. *Surety.* The contractor's surety coverage should be clearly identified, both as to full name of the surety company and home address as well as the name and address of the broker handling the contractor's account. Here again, if there is some question of the contractor being able to perform, a call to the broker or even the surety's bond manager at the home office may clear up any doubts. It is well to keep in mind that if later during construction, the contractor cannot be made to perform in a satisfactory manner, a call to the surety's home office bond manager will usually help.

3. *Insurance.* The contractor's insurer and the broker handling the account should be known. Here again, if some doubt exists, call the broker or, better yet, the insurance company.

4. *Construction ability.* The ability of the contractor to complete the subject project should be thoroughly investigated. Just because the contractor has successfully built some large hotels or high-rise apartments does not necessarily mean that he has the ability to build high-rise health science and hospital buildings. His experience record during the past five years should be carefully examined. What he built 20 or even 10 years ago means little as methods change and so does the contractor's organization.

5. *Completion ability.* The ability of the contractor to meet reasonable completion dates successfully with a minimum of time extensions should be considered. Any history of the assessment of liquidated damages against the contractor on previous projects should not be overlooked. Does the contractor make good time during the construction of the work with his own forces and then drag during the period his subs are finishing the work? How does he get along with his subs? Does he run the subs in finishing or do the subs run his job for him during this period?

6. *Personnel.* It is important to know the office and field personnel who will administer and supervise the construction. Has the contractor skilled personnel and are they available for the subject project? (This information may not be known prior to bidding, but it should be considered in evaluating the responsibility of the contractor after bidding.)

7. *Equipment.* The amount, type, and condition of the contractor's equipment is an important qualification, especially if the contractor is

going to do more than the usual minimum with his own forces. It is also important to know if any of this equipment is liened. Take a look at his yard, and see if he has a pile of rust or if it is well-maintained equipment.

8. *Work load.* Find out what jobs that contractor has on the books to be built during the period of the subject project's construction time. Is the contractor overcommitted? Does he have other work he is going after or to which he is committed during the next year or two? Find out what portion of the project he proposes to construct with his own forces. Does he intend to sub most of it and act as a broker, or is he a competent builder? (The "broker type" should be avoided.)

9. *Client relationship.* The ability of the contractor's organization to work compatibly with the staff of the owner and the architect is most important. Check this with other clients. Does he submit his extra proposals promptly, and do the estimates for these leave a bad taste with either the owner or architect? Is he cooperative in the field, or is he going to build "his way" without regard to the wishes of the owner or architect? Does he follow written instructions faithfully and promptly?

10. *Type of firm.* The bid will tell whether the firm is an individual, a partnership, a corporation, or a joint venture that will submit a proposal. This qualification should be known ahead of the bidding, especially if it is a joint venture. In that case, both ventures should be qualified, and it should be known which will sponsor or run the job.

11. *Safety record.* The safety record, as reflected in the workmen's compensation figures, and in the contractor's acceptance by his bonding company as reflected in his bond rate, indicates whether he is a preferred risk or run of the mill.

Figure 9.1 shows a short prequalification questionnaire that asks for information that a typical client requires.

9.3. Personnel and Equipment

An important way to improve your company's image on the prequalification questionnaire is to list the personnel (item 2) and the equipment (item 8) that will probably be used on the job. This is particularly true for a new or young contractor. It is particularly important to do this when prequalifying with private companies. The individual biographies of personnel enable you to get credit for the 20 years experience the project manager has had with another large successful contractor or the 25 years field experience possessed by your pile-driving foreman. An equipment schedule (detailed to your advantage) demonstrates that your company has the necessary equipment

1. Company name _____
   Address _____
   _____
   Telephone (   ) _____
   Year founded _____
   Contractor's license no. _____ Class _____

2. Officers, years in company, years in construction:
   President _____ _____ _____
   Vice president _____ _____ _____
   Secretary _____ _____ _____
   Treasurer _____ _____ _____
   _____ _____ _____ _____
   _____ _____ _____ _____

3. References: Bank(s) and branch _____
   _____
   Surety company _____
   Office _____

4. List 5 largest projects completed in previous 5 years.

Project	Location	Description	Contract Amount	Year Completed
A. _____	_____	_____	_____	_____
B. _____	_____	_____	_____	_____
C. _____	_____	_____	_____	_____
D. _____	_____	_____	_____	_____
E. _____	_____	_____	_____	_____

5. Financial data
   (1) Cash on hand _____   (14) Accounts payable _____
   (2) Accounts receivable _____   (15) Notes due within 1 year _____
   (3) Inventory at cost _____   (16) Income and withholding tax accrued
   (4) Investments in joint ventures ____   (current) _____
   (5) Work in progress (costs in excess of   (17) Work in progress (billings in excess
       billings) _____   of earnings) _____
   (6) Total current assets _____   (18) Other current liabilities _____
   (7) Land and fixed plant _____   (19) Total current liabilities _____
   (8) Less depreciation(− _____) (20) Long-term notes payable _____
   (9) Equipment owned _____   (21) Capital _____
   (10) Less depreciation(− _____) (22) Retained earnings _____
   (11) Other fixed assets _____   (23) Profit earned _____
   (12) Total fixed assets _____   (24) Less estimated income tax(− _____)
   (13) Total assets _____   (25) Total fixed liabilities _____
                                          (26) Total liabilities _____

FIGURE 9.1. Prequalification Questionnaire

Compute (a) (Item 6* less item 19) × 10 = $ _____
         (b) (Item 13* less arithmetical sum of items 19, 20, and 24) × 4 =
              $ _____
              * To these figures may be added the value of unconditional bank
                guarantee from U.S. or state bank. Attach notarized guarantee.
                Amount: $ _____

6. Contract volume of work on hand $ _____
   Completed to date $ _____

7. Have you ever failed to complete a contract? If so, attach details _____
   Do you have suits pending against you, other than those *fully* covered by
   insurance? _____
   If so, attach details _____

8. List principal items of construction equipment owned.

Number	Type	Make/Model	Age	Condition	Location

9. List three project performed of similar nature to current project on which it is desired to submit a proposal.

Project	Description	Location	Contract Value	Year Completed

10. List of proposed subcontractors and major suppliers.

Type of Subcontractor or Supplier	Name	Location	License #(Subcontractor Only)

11. List of 5 client references.

Company	Address	Name of Individual to Contact	Position

FIGURE 9.1. (*continued*)

12. Citations, awards (with year):

_____
_____
_____

Attested:

                                                          Signed _____
_____   By _____
_____   Title _____
                                                          Affix corporate seal.

FIGURE 9.1. (*continued*)

available, whether it be owned, chartered, or leased. Thus, where a special large piece of equipment is essential for the performance of the job (for example, a dredge), the enclosure of a charter or lease agreement may be as effective as if you owned the item outright.

However, do not get yourself trapped into guaranteeing that certain personnel or equipment will be available for a particular job. This mistake could be serious if the same personnel or equipment are subsequently required on another job or if one of your key persons changes jobs. In both cases, you should qualify as to availability. Such phrases as "as and when required" or "in the judgment of the contractor" can be used to avoid making a firm commitment.

An example of this problem arose on one bridge job when Ben C. Gerwick, Inc., listed a very large derrick barge. The owner, the U.S. Corps of Engineers, refused to let Gerwick move the derrick barge to another project after its main work had been completed, because it would be needed again for a few days several months later. This dilemma was only resolved by submittal of a letter guaranteeing that the derrick barge would be returned to the first job when needed. From that time on, the qualifying phrase "for the periods and times required on this project, as determined by the contractor" was added.

9.4. Financial Data

Almost all prequalification questionnaires concentrate on financial data (item 5), perhaps because it is easiest to quantify. Two common measures of financial strength are net quick (item 5a) and net worth (item 5b).

$$\text{Net quick} = \text{current assets} - \text{current liabilities}$$

A contractor is usually considered to have adequate resources to handle a volume equal to 10 times net quick. This ratio may be increased in special cases where the company has proven expertise.

9.4. Financial Data

$$\text{Net worth} = (\text{current assets} + \text{fixed assets})$$
$$- (\text{current liabilities} + \text{fixed liabilities}).$$

A contractor is usually considered to have adequate resources to handle a volume equal to 4 times net worth.

As an example of how net quick and net worth are viewed by a prospective client, consider Mom and Pop Construction, Inc., of Princeton, New Jersey, with assets and liabilities shown below.

Current assets	$1,000,000
Fixed assets:	
Equipment, land, physical plant, insurance	940,000
Current liabilities	900,000
Fixed liabilities:	
Long-term notes	500,000
Reserve for income tax	100,000

Note: Reserve for income tax can be a long- or short-term liability, depending upon whether it is due this year or in a future year. Here we have assumed it is due in a future year.

$$\text{Net quick} = \$1,000,000 - \$900,000$$
$$= \$100,000$$
$$10 \times \text{net quick} = \$1,000,000$$
$$\text{Net worth} = (\$1,000,000 + \$940,000)$$
$$- (\$900,000 + \$500,000 + \$100,000)$$
$$= \$440,000$$
$$4 \times \text{net worth} = \$1,760,000$$

In this example there is an indication of a possible problem: the fixed net worth is proportionally much greater than the net quick. This is an indication that Mom and Pop Construction, Inc., is not able to make full use of its fixed assets. It needs more quick assets if it is to have a well-balanced operation. This situation is very typical of the construction company that has grown up from a personal operation into a middle-sized business.

As another example, consider the Kash Rich Company of San Francisco, California, with the assets and liabilities as shown below:

Current assets	$1,000,000
Fixed assets	200,000
Current liabilities	600,000
Fixed liabilities	300,000

$$\text{Net quick} = \$1,000,000 - \$600,000$$
$$= \$400,000$$
$$10 \times \text{net quick} = \$4,000,000$$
$$\text{Net worth} = (\$1,000,000 + \$200,000)$$
$$- (\$600,000 + \$300,000)$$
$$= \$300,000$$
$$4 \times \text{net worth} = \$1,200,000$$

In this case, the company has adequate current funds, but its operations are limited by inadequate total assets. This example would be typical of a successful "broker" type of general contractor. However, it is still not maximizing its opportunities.

There exist several ways to vary the financial data on a prequalification questionnaire to the contractor's advantage. The contractor should remember that the accountant normally keeps books on a basis to satisfy the U.S. income tax requirements. He legitimately depreciates equipment at the most rapid rate possible. He legitimately values inventory and real estate at actual cost. He legitimately records profits only when formally earned; for example, the company may be on the completed-contract basis and may record earnings only in the year when the contract is completed.

All these practices, which may be entirely proper from a tax point of view, result in gross understatements of a company's true worth. Therefore, while you must use the accountant to give you the data for a prequalification statement, it is only the marketing executive who can properly prepare the report.

In the case of West Coast Marine Contractors, its fixed assets were seriously understated due to the fact that its large fleet of marine construction equipment was valued at original cost less accelerated depreciation. West Coast Marine Contractors therefore engaged a certified marine appraiser to appraise all its equipment and to certify the present value "to an operating company," all of which of course exceeded both its book value and its value at forced sale by several hundred percent. This "appraised value" was accepted by several major governmental agencies for prequalification purposes. Similarly, the waterfront real estate, which had been originally purchased for practically nothing, was revalued by a certified appraiser to give its current market value.

In the first case, discussed above, where there is inadequate cash to support a balanced operation, several steps are possible. One is to obtain a special bank loan for the project in question, in which the bank makes its rights secondary to the liabilities incurred under a particular contract. The bank gives a letter stating that a specific sum of money will be made available on loan until this specific contract is completed. The agency then treats this as an additional short-term asset, balanced by a long-term liability. The bank will usually require that the money made available by this loan be used only for payments on materials and services and labor used on the specific contract and that the funds not be comingled with the company's other operations.

Another method is to form a partnership or joint-venture agreement between your company (presumably a corporation) and an individual who has substantial liquid and fixed assets. Often this individual may be a major stockholder in the company who wants to see his company's opportunities maximized. Thus, his assets are added to those of the company for purposes of prequalification. He, in turn, remains personally liable under the contract. Therefore, he will have entered into a separate agreement with the company to provide that his assets are not called upon until those of the company have been first exhausted.

Of course, the full joint venture, in which two companies merge their current and fixed resources, is widely used to enable prequalification on a major job. Most governmental agencies require that the new joint venture file a specific prequalification statement in its own right, rather than just accepting the sum of the two individual companies' positions.

The financial qualifications discussed above are directed toward the current volume of work which you are believed capable of handling without undue financial stress. Thus once these totals are determined, they pertain both to incomplete work under contract and to new work on which you desire to bid.

So for any specific project, you may be required to submit a statement as to "uncompleted work on hand." An accountant may interpret this as the original contract value of all work still incomplete, in which case you will seldom be able to bid any new work at all. More realistically, he may value it as the contract value of all incomplete work under contract, less the progress estimates invoiced to date. Even this is in error by the value of the work performed in the present month, so this, too, needs to be estimated.

A less conservative but sometimes justified procedure is to take the estimated cost to complete all uncompleted contracts, plus an allowance for overhead, as representing the value of uncompleted contracts on hand.

The governmental agencies which evaluate these prequalification statements have a certain amount of discretion in their determinations. For example, in the case of a West Coast contractor with a long and successful experience in bridge construction, they often used a multiple of 20 on "net quick," rather than 10, "for bridges only." For any other type of work, such as road work, the factor of 10 was applied. Note the specific wording of the "Rating Formula" at the end of Figure 9.2.

9.5. Subcontractors, Major Suppliers, and References

Item 10 of the prequalification questionnaire requires the listing of proposed subcontractors and major suppliers. In order to maintain your position in future bidding and negotiations with these suppliers and subcontractors, it is best to name several in each category who are of good standing and who are agreeable to working with you on the project or projects in question.

An exception arises when the subcontractor or supplier offers you a unique advantage. Then you may purposely want to tie your two firms together, so as to take full advantage of his unique capabilities or equipment. In that case, it is wise to have full understanding between contractor and subcontractor or supplier as to the terms and conditions under which you will operate if awarded the job.

Suppliers and subcontractors call on the same clients that you do. They are excellent sources of early information. They can give you favorable or unfavorable endorsements. Finally, they must give you proposals and prices, which can either help or hurt you competitively. Therefore, take care to maintain top relations: pay bills promptly when due, treat them fairly, and write a letter of appreciation for good performance or special services.

A final item (item 11) on the form is a list of clients and references. This will normally be required, and some or all references will normally be contacted. Therefore, it is important that your company receive favorable recommendations. The best way to assure this is, of course, to complete all projects in a satisfactory manner. Also, it is good to retain periodic contacts with past clients: they may also be a source of future work.

Before listing a reference, it is always advisable to contact him and obtain his approval. This has a double effect: it prepares him to give you a good endorsement if contacted and it lets him know you are still in business and value your contractor-client relationship with him. Therefore, first get approval by telephone; then write a letter of confirmation and thanks.

9.6. Special Considerations in Prequalification

Even though the prequalification questionnaire has been completely filled out, there are still additional steps that should be taken to stress your complete performance and abilities. Additional forms, amplifying documents (for example, see Figure 9.2), photographs of past work, and brochures can be attached to the prequalification questionnaire. This is why brochures, photos, and personal contact help: they assure you of maximum opportunity in the field in which you specialize. (These items will be discussed in later sections.) Also, a personal visit and follow-up visit can establish an identity for your firm.

Finally, check to make sure your prequalification has been received and filed in the proper place. With most large companies, this is the purchasing department, although governmental agencies often have separate contracts departments.

STATE OF CALIFORNIA

CONTRACTOR'S STATEMENT OF EXPERIENCE AND FINANCIAL CONDITION

To be filed with the State of California by Contractors proposing to bid on State Projects in accordance with provisions of **Article 4** of Chapter 3 of Part 5 of Division 3, Title 2, of the Government Code

ISSUE PREQUALIFICATION TO:

NAME OF FIRM _____

ADDRESS _____
      (Street or P.O. Box)

_____
(City)   (State)   (Zip Code)

TELEPHONE: AREA CODE _____ _____

DATE OF FINANCIAL STATEMENT _____

printed in CALIFORNIA OFFICE OF STATE PRINTING

T__ GS__ WR__ AD__

FIGURE 9.2. Contractor's statement of experience and financial condition.

INSTRUCTIONS

1. *Frequency of Submission.* A new statement will be required only once each year, unless a statement is specifically requested by the State of California, or unless a substantial decrease in the contractor's financial resources occurs subsequent to the date of a statement submitted. Otherwise, it will be optional for a contractor to file new statements at more frequent intervals; in which case, the rating in effect will be based on the latest statement on file. In the event the State of California requests submission of a new statement and the contractor does not comply with the request within thirty days, the rating based on the statement on file may, at the discretion of the State of California, be considered no longer in effect.

2. *Effective Period of Ratings.* When the financial information shown in the statement on file becomes a year old, the contractor should submit a new statement. Unless a new statement is specifically requested, prequalification ratings based on statements of financial condition on file will remain in effect no longer than fifteen months after the date of the financial condition of the statement. A 30 days extension may be granted at the Contractor's or Accountant's written request, giving a justifiable reason.

3. *Age of Financial Information.* The State of California reserves the right to reject statements in which the financial condition shown is as of a date six months or more prior to the date of filing. In no event will a statement be accepted if the financial condition shown is as of a date more than one year prior to filing.

4. *Number of Copies.* Only one copy of the statement is required even though the contractor may desire to qualify to perform work for more than one Department.

5. *Data Required.* All applicable portions of the form should be completed, with schedules being attached if the space provided does not suffice. Failure to include the information called for may result in a greatly reduced rating or no rating at all.

It will be acceptable, in lieu of completing the specific schedules in the financial portion of the statement, for the contractor to submit the customary accountant's report and schedules, provided that they include all of the information specifically requested in the form.

It is essential that construction experience of the contractor for the prior three years be shown, as such experience is considered in establishing prequalification ratings.

6. *Affidavits.* The appropriate affidavit must be completely executed or the statement will be returned. Where a consolidated statement is submitted to obtain joint prequalification of several organizations, an appropriate affidavit must be executed for each entity in the combination.

7. *Accountant's Certificate.* The certificate of a Certified Public Accountant or Public Accountant will be required in all cases. A suggested form of certificate is included which may be used if appropriate. However, it will be acceptable for the accountant to submit a certificate in his own words, including such qualifications as may be necessary in view of the scope of this assignment; provided that such qualifications shall not be so extensive as to nullify the value of the statement or its usefulness to the State of California.

Bearing in mind that working capital and net worth are important factors in determining the prequalification rating of a contractor, the accountant will perform a valuable service for his client and at the same time assist the Department if he will furnish, by supplementary schedules or as a part of his certificate, any information not specifically called for by the statement which in his opinion might properly be taken into consideration.

In the event that the contractor's job income and expenditures are accounted on a completed contract basis and the balance sheet includes an item reflecting the excess of costs to date over billings to date, or vice versa, the elements of "Accumulated Costs" and "Billings to Date" must be shown in support of the balance sheet item.

8. *Licensing.* Attention is directed to Section 14311.5 of the Government Code, amended by Chapter 1050, Statutes of 1967, shown under the "Prequalification Law" on the inside of the back cover, relative to contractors' licenses on all state projects where federal funds are involved. While in such cases no bid submitted or contract thereafter awarded will be invalidated by failure of the bidder to be properly licensed in accordance with the laws of this state, payment will be dependent upon proper licensing, and any bidder or contractor not so licensed will be subject to all legal penalties imposed by such laws, including but not limited to any appropriate disciplinary action by the Contractors State License Board.

FIGURE 9.2. (*continued*)

STATE OF CALIFORNIA

CONTRACTOR'S STATEMENT OF EXPERIENCE AND FINANCIAL CONDITION

PLEASE CHECK THE DEPARTMENT(S) WITH WHICH
YOU WISH TO ESTABLISH A PREQUALIFICATION RATING.

☐ **DEPARTMENT OF TRANSPORTATION**
HIGHWAY CONSTRUCTION

☐ **DEPARTMENT OF GENERAL SERVICES**
BUILDING CONSTRUCTION

☐ **DEPARTMENT OF WATER RESOURCES**
WATER PROJECT CONSTRUCTION

PREQUALIFICATION LAW

GOVERNMENT CODE SECTIONS:

14312. The questionnaires and financial statements are not public records and are not open to public inspection.

14313. The department shall furnish to each bidder a standard proposal form, which, when filled out and executed may be submitted as his bid. Bids not presented on forms so furnished shall be disregarded. The department shall not furnish proposal forms to any person who is required to submit and has not submitted a questionnaire and financial statement for prequalification at least FIVE days prior to the date fixed for publicly opening sealed bids and been prequalified for at least ONE day prior to that date.

Information relating to filing this statement may be obtained at:

DISBURSING OFFICE
* DEPARTMENT OF TRANSPORTATION
P. O. BOX 1139
1120 N STREET, ROOM 5101
SACRAMENTO, CALIFORNIA   95805
AREA CODE (916) 445-8875
Please use street address if sending by special courier or express mail.

* Department of Transportation acts as agent for the other departments in processing the Prequalification Statements.

FIGURE 9.2.   (*continued*)

CONTRACTOR'S STATEMENT OF EXPERIENCE

☐ A Corporation
☐ A Co-partnership
☐ An Individual
☐ Combination

* **NAME**..
(Name Must Correspond With Contractor's License in Every Detail)

PRINCIPAL OFFICE..
(Street or P.O. Box)   (City)   (State)   (Zip Code)

The signatory of this questionnaire guarantees the truth and accuracy of all statements and of all answers to interrogatories hereinafter made

1. Are you licensed as a Contractor to do business in California?........... License No........... Type...........

    Classification (Type) of Specialty Contractor:...........

2. How many years has your organization been in business as a contractor under your present business name?
....

3. How many years experience in........... (Type) ....construction work has your organization had:

    (a) As a general contractor?...........   (b) As a subcontractor?...........

4. Show the projects your organization has completed during the last three years in the following tabulation:
    To assure maximum consideration for your prequalification rating, be specific as to the nature of the work your firm actually performed.

YEAR COMPLETED	TYPE OF WORK	VALUE OF WORK PERFORMED	LOCATION OF WORK	FOR WHOM PERFORMED

* Except as provided in Section 14311.5 of the Government Code, a contractor prequalified through the submission of a Statement of Experience and Financial Condition who wishes to bid on projects handled by the State of California must be licensed under the California "Contractors' License Law". The licensing must correspond with the prequalification as to type of organization; i.e., a contractor licensed as a corporation must be prequalified as a corporation, a contractor licensed as a copartnership must be prequalified as a copartnership, etc. Where the prequalification is in the name of and based on a combination of such organizations, then the combination must be licensed as such and any bids based on such prequalification must be in the name of the combination so prequalified. The license, or licenses, held by a contractor must authorize the type of work on which he requests permission to bid. Corporations not incorporated in the State of California must take the necessary steps to permit doing business in the state.

PLEASE INCLUDE THREE COPIES OF ANY ATTACHMENTS TO THIS PAGE

FIGURE 9.2. (*continued*)

5. Have you or your organization, or any officer or partner thereof, failed to complete a contract? _____ If so, give details
_____
_____

6. If you have a controlling interest in any firms presently prequalified with the State of California, show names thereof
_____
_____

7. To determine status of assets, in what other lines of business pertaining to this Financial Statement do you have a financial interest?_____
_____

8. Name the persons with whom you have been associated in business as partners or business associates in each of the last five years _____
_____
_____
_____

9. What is the construction experience of the principal individuals of your present organization?

INDIVIDUAL'S NAME	PRESENT POSITION OR OFFICE IN YOUR ORGANIZATION	YEARS OF CONSTRUCTION EXPERIENCE	MAGNITUDE AND TYPE OF WORK	IN WHAT CAPACITY

If a corporation, answer this:	If a copartnership, answer this:
Capital paid in cash, $_____	Date of organization_____
When incorporated_____	State whether partnership is general, limited or association
In what State_____	
President's name_____	Name and address of each partner:
Vice President's name_____	
Secretary's name_____	
Treasurer's name_____	

WERE YOU PREQUALIFIED LAST YEAR? YES____ NO____

PLEASE INCLUDE THREE COPIES OF ANY ATTACHMENTS TO THIS PAGE

FIGURE 9.2. (*continued*)

WHERE PREQUALIFICATION IS BASED ON A COMBINATION OF ORGANIZATIONS, THE APPROPRIATE AFFIDAVITS MUST BE EXECUTED FOR EACH MEMBER OF SUCH COMBINATION.

AFFIDAVIT FOR INDIVIDUAL

_____
(Name of individual)

doing business as _____
(Name of firm, if any)

certifies and says: That he is the person submitting the statement of experience and financial condition; that he has read the same, and that the same is true of his own knowledge; that the statement is for the purpose of inducing the State of California to supply the submittor with plans and specifications, and that any depository, vendor, or other agency therein named is hereby authorized to supply said State of California with any information necessary to verify the statement; and that furthermore, should the foregoing statement at any time cease to properly and truly represent his financial condition in any substantial respect, he will refrain from further bidding on State work until he shall have submitted a revised and corrected statement.

I certify and declare under penalty of perjury that the foregoing is true and correct.

Subscribed at _____, _____, State of _____,
　　　　　　　　(City)　　　　　　　　(County)

NOTE: Statement will be returned unless affidavit is complete including the date of signature.

on _____ 19____
　　　　(Date)

_____
(Applicant must sign here)

AFFIDAVIT FOR CO-PARTNERSHIP

_____, *certifies and says:* That he is a partner of the partner-
(Name of partner)

ship of _____ ;
(Name of firm)

that said partnership submitted the statement of experience and financial condition; that he has read the same and that the same is true of his own knowledge; that the statement is for the purpose of inducing the State of California to supply the submittor with plans and specifications, and that any depository, vendor, or other agency therein named is hereby authorized to supply said State of California with any information necessary to verify the statement; and that furthermore, should he foregoing statement at any time cease to properly and truly represent the financial condition of said firm in any substantial respect, they will refrain from further bidding on State work until they shall have submitted a revised and corrected statement.

I certify and declare under penalty of perjury that the foregoing is true and correct.

Subscribed at _____, _____, State of _____,
　　　　　　　　(City)　　　　　　　　(County)

NOTE: Statement will be returned unless affidavit is complete including the date of signature.

on _____ 19____
　　　　(Date)

The foregoing statement and affidavit are hereby affirmed.

_____
(Member of firm must sign here)

_____　　_____
(Remaining members of firm sign here)　　　　　(Name of firm)

FIGURE 9.2. *(continued)*

AFFIDAVIT FOR CORPORATION

_____, certifies and says: That he is_____
      (Name of officer)   (Official capacity)

of the_____,
                    (Name of firm)

the corporation submitting the statement of experience and financial condition; that he has read the same, and that the same is true of his own knowledge; that the statement is for the purpose of inducing the State of California to supply the submittor with plans and specifications, and that any depository, vendor, or other agency therein named is hereby authorized to supply said State of California with any imformation necessary to verify the statement; and that furthermore, should the foregoing statement at any time cease to properly and truly represent the financial condition of said corporation in any substantial respect, it will refrain from further bidding on State work until it shall have submitted a revised and corrected statement.

I certify and declare under penalty of perjury that the foregoing is true and correct.

Subscribed at_____, _____, State of_____,
          (City)   (County)

NOTE: Statement will be returned unless affidavit is complete including the date of signature.

on_____ 19_____.
       (Date)

_____
(Officer must sign here)
NOTE.—Use full corporate name and attach corporate seal.

COMMENTS

FIGURE 9.2. *(continued)*

CERTIFICATE OF ACCOUNTANT

I (WE) HAVE EXAMINED THE FINANCIAL STATEMENT OF
_____AS OF_____,
MY (OUR) EXAMINATION WAS MADE IN ACCORDANCE WITH GENERALLY ACCEPTED AUDITING STANDARDS, AND ACCORDINGLY INCLUDED SUCH TESTS OF THE ACCOUNTING RECORDS AND SUCH OTHER AUDITING PROCEDURES AS WE CONSIDERED NECESSARY IN THE CIRCUMSTANCES.

IN MY (OUR) OPINION, THE ACCOMPANYING FINANCIAL STATEMENT INCLUDED ON PAGES_____ TO _____, INCLUSIVE, SET FORTH FAIRLY THE FINANCIAL CONDITION OF

_____

AS OF_____, IN CONFORMITY WITH GENERALLY ACCEPTED ACCOUNTING PRINCIPLES APPLIED ON A BASIS CONSISTENT WITH THAT OF THE PRECEDING YEAR.

OR

I (WE) HAVE REVIEWED THE ACCOMPANYING FINANCIAL STATEMENT OF
_____AS OF_____
ALL INFORMATION INCLUDED IN THE FINANCIAL STATEMENT IS THE REPRESENTATION OF THE MANAGEMENT (OWNERS) OF THE COMPANY.

A REVIEW CONSISTS PRINCIPALLY OF INQUIRIES OF COMPANY PERSONNEL AND ANALYTICAL PROCEDURES APPLIED TO FINANCIAL DATA. IT IS SUBSTANTIALLY LESS IN SCOPE THAN AN EXAMINATION IN ACCORDANCE WITH GENERALLY ACCEPTED AUDITING STANDARDS, THE OBJECTIVE OF WHICH IS THE EXPRESSION OF AN OPINION REGARDING THE FINANCIAL STATEMENTS TAKEN AS A WHOLE. ACCORDINGLY, I (WE) DO NOT EXPRESS SUCH AN OPINION.

BASED ON MY (OUR) REVIEW, WITH THE EXCEPTION OF THE MATTER(S) DESCRIBED IN THE FOLLOWING PARAGRAPH(S), I AM (WE ARE) NOT AWARE OF ANY MATERIAL MODIFICATIONS THAT SHOULD BE MADE TO THE ACCOMPANYING FINANCIAL STATEMENTS IN ORDER FOR THEM TO BE IN CONFORMITY WITH GENERALLY ACCEPTED ACCOUNTING PRINCIPLES.

_____
(Accountant must sign here)

_____
(Print name or firm)

Telephone: _____   License No. _____
☐ Public Accountant
☐ Certified Public Accountant

Special note to Accountant:
The above Certificate of Accountant must not be made by any individual who is in the regular employ of the individual, co-partnership or corporation submitting this statement; nor by any individual who is a member of the concern if his financial interest is over 10%.

THE CERTIFICATE OF A LICENSED ACCOUNTANT WILL BE REQUIRED IN ALL CASES.

FIGURE 9.2. *(continued)*

CONTRACTOR'S FINANCIAL STATEMENT

NAME..

Condition at close of business.. 19......

ASSETS	DETAIL	TOTAL
Current Assets		
1. Cash..		
2. Notes receivable................................		
3. Accounts receivable from completed contracts...........		
4. Sums earned on incomplete contracts...................		
5. Other accounts receivable............................		
6. Advances to construction joint ventures...............		
7. Materials in stock not included in Item 4.............		
8. Negotiable securities................................		
9. Other current assets.................................		
TOTAL...		
Fixed and Other Assets		
10. Real estate...		
11. Construction plant and equipment....................		
12. Furniture and fixtures..............................		
13. Investments of a non-current nature.................		
14. Other non-current assets............................		
TOTAL...		
TOTAL ASSETS..		
LIABILITIES AND CAPITAL		
Current Liabilities		
15. Current portion of notes payable, exclusive of equipment obligations and real estate encumbrances..........		
16. Accounts payable....................................		
17. Other current liabilities...........................		
TOTAL...		
Other Liabilities and Reserves		
18. Real estate encumbrances............................		
19. Equipment obligations secured by equipment..........		
20. Other non-current liabilities and non-current notes payable........		
21. Reserves..		
TOTAL...		
Capital and Surplus		
22. Capital Stock Paid Up...............................		
23. Surplus (or Net Worth)..............................		
TOTAL...		
TOTAL LIABILITIES AND CAPITAL...........................		
CONTINGENT LIABILITIES		
24. Liability on notes receivable, discounted or sold........		
25. Liability on accounts receivable, pledged, assigned or sold........		
26. Liability as bondsman...............................		
27. Liability as guarantor on contracts or on accounts of others........		
28. Other contingent liabilities........................		
TOTAL CONTINGENT LIABILITIES............................		

NOTE.—Show details under main headings in first column, extending totals of main headings to second column.

FIGURE 9.2. *(continued)*

DETAILS RELATIVE TO ASSETS

1 Cash:
(a) On hand .... $ ....
(b) Deposited in banks named below .... $ ....
(c) Elsewhere—(state where) .... $ ....

NAME OF BANK	LOCATION	DEPOSIT IN NAME OF	AMOUNT

2* Notes Receivable:
(a) Due within one year .... $ ....
(b) Due after one year .... $ ....
(c) Past due .... $ ....

RECEIVABLE FROM	FOR WHAT	DATE OF MATURITY	HOW SECURED	AMOUNT

Have any of the above been discounted or sold? .... If so, state amount, to whom, and reason ....

3* Accounts receivable from completed contracts exclusive of claims not approved for payment $ ....

RECEIVABLE FROM	TYPE OF WORK	AMOUNT OF CONTRACT	AMOUNT RECEIVABLE

Have any of the above been assigned, sold or pledged? .... If so, state amount, to whom, and reason ....

4* Sums earned on incomplete contracts, as shown by engineers' or architects' estimates .... $ ....

RECEIVABLE FROM	TYPE OF WORK	AMOUNT OF CONTRACT	AMOUNT RECEIVABLE

Have any of the above been assigned, sold or pledged? .... If so, state amount, to whom, and reason ....

* List separately each item amounting to 10 per cent or more of the total and combine the remainder.

FIGURE 9.2. (*continued*)

DETAILS RELATIVE TO ASSETS (Continued)

5* Accounts receivable not from construction contracts .... $ ....

RECEIVABLE FROM	FOR WHAT	WHEN DUE	AMOUNT

What amount, if any, is past due? .... $ ....
Assigned, sold, or pledged .... $ ....

6 Advances to construction joint ventures .... $ ....

NAME OF JOINT VENTURE	TYPE OF WORK	AMOUNT

What amount, if any, has been assigned, sold, or pledged? .... $ ....

7 Materials in stock and not included in Item 4
    (a) For use on incomplete contracts (inventory value) .... $ ....
    (b) For future operations (inventory value) .... $ ....
    (c) For sale (inventory value) .... $ ....

DESCRIPTION	QUANTITY	VALUE		
		FOR INCOMPLETE CONTRACTS	FOR FUTURE OPERATIONS	FOR SALE

What amount, if any, has been assigned, sold, or pledged? .... $ ....

8** Negotiable Securities (List non-negotiable items under Item 13)
    (a) Listed—Present market value .... $ ....
    (b) Unlisted—present value .... $ ....

ISSUING COMPANY	CLASS	QUANTITY	BOOK VALUE		PRESENT VALUE (ACTUAL OR ESTIMATED)	
			UNIT PRICE	AMOUNT	UNIT PRICE	AMOUNT

Who has possession? ....
If any are pledged or in escrow, state for whom and reason ....

Amount pledged or in escrow .... $ ....

* List separately each item amounting to 10 per cent or more of the total and combine the remainder.
** IMPORTANT: Items listed under this heading will be given no consideration as working capital unless actual or estimated market value is furnished.

FIGURE 9.2. *(continued)*

DETAILS RELATIVE TO ASSETS (Continued)

9 Other current assets
Bid deposits, prepaid expenses, cash value of life insurance, accrued interest, etc. $ ....

DESCRIPTION	AMOUNT

10* Real estate { (a) Used for business purposes .... $ ....
Book value { (b) Not used for business purposes .... $ ....

LOCATION	DESCRIPTION	HELD IN WHOSE NAME	VALUE

11* Construction plant and equipment .... $ ....

11A What is your approximate annual income from rental of equipment owned by you, exclusive of such income from associated concerns having same ownership .... $ ....

12* Furniture and fixtures .... $ ....

13 Investments of a non-current nature .... $ ....

DESCRIPTION	AMOUNT

14 Other non-current assets .... $ ....

DESCRIPTION	AMOUNT

TOTAL ASSETS $ ....

* Show book value (cost less depreciation) unless an appraisal schedule prepared by an *independent* appraiser is attached; in which case appraised value may be shown.

FIGURE 9.2. (*continued*)

DETAILS RELATIVE TO LIABILITIES

15 | Current Portion of Notes Payable, exclusive of equipment obligations and real estate obligations .... $ ....

TO WHOM PAYABLE	WHAT SECURITY	WHEN DUE	AMOUNT

16* | Accounts Payable: (a) Not past due .... $ ....
(b) Past due .... $ ....

TO WHOM PAYABLE	FOR WHAT	WHEN DUE	AMOUNT

17 | Other current liabilities .... $ ....
Accrued interest, taxes, insurance, payrolls, etc.

DESCRIPTION	AMOUNT

18 | Real estate encumbrances .... $ ....

19 | Construction Equipment obligations secured by equipment: { (a) Total payments due within six months $ ....
{ (b) Total payments due after six months $ ....

TO WHOM PAYABLE	HOW PAYABLE **	AMOUNT

20 | Other non-current liabilities and non-current notes payable .... $ ....

DESCRIPTION	FOR WHAT	WHEN DUE	AMOUNT

21 | Reserves .... $ ....

DESCRIPTION	AMOUNT

22 | Capital stock paid up: (a) Common .... $ ....
(b) Preferred .... $ ....

23 | Surplus (or Net Worth) .... $ ....

TOTAL LIABILITIES AND CAPITAL $ ....

* List separately each item amounting to 10 per cent or more of the total and combine the remainder.
** In this space show amount and frequency of installment payments.

FIGURE 9.2. (*continued*)

STATE OF CALIFORNIA

GENERAL STATEMENT OF BANK CREDIT

(Date)

Disbursing Office
Department of Transportation
P.O. Box 1139
Sacramento, California 95805

Gentlemen:

In connection with the prequalification of

---, a contractor
(Name of Contractor)

under Sections 14310 et seq. of the Government Code to perform contracts with the Departments of the State of California, we hereby declare that said contractor has been extended a line of credit in a total amount not exceeding $ ...., and that such credit will not be withdrawn or reduced without notice to the Department of Transportation.*

This letter is signed with the understanding that it is a document to be used by the State of California for the Department of Transportation, Department of General Services and/or the Department of Water Resources only for the purpose of determining the financial resources of said contractor available for use in performing work under contracts which may be awarded to him by the Departments during the term of his prequalification.

This General Statement of Bank Credit will **EXPIRE** with the Contractor's Statement of Experience and Financial Condition for which the line of credit was issued.

(Name of Bank)

(Address)

By-----------------------------------

(Title)

* Department of Transportation acts as agent for the other departments in processing the Prequalification Statements.

PLEASE NOTE: The above form may be used to augment your Working Capital and completed by your bank, or if they prefer, one with substantially the same provisions may be issued on their own letterhead.

FIGURE 9.2. *(continued)*

ADDITIONAL INFORMATION REQUIRED FOR PREQUALIFICATION WITH THE

DEPARTMENT OF WATER RESOURCES

General Contracting—Heavy			*Other Contractors	
TYPE OF WORK	MARK WITH (X)			
1 All Classes of Construction		D E S I G N	M A N U F A C T U R E	GENERAL CATEGORY
2 Clearing and Grubbing				
3 Excavation				
4 Tunnel				
5 Levee and Flood Control				
6 Sewer and Water Lines				
7 Road Work				Complete Questionnaire Page 14
8 Bridge				
9 Grading and Paving			Electrical	
10 Pipeline				
11 Marine			1 Motors and Generators	
			2 Transformers	
Building Construction			3 Power Circuit Breakers	
1 General			4 Switchgear	
2 Structural			5 Control Boards	
3 Electrical			6 Cables 230KV	
4 Mechanical			7 Station Batteries	
5			8	
Drilling			Mechanical	
1 Diamond Core			1 Pumps	
2 Wells			2 Turbines	
3 Foundations			3 Governors	
4			4 Gates	
			5 Valves	
			6 Cranes and Gantries	
			7	
			Electronic—Complete Page 14	

* Other Contractors are contractors that design, manufacture and/or fabricate equipment off-site and install it at the jobsite under "furnish and install" contracts.

FIGURE 9.2. (*continued*)

DEPARTMENT OF WATER RESOURCES
Questionnaire for "Other Contractors"
(Design—Manufacture—Fabrication—Installation)

INSTRUCTIONS: All firms who intend to participate in design, manufacturing, fabricating, and installing contracts will be required to submit the following data. Such data should be provided on a separate form attached to this statement using the same paragraph numbering as shown below.

For each General Category checked, a separate set of answers to these questions will be required. This questionnaire will be submitted only once and will be updated as required by the State of California. Additional information may be required for particular projects.

A.—Design*
    1. Submit an Organization Chart of your firm.
    2. Describe the major design projects completed by your firm in the last five years. (Describe system or equipment designed, owner, where equipment is installed, and approximate cost of project covered by this design.)
    3. List consulting firms engaged by your company within the past five years. List the specific projects on which they were employed and the nature of their services.
    4. Give the location of your design offices.
    5. List the number of engineers, technicians and draftsmen employed by your company and submit a brief resume of the experience of your supervisory personnel.
    6. List any additional information you believe is important.

B.—Fabrication or Manufacture
    1. Describe the major contracts completed in the last five years. (Describe equipment and ratings, location of installation, owner, and contract amount.)
    2. How long has your firm been engaged in this type of manufacturing or fabrication?
    3. Give location of major plant facilities.
    4. Briefly describe each plant's capabilities, type manufacturing, crane sizes, size and description of major machine tools, etc. Pictures are desirable.
    5. List the firms you normally engage to perform work which is not within your capability. (Painting, castings, galvanizing, machining, etc.)
    6. List sources or suppliers of major components.
    7. List additional information you believe is important.

C.—Quality Control
    1. Describe in detail your quality control program.
        a. What is your procedure to insure that all articles furnished to you have met all specification requirements?
        b. Describe the material testing facilities which are available in your plant and those used outside your plant.
        c. Describe how inspection records are maintained and what records are available to the Department.
    2. Submit your quality control manual.

D.—Laboratory and Testing Facilities
    1. Describe laboratory and model testing facilities and give their location.
    2. Describe equipment testing facilities and give their location.
    3. What testing, if any, is normally contracted out? Describe and list firms engaged.

E.—Field Installation and Service
    1. Describe how you install or erect the product you manufacture, i.e., subcontract, furnish erection engineers, install by own crews, etc.
    2. Describe the training program regarding operation and maintenance which is available to the Department.
    3. Location of maintenance shops and parts storage warehouses.

* An affiliate company may be used to fulfill any portion of the design requirements. However, if an affiliate is used, an affiliate agreement shall be submitted for approval. An outline of the information required to be in the affiliate agreement may be obtained from the Department of Water Resources.

Electronic Contractors

A Special Electronics Questionnaire will be furnished by the Department of Water Resources upon request. This questionnaire will be submitted only once and will be updated as required by the Department of Water Resources.

FIGURE 9.2. (*continued*)

ADDITIONAL INFORMATION REQUIRED FOR PREQUALIFICATION WITH THE
DEPARTMENT OF TRANSPORTATION

SAFETY RECORD

This page *must be completed* for contractor's wishing to be prequalified with the Department of Transportation. It is not required for Water Resources and General Services. See Section 14310.4 of the Government Code which became effective January 1, 1978.

Contractor's Name _____

Address _____ Phone _____

_____

A. Insurance Experience Modification _____% as of _____ (date) (This information is available from your Workman's Compensation Insurer.)

   If no rating, are you self-insured? Yes ☐   No ☐

   Please explain if no rating and not self-insured. _____
   _____
   _____

B. The Department of Transportation is primarily interested in your safety records in regard to public works contracts performed in California. Should you use data collected from a wider range of projects to complete Part C, please briefly describe the type of work and geographic area included. Questions should be directed to Linn Ferguson (916-445-7958) or Ralph Haverkamp (916-445-4279).

C. **CONTRACTOR'S INDUSTRIAL SAFETY RECORD
    FOR LAST 3 COMPLETE YEARS**

	19__	19__	19__
*1. No. of fatalities			
*2. No. of lost workday cases			
*3. No. of lost workdays			
4. Total Man-Hours worked			

* The information required for these items is the same as required for columns 1, 3 and 4, Log and Summary—Occupational Injuries and Illnesses, CAL/OSHA No. 200.
  Remarks: _____
  _____

D. You may attach any additional information or explanation of data which you would like taken into consideration in evaluating the safety record.

A SUBSTANDARD SAFETY RECORD MAY BE CAUSE FOR DETERMINATION BY THE DIRECTOR THAT A CONTRACTOR IS NOT A RESPONSIBLE BIDDER. THIS DETERMINATION REQUIRES A HEARING BY THE DIRECTOR WHERE THE CONTRACTOR MAY PRESENT EVIDENCE OF HIS SAFETY PERFORMANCE.

81357-500 9-80 10M CAM ⊕ OSP

FIGURE 9.2. (*continued*)

BIDDING INFORMATION

1. Any contractor who has prequalified and who desires plans and proposal forms should make his request in writing or he may request plans and proposal forms by telephone to the Department advertising the work. The status of the contractor's work on incomplete contracts with the appropriate Department must be shown on that form.

2. Two or more contractors who have prequalified by filing separate statements and who wish to combine their assets for bidding on a single project may do so by filing an affidavit of joint venture in the form approved by the State of California, but such affidavit will be valid only for the specific project mentioned therein. Should the contractors desire to continue to bid jointly, a joint prequalification statement should be filed. Attention is called to the "Contractors' License Law" with respect to the license requirements for joint bids.

RATING FORMULA

Prequalification ratings in the Department of Transportation and Department of Water Resources are based on ten times working capital or four times net worth, whichever is smaller; subject to adjustment upon consideration of experience, equipment and performance factors. Ratings in the Department of General Services are based on ten times working capital; subject to adjustment upon consideration of experience and performance factors.

Working capital may be augmented by submission of Statement of Bank Credit in a form prescribed by the Department and net worth may be augmented by submitting appraisals of fixed assets prepared by independent appraisers.

PREQUALIFICATION LAW

14310. The department may, and on contracts the estimated cost of which exceeds three hundred thousand dollars ($300,000) the department shall, require from prospective bidders answers to questions contained in a standard form of questionnaire and financial statement including a complete statement of the prospective bidder's financial ability and experience in performing public works. When completed, the questionnaire and financial statement shall be verified under oath by the bidder in the manner in which pleadings in civil actions are verified.

14311. The department shall adopt and apply a uniform system of rating bidders, on the basis of the standard questionnaires and financial statements, in respect to the size of the contracts upon which each bidder is qualified to bid. When bids for more than one project are to be received at the same bid opening, the department may permit a bidder to submit bids for each project within such bidder's prequalification rating, even though such rating is insufficient to permit the bidder to be awarded the contract for each project bid upon.

In no event shall any bidder be awarded a contract if such contract award would result in the bidder having under contract work for which prequalification is required in excess of that authorized by his prequalification rating. In determining whether an award of a contract would result in a bidder having under contract work in excess of that authorized by his prequalification rating, the department may use its estimated cost of such contract rather than the amount of the bidder's bid. If the department determines that a bidder would be awarded the contract for two or more projects but cannot be awarded the contract for all such projects because of the inadequacy which of the bids of such bidder are to be accepted and the contract awarded thereon and which of the bids of such bidder are to be disregarded. In making its decision the department shall be guided by the combination of contract awards which will result in the lowest total cost for the projects involved.

14311.5. In all state projects where federal funds are involved, no bid submitted or contract thereafter awarded shall be invalidated by the failure of the bidder or contractor to be properly licensed in accordance with the laws of this State, nor shall any such contractor be denied payment under any such contract because of such failure; provided, however, that the first payment for work or material under such contract shall not be made by the State Controller unless and until the Registrar of Contractors certifies to him that the records of the Contractors State License Board indicate that such contractor was or became properly licensed between the time of bid opening and the making of the certification. Any bidder or contractor not so licensed shall be subject to all legal penalties imposed by such laws, including but not limited to any appropriate disciplinary action by the Contractors State License Board, and the department shall include a statement to that effect in the standard form of prequalification questionnaire and financial statement.

14312. The questionnaires and financial statements are not public records and are not open to public inspection.

14313. The department shall furnish to each bidder a standard proposal form, which, when filled out and executed may be submitted as his bid. Bids not presented on forms so furnished shall be disregarded. The department shall not furnish proposal forms to any person who is required to submit and has not submitted a questionnaire and financial statement for prequalification at least FIVE days prior to the date fixed for publicly opening sealed bids and been prequalified for at least ONE day prior to that date.

FORM DSB-70 (REV. 1-81)

FIGURE 9.2. (*continued*)

10
The Proposal

10.1. The Basic Proposal Format

The proposal letter normally has a basic format—introduction, body, conclusion, and appendixes.

The *introduction* contains a description of what the project is and gives a reference to the bid invitation or plans and specifications to be bid on. Also, the introduction will contain a basic interpretation of the entity of the bidder—for example, "the Adam Smith Company, a general building contractor incorporated under the laws of the state of Ohio, whose address is . . ."

The *body* defines various legal conditions that have not been defined in the bid proposal or invitation to bid. Typical conditions to be defined include schedule, price, and payment terms and exceptions or qualification to the specifications (stipulations) and contractual clauses, such as insurance required and work to be performed by the owner (for example, obtain the building permit). Since there is only limited room for sales arguments in the proposal body, normally only the key sales arguments are presented. For example, if schedule is the key issue, there might be a guarantee to finish the project three months ahead of the owner's proposed schedule. Typical sales arguments might include a statement about proper personnel, special sources of material, favorable contract terms, high quality, or an attractive financing package.

The *conclusion* is normally a short paragraph that ends the proposal with something that is pleasant and positive. A typical ending is, "We trust that this proposal will meet with your satisfaction. I will call next week to see if you have any questions."

Since the actual proposal is purposefully relatively short, additional material may be included in the *appendixes*. For example, the appendixes may

184   *The Proposal*

contain a four-page bid sheet with unit prices for overruns or underruns. Too much detail in the proposal proper disrupts the continuity when it is read by the client and detracts from the key elements of the proposal.

10.2. Exceptions and Stipulations

Many bid documents contain a clause or clauses that are inequitable, objectionable, or technically wrong from the contractor's point of view. In the proposal the contractor often finds it necessary to take exceptions to certain parts of the specifications and qualify his bid or add stipulations to his bid.

The contractor would like to use some method or device to remove the clause without being disqualified. It is most effective not to restate the bad clause: this just reinforces it and allows the owner's engineer to make a detailed comparison of the old clause and a new clause. This also forces the person to tell his boss that he made a mistake if the old clause is thrown out. The contractor should "suggest" a new clause that can be modified by the owner. Then the owner's engineer and lawyer can have the last word. For example, consider an objectionable clause that states that use of jetting in the pile-driving operation is not permitted. A positive approach might be as follows: "Our studies indicate that limited jetting may be required to obtain the specified penetration of the piles. Therefore, we suggest that a clause similar to the following be considered."

All risks should have been identified by the contractor during the bidding period, and unusual or highly objectionable risks should have been pinpointed. Since a client will not welcome qualifications or stipulations on risks, clauses should not be challenged unless absolutely necessary. For example, consider an $11 million foundation contract where there exists a possible 10-in. water pipe that may need to be moved at a cost of $15,000. The contractor might feel justified in stipulating that the owner will pay to have the pipe relocated if it is encountered; however, this stipulation could be a source of irritation to the owner. The contractor would be better off to include $7500 in his price as a contingency and not use the stipulation.

Some risks can be totally unacceptable. For example, on a Ben C. Gerwick, Inc., pile job for a major public utility, the utility wanted Gerwick to pay for any downtime of its computer in the building next door, caused by vibrations due to the pile driving. The downtime could cost several million dollars. No insurance company would take such a risk. Ben C. Gerwick, Inc., refused to accept such a risk. Instead, a procedure to minimize the vibrations from the pile driving was mutually agreed on by Gerwick and the utility. As long as Gerwick followed this procedure, it was contractually exempt from the risk.

When you do take exceptions, take exceptions on the key risk item. If you take too many exceptions, the owner is quite likely to throw out your bid. This very case happened to a California building contractor on a large

Southern California contract. The contractor's lawyer, who was new to the business, dictated changed wording to 11 contractual clauses. The owner decided that the contractor was nonresponsive: he had changed too much.

On the items you do take exceptions to, offer a reasonable contractual clause to substitute. For example, on a Nigerian oil company contract, a contractor we will call Overseas Construction was asked to take complete responsibility for any pollution that occurred during the contract period, regardless of cause or source. Overseas Construction qualified its proposal to take only the responsibility for pollution caused by its negligence and refused to take the responsibility for an oil spill that could be caused by another contractor or by the very nature of the required work or by geological conditions.

10.3. How to Avoid Late Payments

In the construction business there exists the constant problem of late payments. A late payment can damage the contractor's cash flow and can force the contractor to procure a short-term note, which adds additional financing costs to the job. Therefore, contractors endeavor to insert a clause in the proposal to avoid this problem. One clause that has been tried in the past is that the owner will be charged an interest rate of 1 percent per month for a late payment. This can backfire. An owner who needs additional financing for his own operations is likely to take you up on your offer to lend him money at 1 percent instead of a bank loan for his own financing. Some court decisions have shown this practice to be legal—the contractor has already agreed that the owner can be late, provided he pay interest. Most contractors do not want to go into the banking business. Further, in practice, no client will really pay a late-payment penalty charge. There is always a good excuse as to why the payment was not made on time. For example, very few contractors' invoices are absolutely perfect; there is normally a small loophole in the bill that can be used to excuse the late payment. ("It was sent to the wrong department." "It did not include full substantiation.")

A major automobile company had a policy of always paying its progress payments 30 days late. It would find a $6 error in a discount that had to be corrected before payment could be made. Or "Item 12" would be unclear, and it would ask for substantiating data. There was always something that would prevent the payment from being made on time. Since the company would be making a total of perhaps $100 million of payments to all its contractors in the United States each month, this policy had the effect of letting the company have $100 million more working capital interest free. If a generally ethical corporation like one of the major automobile companies uses this policy, think what unethical owners may try to do, particularly in times of high interest rates!

One effective way to help the late-payment problem is to offer, during contract negotiations, a lump-sum price incentive if the owner makes a cash advance each month, sufficient to cover the expenditures during the month. If the owner agrees to cash advances, the contractor can then offer a discount equal to the financing charges saved by the contractor. This plan can be a very attractive marketing point.

Also, a clause should be inserted that requires the payment of all but the amount in dispute in the case of inadequate substantiation or minor errors. This way a $1,100,000 progress payment will not be held up by a $50,000 dispute—the contractors will receive the $1,050,000 that is not in dispute.

It is common practice for an owner to hold back the full 5 or 10 percent retention until all bills, claims, and punch lists are fully settled. In the case of machinery installations, particularly, this may take many months. The contract should provide for release of the retention upon occupancy and/or beneficial use by the owner or upon substantial completion by the contractor. The owner should then retain only enough to cover disputed items and items needing correction.

10.4. Technical and Contractual Alternatives

Technical alternatives for the proposal can be handled in a number of ways—prior approval, undisclosed (for further negotiations), or disclosed with full details.

Consider the case where contractor A has conceived of a technical alternative to use lightweight concrete by which he can offer the owner a savings of $30,000. The bids are as follows:

Contractor A	Contractor B
$3,722,000	$3,716,000

Contractor A will deduct $30,000 from his bid if the technical alternative of lightweight concrete is used. Therefore, contractor A's bid would be $3,692,000. Contractor B has offered no technical alternative.

In this case the owner may possibly engage in some "ethical" shopping and negotiate with contractor B who has the low bid on the main proposal. He will ask him for his deduction if he substitutes lightweight concrete. Quite likely, contractor B will deduct $30,000 or more from his bid by using contractor A's technical alternative. The owner thus gets to take advantage of both the technical innovation and the low bid. Contractor A has come up with a free technical alternative for contractor B. What can contractor A do?

Contractor A could have submitted his proposal in a different manner. Perhaps the best way would have been to submit a bid price of $3,692,000 and state that it is based on the technical alternative of using lightweight concrete. Contractor A probably will wisely elect not to state the additional price required for using regular concrete. This has proved successful in many cases.

Another way would have been to submit a bid price of $3,722,000 and then tell about the technical alternative and simply say that it will produce a significant cost savings. If the owner is interested, he will at least talk to contractor A. Once contractor A is in discussions, he can use his personal powers of persuasion to augment his reduction in price.

There is no one right way to present a technical alternative. Therefore, you need to try to get to know the owner and then guess how he will react.

Asking for approval ahead of time may lead to the owner's denial (in which case you have lost bargaining power) or to his issuance of an addendum to all bidders, giving them the chance to bid on your alternative.

Besides technical alternatives, you can suggest contractual alternatives that are bait for follow-up talks. You suggest a contractual alternative that "offers the promise of significant savings to the owner." This is particularly effective if you judge that all bids will be over his budget.

10.5. How Not to Sell a Proposal

Now let us consider a case where a proposal turned out to be a comedy of errors. F & C Construction Company was asked by Minerals Mining Company to submit a proposal for a large ocean-terminal project in Australia. F & C Construction had previously worked on the job site and was in an excellent position to be awarded the contract. In fact, only three companies were asked to submit proposals, and the owner would accept almost any type of reasonable contract—cost plus, target estimate "or lump sum." F & C's proposal contained five major errors.

1. F & C's estimated price was about $24,600,000, and the competitors' bids were about $23,800,000 and $23,750,000. The owner assumed that F & C had the high price and used this assumption throughout the selection process. Actually, F & C's bid was in U.S. dollars, and the competitors' bids were in Australian dollars. At the time a U.S. $1.00 equaled Australian $1.25. Therefore, F & C's price was actually low; however, the owner did not know this because F & C did not make this clear. Nationalism runs high in Australia, as in most countries. Regardless of what currency you will later specify for payments and regardless of how you later may wish to cover currency reevaluation, it is poor tactics to quote in a foreign currency, especially if the numerical value is higher, as in the case quoted. In this case, U.S. was "foreign."
2. The project had a 14-month completion schedule; the owner had made commitments to load ships in 14 months from the starting date. F & C's proposal included a provision for payment for main office overhead of $16,000 per month for a minimum of 18 months. This indicated to the owner that it was not even attempting to meet the owner's 14-month schedule.

3. The only equipment (a derrick barge) in the South Pacific of large enough capacity to perform the work belonged to F & C. In its proposal F & C agreed to furnish the equipment at a lump sum of $2,500,000 for equipment rental over an 18-month period. F & C also inserted a clause that this equipment was definitely not available to be rented to another contractor and that if it were not used on this job, it would be moved to the Arabian Gulf. The threat had a reverse effect. Minerals Mining Company designed and built its own fully adequate crane for $2,200,000 (and sold it for about half that after the project was completed). Few executives in the decision-making role got where they are by backing down to threats.
4. F & C had an excellent project manager who had worked for the owner before, and who would be the preferred choice of the owner to have on this job. In the proposal it was stated that this individual would be assigned to the job. However, F & C did not ask or tell this project manager about this. Since one of the owner's key executives knew this man well, he telephoned him to say that he had heard that he (the project manager) might be coming down. The "assigned" project manager replied, "Not on your life; I am staying in the United States for the next two years." Well, so much for that sales argument.
5. The competition showed more insight concerning the job than F & C, who had been working there for two years. F & C just seemed to take the job for granted. The owner asked for a detailed work program. F & C wrote back and rather bluntly refused the request, saying it would furnish the schedule after award. The competition, of course, did furnish detailed schedules.

It is not surprising that F & C was not awarded this contract.

As another example, consider the following pricing on a segregated lump-sum bid.

Item 1	$100,000
Item 2	200,000
Item 3	100,000
Item 4	111,472
Bid Total	$511,472

Most owners will react negatively to this pricing. They will feel that for items 1, 2, and 3 the contractor did not even do a cost estimate—he just threw out some numbers. This probably is an incorrect assumption. What probably happened is that the contractor made a detailed estimate of the total job, came up with a bid of $511,472, and then just apportioned it back on an approximate basis to the bid items. How much more effective would be a schedule of prices such as this:

Item 1	$ 98,200
Item 2	206,500
Item 3	87,750
Item 4	119,022
Bid Total	$511,472

10.6. Legal Requirements

The laws of the various states differ as to the requirements for qualification to do business in that state, that is, to become "domesticated" within that state. A few states require that a corporation be qualified prior to submission of a proposal; otherwise, the contract that results from that proposal is not valid, for the corporation has no status in the eyes of the law. Among other things, this means that you cannot sue in the courts to recover a claim, no matter how legitimate. Most states require that you become domesticated prior to or concurrently with the signing of the contract. In any event, check with your lawyer!

Most public works and governmental contracts require that certain data and documents be submitted with the bid. These may include a bid bond, an affidavit of noncollusion, a certification of compliance with the laws concerning solicitation of minority subcontractors, and a board of directors' resolution authorizing the individual who signs the bond to commit the corporation. Data may be required, such as a list of equipment to be used, a statement as to when work will commence, a schedule date for completion, and even the name of the project manager. Many local public works contracts require a listing of subcontractors and sometimes even their prices!

Failure to completely and responsively fill out the above information can cause the contractor to lose the job. First, it may disqualify him. Usually, however, the agency reserves the right to waive informalities such as these. In today's climate, however, the second bidder may join with a subcontractor or a minority contractor, to legally challenge the award. The net result is often that all bids are rejected.

Under such a challenge, the contractor may find himself in the undesirable position of having been low bidder, with his price out for everyone to shoot at, but with bids rejected. In such an impossible situation, about all he can do the second time around is to raise his price! That has worked, believe it or not. Much more often, he loses the job. Since a contractor is only successful bidder on 1 out of 6 or 10 or more jobs, he cannot afford to have that one job thrown out because of a careless oversight.

Experienced contractors assign a man to assemble and fill out all parts of the bid documents several days ahead of time. He flags any blanks requiring a last-minute insertion. This minimizes the chance of missing a key blank in the last-minute rush.

Turner Construction Company
44 Montgomery Street, San Francisco, Calif. 94104
Telephone (415) 391-1310

Turner

January 29, 1980

Mr. XXXXXXXXX
Director/Bay Area Projects
XXXXXXXXXXXXXXXX
XXXXXXXXXXXXXXXX
San Francisco, California

Dear Mr. XXXXXXXX:

We are pleased to submit this two-part brochure outlining our capabilities to provide Construction Consulting/Management services for your proposed Office Building Complex.

The Turner San Francisco Territory office was founded in 1968 and now has a staff of over 100 career employees (including jobsite staff). From this office, we have contracted for more than $700 million of new building construction. On substantially all of these projects, we were selected at the outset of design to work on the Owner/Architect team to ensure that the building that evolves represents the most value for the Owner's construction dollar.

This office is fully self-contained; all preconstruction and construction phase services on your project will be performed locally by existing staff. These staff members are familiar with the Turner Project Management Control System and have worked together on other projects in the Bay Area.

Your project will be under the direct supervision of Paul A. DeMange, a vice president of the company who will serve as the Project Executive. Mr. DeMange has 27 years of construction experience with Turner. He has worked successfully with your Architects on a previous project.

We are also assigning Walter B. Denmead, one of Turner's senior project managers. He joined Turner 15 years ago and has spent the last 7 years in the Bay Area managing projects ranging in size up to $40 million. The other staff members proposed are all currently assigned to the San Francisco office.

FIGURE 10.1. Turner Proposal.

Mr. XXXXXXXX
January 29, 1980
Page 2

Turner is a member of the Associated General Contractors of California and, as such, is a signator to the trade asgreements with the various unions which will be performing the actual construction work on your project. This important access to local labor will facilitate the resolution of the costly jurisdictional disputes that invariably arise among the trades in our business.

Your project master plan projects a significant facilities growth over the next decade. This Turner office is a well-established permanent operation and we have proposed a staff which we believe can see this project through to completion.

Turner has the experience, the local resources, and the management systems necessary to become an effective member of the Owner/Architect/Contractor team. We look forward to working with you.

Sincerely,

J. D. Quinn, Jr.
Contract Manager

JDQ/mkp

FIGURE 10.1. *(continued)*

Turner

APPENDIX INDEX

1. Estimating and Budget Control

    Preliminary Contract Data Sheets
    Turner Building Cost Index
    Cost Comparison Sheet
    Project Cost Model
    Final Estimate Example

2. Schedule Control

    Minutes of Meetings
    Resource Allocation Control System
    Project Scheduling System
    Market Conditions Report

3. Cost Control System

    Contract Budget Report (Summary Budget)
    Change Estimate Report
    Indicated Outcome Report

4. Experience

    San Francisco Territory Projects
    Low-rise Office Buildings
    Construction Management Experience
    Turner Cities

5. Reserve Staff

FIGURE 10.1. (*continued*)

Turner

INDEX

1. Budget Control

2. Scheduling and Control of Progress

3. Quality Control

4. Project Staff

    Resumes
    Territory Staff

5. Turner Project Management Control System

    Project Scheduling System
    Resource Allocation Control System (RACS)
    Cost Control

6. Experience

7. Reimbursable Costs

8. Financial Information

    Annual Report

9. References

    Commendation Letters

FIGURE 10.1. *(continued)*

DOWNRIVER CONSTRUCTION COMPANY
Sample Proposal Division
2444 Virginia Street, Suite 416
Berkeley, California 94720

August 12, 1980

Westernwood Manufacturing Company
127 Champs-Elyseés
New Orleans, Louisiana 12345

    Attention: Mr. R. L. Sherwood, Manager of Procurement

    Subject: Foundation Structures and Cooling Water Circulation System, Baton Rouge Plant Expansion Project

    Ref.:   (a) Your Inquiry 7-156 dated 10 December 1980.

Gentlemen:

    We are pleased to submit our proposal for performing the construction work for the above project in accordance with the plans, specifications, and contract provisions attached to your letter of inquiry (reference (a) above).

    Our lump-sum price for all work, excluding the area fill and embankment, is $7,144,300. For the area fill and embankment, we quote the unit price of $2.47 per cu yd.

    We propose to comply with your interim and final scheduled dates for completion, provided the work is awarded and the notice to proceed is received by February 15, 1981. Further, we request that all shop drawings be acted upon and received back by us within 10 days of the date of delivery to your resident engineer. While we recognize that the time is short, you have set an extremely tight schedule and, thus, the above provision becomes necessary.

    We understand that the environmental impact statement has been submitted and approved and that you expect to obtain approvals for work on the levee from the Levee District and the Corps of Engineers by early January. You have stated that all other permits will be procured by you prior to actual start of work.

    Should there by any delay due to action of a governmental agency prior to or during performance of the contract, other than that solely within our control, then an appropriate time extension is to be issued. Further, if such a delay occurs after work is commenced, we are to be paid for our out-of-pocket job overhead costs, including rental of equipment on job site at 50 percent of A.E.D. rates, plus an agreed figure to cover main office costs directly associated with the project.

    With regard to the area fill and embankment, your specifications provide for "measurement in place" but do not give any details as to how the quantity will be measured. Since considerable, but variable, settlement may take place, we propose the use of flat plate settlement markers on a 100-ft grid spacing, to delineate the undersurface of the fill.

    Your standard form of contract, as attached to your formal inquiry, contains several clauses which we feel are not fully appropriate to this particular project. We should like the opportunity to meet with you to arrive at mutually acceptable modifications.

    In particular, we should like to have a clause covering escalation in prices of structural steel and electrical equipment. As an alternative, a provision for advance procurement and payment is acceptable to us.

FIGURE 10.2. Downriver Construction Company proposal.

Final payment is to be made within 10 days after completion and acceptance or beneficial occupancy by you. If there remain any minor incomplete or unsettled items, the retention or withhold should be limited to their reasonable value.

Due to current Corps of Engineers' construction immediately upstream of your project, there exists the risk of severe flooding during the construction period. We have been unable to obtain insurance covering this specific risk, since the ability to recover from the government is not certain even if it is under its control or is its fault. We believe this is a matter of joint concern for both you and ourselves during construction and during early operations. We should like the opportunity to discuss this with you, as we believe a protective dike or wall along the northern boundary may be warranted as an additional item to the contract (extra work), in order to minimize this risk.

We are in a particularly favorable position for your project since we are just now completing a somewhat similar project (the White River pulp mill) upstream from your project. We are licensed and qualified for work in the state of Louisiana.

We trust that you will find this proposal acceptable and that we will receive your order for this challenging and important project.

Very truly yours,

DOWNRIVER CONSTRUCTION COMPANY

by _____
R. W. Wester, President

FIGURE 10.2. (*continued*)

INTERNATIONAL CONSTRUCTION
Berkeley, Rome, Paris

April 27, 1981

Chip House
101 Other Drive
Silicon Valley, California 94040

Attention: Mr. George P. Burdell

Gentlemen:

In response to your request of March 30, 1981, for a quotation, we are pleased to submit the following proposal for construction management services in connection with your planned Silicon Valley facility.

During the planning and design phase of the project, we propose to:

Consult with, advise, assist, and make recommendations to you and your architect on all aspects of the planning for construction.

Review the architectural, civil, mechanical, electrical, and structural plans and specifications as they are being developed, and advise and make recommendations with respect to such factors as construction feasibility, possible economies, availability of materials and labor, time requirements for procurement and construction, and projected costs.

Prepare a budget estimate at the completion of schematic plans and an updated estimate at the completion of design development documents. We will continue to review and refine these estimates as the design develops, and we will advise you and your architect if it appears that the budgeted targets for the project will not be met.

Prepare detailed cost estimates of each of the bid packages prior to receipt of bids from the trade contractors.

Recommend for purchase and expedite procurements of long-lead delivery items.

Prepare a progress schedule for all project activities by you, your architect, and the trade contractors.

Administer competitive bidding by various trade contractors, analyze bids received, recommend awards to you, and contract with the various trade contractors.

During the construction phase of the project, we propose to:

Maintain a competent, full-time supervisory staff at the job site to coordinate and provide construction of the work and the progress of the trade contractors.

Provide a quality control program to enforce compliance by the contractors with the contract documents.

Expedite and process all shop drawings, samples, catalogs, and other project papers.

Establish and maintain effective programs relating to safety, job-site records, labor relations, and progress reports.

FIGURE 10.3. International Construction Company proposal.

Provide and maintain a detailed schedule for progress control, and provide reports to you concerning construction progress.

Review and process all applications for payment by trade contractors and material suppliers.

Review, make recommendations for, and process any requests for changes in the work, and maintain records.

Establish and maintain a detailed cost control system which will provide current information on cash flow projectives, construction costs, fees, change orders, and contingency.

Perform general conditions items of work as required.

Perform any portions of the construction work which you may request.

We propose to provide these services under a mutually satisfactory contract providing for International Construction's reimbursement for the actual direct cost of its field staff, office expenses, and any general conditions work undertaken, plus a fixed fee. We propose that the fee be fixed as a lump sum at the time the construction budget is agreed upon. This fee will not vary unless there is a change in the scope of the project. The fixed fee would include the costs for the services of the corporate officers of the company, the estimating and purchasing departments and the project development coordinator. Also included within the fee would be the administrative overhead expenses related to the Berkeley and other corporate offices and our profit. All other costs would be reimbursable at direct cost with no markup for overhead or profit.

We propose to utilize a technical staff composed of John E. Internation, Jane S. Smith, B. C. Reagan, and a job accountant as indicated in the attached brochure. Based upon the use of this staff, the scope of the work as outlined, and the nine-month construction schedule shown in the provided documents, we estimate that the total cost of the reimbursable expenses, exclusion of general conditions work, will be $250,000. Also, based upon the present project scope, we propose that the fee be fixed at the rate of 3 percent of the budgeted construction cost.

Based upon our extensive experience in construction work in this area and upon our "track record" as construction manager for many leading projects, we believe we are uniquely qualified to undertake this work for you. We are confident that the staff we have proposed for the project is fully capable of providing you with an efficient, timely construction program.

We are pleased to have the opportunity to present this proposal, and we look forward to hearing from you.

Very truly yours,

Contract Manager

FIGURE 10.3. *(continued)*

198   The Proposal

Ben C. Gerwick, Inc., was once bidding on its largest project, which at today's prices would be $140 million. There were three alternates, on only one of which it seriously wished to bid. Pasted in the upper left-hand corner was a copy of the bid advertisement. It covered a note saying that the bidder must bid on all three alternates, but the estimator, in his hurry and concentration on pricing, did not lift the advertisement to read the note underneath.

Gerwick bid on one alternate only. It was low by a hair. The owner verbally stated that he would award. The second bidder filed a legal protest that Gerwick had not bid on the other two alternates. It took two weeks for the board to get a legal opinion allowing it to award to Gerwick. By this time, faced with environmental objections, the board of commissioners had had second thoughts; it rejected all bids and completely redesigned the project!

Figures 10.1, 10.2, and 10.3 show the various aspects we have been discussing in actual form.

10.7. After the Proposals Have Been Submitted

A construction company will typically have a number of proposals outstanding, which may or may not result in contracts. This places the contractor in a difficult position: although he desperately wants a percentage of them to result in contracts, he would be overloaded if they should all be awarded. This is particularly serious if he has one very large proposal outstanding, with perhaps partial encouragement. This may prevent him from bidding on other large projects, and yet he could end up with nothing.

J. H. Pomeroy Company was once low bidder on the same day on a very large harbor project and on a missile silo complex. It had to stop bidding on all other projects as it had no more financial capabilities or personnel resources. Four weeks later, all bids were rejected on the harbor project. Two weeks after that the missile silo bids were also rejected, and the government entered into negotiations with an outside contractor who had not even bid the first time.

This not only cost Pomeroy two of its largest-ever jobs but also prevented it from bidding in the meantime.

Private work may present even greater problems in this regard because bids are normally not publicly announced. It is, of course, important to stay in contact with the client and to endeavor to engage him in detailed discussions. From these, you may be able to learn where you stand. You may be able to get the client to work with you to resolve problems which have arisen from the bids.

Unfortunately, many large companies have trained their personnel to ask questions but give out no information.

This can be a highly critical period for marketing. Your company has gone through the lengthy process of convincing the client to place you on the bid list. You have prequalified, prepared an estimate, and submitted a

TABLE 10.1. Emkay Summary of Active Proposals

SUMMARY OF ACTIVE PROPOSALS
EMKAY—NORTHERN CALIFORNIA

Prepared by: Shaw
Date: August 17, 1981
Page Number: 1

Bid Due Date	Priority	Type of Estimate/ Bid or Negotiated	Project/Location/ Type of Work and Scope	Proba-bility of Award	Estimated Amount, in $1000	Estimator/ Proposal Manager	Owner/ Agent	Comments
August 21, 1981	A	Neg. CPFF	Hayward—Diablo XV 27,000 S.F.	...	500	Sample	Bedford	Budget
...	A	Neg.	Montebello Phase I Dirt and Underground Oakland, California	80%	929	Sample/ Shaw	I.C.D.	Given to owner
...	...	Neg. CPFF	Soscol III Napa, California	80%	...	...	Bell	Awaiting architect Final drawings
...	B	Neg. CPFF	Bier Gunderson 82,000 S.F. Interior Oakland, California	50%	1,500	Sample/ Shaw	Fleischer	Quoted, meet with owner and architect week Aug. 25 for design schedule.
August 18, 1981	B	Neg. CPFF	Pleasanton Park 9 Industrial Buildings	...	...	Sample/ Shaw	Reynolds & Brown	Preliminary budget
...	B	Neg.	Worldco Sutter & Franklin, San Francisco 11-story; 90,000 S.F. Office	25%	...	Shaw	Tony Chan	Architect working on preliminary drawings

TABLE 10.1. (*continued*)

SUMMARY OF ACTIVE PROPOSALS
EMKAY—NORTHERN CALIFORNIA

Prepared by: Shaw
Date: August 17, 1981
Page Number: 1

Bid Due Date	Priority	Type of Estimate/ Bid or Negotiated	Project/Location/ Type of Work and Scope	Probability of Award	Estimated Amount, in $1000	Estimator/ Proposal Manager	Owner/ Agent	Comments
...	B	Neg.	Lennon Lane Grove Office Complex Walnut Creek, California	...	4,000	Sample/ Shaw		
...	B	Neg.	Urban West Office Park Pleasant Hill, California 500,000 S.F. Office	25%	20,000	Shaw	Urban West	Arch-Kaplan McLaughton Diaz. No decision on G.C.
October 1, 1981	...	Design Build	San Rafael Parking Garage 3 level; 141,000 S.F.	...	...	Shaw	City of San Rafael	

proposal. Do not let up now. Chances are the client has received several bids which are very close. As a private owner, he can select whichever one he thinks is in the best interests of his company. Qualitative considerations, such as ability to work with his people cooperatively and safely, reputation for meeting schedule, or demonstrated ability to establish good labor relations, may be more important to the client than minor differences in price.

To assist in evaluating the total picture of outstanding proposals, many companies keep a record as shown in Table 10.1.

11

Salesmanship and Advertising

11.1. What Influences an Owner's Decision to Call You or to Award You a Contract?

Many contractors, particularly the larger ones, put on what is generally referred to as a "dog and pony show." Usually this is not done on initial contact with the owner but is a major presentation made after determining his needs and showing how your company can fill them. Generally, this includes three to eight participants from the contractor's firm, most of whom are in management and operations.

It usually involves presentation of brochures, including a tailor-made proposal and sales pitch aimed at this specific owner and project. Frequently, use is made of such aids as large charts, models, slides, and even sound movies.

Whether you put on a fully orchestrated dog and pony show, simply call on the owner by yourself or call accompanied by your chief engineer and general superintendent, there are certain essential points you should decide in your sales effort.

It is very important to convey to the owner that your company is interested in his unique problem. Along this line, it is essential that you, or one of your sales team, really does understand his particular problem. If he senses that you do not understand his problem, he will have serious doubts that your company can provide the solution, no matter how elaborate a performance you may put on.

You should demonstrate to the owner that your company has the capacity, depth, and experience to perform in a fully professional manner. This can involve a number of techniques. For example, you can cite experience on similar projects showing the owner's name, location, year performed, and difficulties encountered. You can cite your record of on-time performance, particularly if time is important to this owner.

It is important to give evidence of your ability to complete work within a budget. Cite repeat contracts with same owners, plus your ability to compete in the marketplace.

Be sure that the owner is satisfied that you have experienced, professional people in your company who will be committed to his specific job. Have them with you, if at all possible; otherwise, provide him with biographies of these key people, showing their educational background and professional track record. If the proposed project will require special or unique equipment, such as special pile-driving hammers, drill rigs, derrick barges, slurry wall equipment, or prestressed concrete facilities, it is especially important to show photos of the equipment and assure the owner that this equipment is nearby and can be committed to his job. If your company has in-house engineering capability with capacity to cope with the owner's problem, you have potentially great appeal.

Perhaps you can "fast track" the project, saving the owner from two to five months on beneficial occupancy. You may provide single responsibility so as to relieve the owner of the hassle of engaging an architect-engineer. Then he is able to deal with one party for both design and construction. Do not push turnkey if the owner has his own engineering department or has commenced negotiations for his own architect-engineer. It may not be ethical, and in any event, you may have to live with these people on future jobs.

Reassure the owner of the financial stability of your company if he does not already know you. Usually, a recent financial statement will suffice. Most private owners assume that you have or can get proper financing, unless it is a very large job. A list of satisfied clients and references could be included. Some owners will accept these on their face value; others will check them out, so be sure the clients on your list are satisfied and will say something good about your company. If possible, give names of people to contact for references, along with their title or responsibility. A good point to make with the owner is the ability of your company to be innovative and to adapt its unique capabilities to the solution of this particular problem.

The highest form of salesmanship is nothing more than service.
Forbes Epigrams

11.2. Client Relationships on a Social Basis

Why do you want to establish a relationship with the potential client on a social basis? The reason, of course, is so that you can facilitate communications.

In marketing it is good to be on a first-name basis with the client. However, it is advisable not to be overly friendly. Excessive familiarity may backfire.

As another note on social etiquette, if you have gone out to dinner with your client the night before, particularly if the social affair involved liquid refreshments, do not make too much, if any, reference afterward. It is bad manners to refer to it in the office the following day. Because heavy social engagements can as often backfire as succeed, some highly successful contractors make sure their entertainment is conducted in good taste. This has the effect of saying: "I respect you and think of you as a fine, upstanding, and ethical person."

Spouses are sometimes involved in entertaining a client—for example, going out to dinner with the client and his wife. They can be a tremendous asset in establishing a good personal relationship with a client. What are some of the things that can be done? Nothing is more effective than if your wife tells the client's wife, "My husband really likes working with your husband. He thinks he is one of the most brilliant engineers he has ever worked with." This is tremendously effective because it is going to get back to your client. But it is only effective if there is at least a strong element of truth in it.

11.3. Projecting the Right Image through Dress

There is some truth to the cliché, "Clothes make the [marketing] man." A paperback book along these lines is *Dress for Success* by John T. Molloy. Reading this book is a must for anyone in marketing, with perhaps one exception—the person in marketing who is at the very top of his profession; he probably already knows everything in the book. Molloy would not claim he could make an idiot into a marketing genius; however, he does suggest that improper dress can keep a good man from reaching the top. What is proper dress?

As Molloy states in his book:*

The clothing combinations I recommended for business wear do not come from the pages of men's fashion magazines. Neither are they determined by my own personal opinions. They are the result of research and testing, and they reflect the conscious and unconscious reactions of a valid cross section of the American public. They work because the American public says that these clothes, in these combinations, project a look of good taste, of credibility, and of upper-middle class success.

In the book, Molloy tells what is proper dress for you. It depends on the situation, your age and physical size, the geographical region, the level at which you are marketing, and other variables. There is no room to describe what clothing combinations are right for you in this text—it takes a book the size of Molloy's to do so.

* John T. Molloy, *Dress for Success* (New York: Warner, 1975), p. 93.

Now there are times when one wishes to project another image. One of the authors was involved in a negotiation for an important project in which everyone recognized that the critical item was the heavy rigging under extremely difficult conditions. Although he brought in his engineers who presented detailed sequential drawings of the hoisting and rigging operations, it was clear that the client's staff was unconvinced. Then he asked his rigging superintendent to attend. The superintendent, realizing he was going to be meeting executives of a large public utility, came to the office in suit and tie. In a flash of intuition, the author asked the superintendent to change back to his work clothes; clean and neat but befitting a rigging superintendent on the job.

It was obvious in the subsequent meeting that here was a man who could and would deliver. The client's executives paid no attention to the engineers but, instead, directed all their attention to the superintendent.

They then awarded the contract with one stipulation: the rigger was to be personally in charge of all critical operations.

This is only one of many examples. Famous trial lawyers study the appearance of their witnesses with care and require them to dress according to the image they wish to create.

In one arbitration of a dispute between a subcontractor and a general contractor, the subcontractor showed up in badly worn work clothes, the general contractor in a stylish business suit. The award to the subcontractor was explicitly stated to be based, among other things, on the fact that he was a poor, small, hard-working subcontractor, unable to match wits with a clever, sophisticated general contractor.

11.4. The Expense Account and Its Abuse

It is common knowledge that people who are entertaining for business are on an expense account. Many jobs in private industry include an expense account allowance. Depending on the business, it is often necessary to entertain, for example, the president of the company or the resident engineer. However, be careful about excessive use of the expense account. Try to think as to what is appropriate for the situation.

A marine contractor had a job for a West Coast port, and it was going pretty well. The Beavers, the honorary society for western contractors, were having their annual dinner. From the point of view of the contractors, this was one time of the year when they got together, went black tie, and had a fairly lavish dinner (with no shortages of liquid refreshments). In general, the contractors see nothing wrong with this affair. Now, the contractor invited the resident engineer from the port job to this night of festivities as his guest. The contractor and the resident engineer had gotten along just fine, and they were ahead of schedule. However, for the next six months the resident engineer crucified the contractor. There was nothing which the contractor could do right. Finally, toward the end of the job the resident engineer was letting up a little. One day he was talking to the contractor, and he volunteered the reason for the change. He stated, "I always thought you guys were doing a good job and just barely making a profit. However, when I saw that thing the Beavers put on and all those thousands of dollars spent, I realized that you were rolling in money. I thought I should get a little of it back for the port." This particular man had never been entertained in this way before; and he was overly if unfavorably impressed. This was a case where entertainment backfired 100 percent. Therefore, watch out for excessive entertainment that is out of line with a particular situation.

On the other hand, when you are dealing with somebody, such as an executive of an oil company, who is used to quite a bit of entertainment, you have to be willing to take him out to dinner occasionally. This is what he's used to and what he expects. However, he may be as glad as you are if the evening ends with a drink and a good dinner and does not go on until late at night. Therefore, with respect to entertainment, you need to evaluate the situation and do what seems to fit the circumstances and the person.

Most major private companies and governmental agencies prohibit gifts both ways. They prohibit their employees from accepting a gift, and they prohibit the contractor or engineer from giving a gift.

In order not to make this policy too blunt and in order not to offend people, some companies will send out a letter to this effect:

Christmas season is again approaching. We, of course, are happy to see another year completed of mutually rewarding relationships; and we hope they will continue through the new year.

Our company policy prohibits our employees from accepting any gifts. If you do wish to make a contribution, we suggest that you make it to your favorite charity. This will certainly be appreciated by us and by them.

This is an excellent way to handle the problem of gifts. A note to the company involved telling them that you appreciated their note and that you have made a gift to such and such a charity may be quite effective!

There are still many cases today when holiday gifts are a tradition. For example, in shipbuilding, the naval architect will often personally deliver champagne to his client at Christmas time. However, today it is usually one bottle and no longer a case.

The most proper and safe remembrance is a Christmas card, especially if it contains a handwritten note.

The whole problem with gift giving is its abuse. With a gift, someone appears to be trying to buy a favor from someone else. In one of the authors' experience as a contractor, he would run into situations where exorbitant demands for "gifts" would be placed on him. The policy of refusing the demand in a nice, polite, face-saving manner almost always worked. (The problem is discussed in Chapter 19.)

11.5. The Use of a Preliminary Design or a Budget Estimate as a Sales Tool

Sometime during or after the second meeting with a client, you may feel sufficiently well entrenched to suggest that your company undertake a preliminary design, so as to produce the requested budget estimate. This can be implemented in several ways. Sometimes the owner is willing to pay outright for your engineering costs in this regard. However, he may then feel free to later go to a competitor and try for a better solution if he has paid for this initial effort. Alternatively, you may decide to perform the preliminary engineering and make the estimate without charge, gambling that you can come up with an acceptable design and budget and can later negotiate a contract. Some contractors have found that this is not as costly as it sounds, provided they confine such work to the fields in which they specialize and thus have prior experience and competence.

Assuming that you have reached the point where you are performing a preliminary design and budget estimate for the owner, you have a unique opportunity to steer the design in a direction which will favor the equipment, facilities, people, and skills peculiar to your company. This is a great way to give your company a running start over your competition, but neither do it at the expense of the owner nor if the end product compromises in any way competent engineering, workmanship, utility, or durability. For example, one job called for 12-in. square piles. The successful contractor's yard was set up to manufacture 14-in. piles at a lower cost than 12-in. piles. In this case it was able to sell the owner on using 14-in. piles for a higher-load capacity, and the owner obtained a more efficient foundation at a lower cost.

11.6. Use of Client Involvement to Sell Your Ideas

The following is an example of a successful sales approach, that of client involvement. A contractor we will call White Construction Company had been trying to persuade the bridge department of the state to make better use of prestressed concrete girders for long spans. White was encountering great difficulty. The state was always finding some excuse for using either structural steel or cast-in-place concrete. The state had used structural steel in the past, and it was very familiar with it. Many times White had gone through the procedure of drawing up a detailed solution using prestressed concrete girders instead of the structural steel which had been specified by the state. In every case the alternative solution had been rejected. The state's engineer would provide one reason after another. For example, he "would feel that the structural steel weighed less and it would not load the foundation so heavily." Or "the lateral response to an earthquake will be too great due to the heavier structure if prestressed concrete is used." Finally, a major project came up. White was particularly anxious to convert that project to prestressed concrete, since the project already called for prestressed concrete approaches. However, the long main span was designed in structural steel. The structural steel main span did look awkward; there was no doubt that it was aesthetically displeasing. Since a prestressed concrete girder would obviously be more aestheticaly acceptable, White knew it had at least one point on its side. White Construction went through its usual procedure of redesigning the bridge in prestressed concrete. It engaged a skilled professional engineering designer to make a full structural analysis of the girder and to produce a nice set of plans. The added deadweight of the redesigned structure turned out be very modest because only the girders had changed—the concrete deck remained the same. The added effect of the additional weight on the earthquake loading was minimal; only a small amount of additional reinforcement was needed for each column. Since the incremental changes in weight and seismic moment were small, only two additional piles per pier were required. After gathering all of this ammunition, White was ready to go to the state and prove that prestressed concrete should be used. One of White's people stated; "It's no use. They will never buy it. I do not know what excuse they will use this time, but we are asking them to admit they made a mistake, or at least were not up to date. We are asking them to admit they did not use enough imagination and did not take enough time to perform the adequate calculations." Therefore, White used another approach. White took the engineer's nice drawings and traced them freehand on a piece of paper without a border. The drawings looked as though he had just sketched them out. However, the dimensions were pretty close to what was right. He then presented this material informally to the state's engineer and suggested that perhaps some of its people could look at this new idea as a possible way to use prestressed concrete for the main span. White told the state's engineer that he assumed the state had probably wanted to use prestressed concrete in the structure because it would be aesthetically more pleasing but realized he was concerned about the additional deadweight and the seismic behavior of the structure. He humbly stated that the people in his company were, of course, not professional designers; however, they had tried to do some crude figures on what might be required. At this point he gave them "approximate" values for the additional steel and piles required. White was just asking the state's engineer to check over the calculations, suggesting that it would be an interesting exercise even if the state did not use this design. The chief engineer of the state replied, "I am not going to change the design. That is out of the question; we never change a design once it is released. But, I will tell you what I will do. I will check your figures so you and we will know for the future. This kind of approach is quite interesting."

This case turned out to be the first time the state ever did change the design after it had been released. It issued a complete revision showing all prestressed concrete. White Construction was low bidder and was awarded the job, and the bridge was built according to the design it had submitted in the redrawn sketch. Further, the state subsequently used the same design on its next large bridge.

11.7. The Psychology of Closing Sales

Understanding "buyer's remorse" is essential to the psychology of closing sales. People generally experience buyer's remorse at and after the closing of any sale whether it be the sale of a major project service or a typical consumer item. Consider the case of a man deciding which cigar to buy at the drugstore. If he buys the more economical cigar, he can experience buyer's remorse and wish he had bought the more expensive, better-tasting one. If he buys the more expensive cigar, he can experience buyer's remorse and wish he had not been so extravagant. Using the psychology of closing sales, the smart merchant will make a reassuring comment about whatever choice the man had made. This same principle should be followed in making a sale of a major project service. For example, after you finish negotiating a contract to build a major building, a reassuring comment that lets the client know that "he has just negotiated an extremely favorable price" or that "his timing was just right to assure a timely delivery of structural steel" may keep a deal from falling through the next day.

11.8. Advertising: Its Use and Misuse

Advertising has proved to be an extremely successful method of promoting product recognition, company standing, and, in the case of individual consumer-oriented products, direct sales as well. However, with engineering and engineered construction, different factors are involved.

First, from a strictly sales point of view, the potential customers are limited in number and usually a highly specialized and sophisticated group. Thus the medium used and the message need to be geared to their interests.

Advertising is a relatively expensive aid to marketing. Careful evaluation of the cost per potential customer will usually limit the media employed to business-oriented and technical journals that cater to specific groups or to direct mail. You can spend too much on advertising. It can cost as much money as another salesperson.

Since the average client is flooded with technical letters, journals, and advertisements, the message must be carefully selected so as to attract attention, create a favorable impression, and encourage retention. A clever and effective advertising program will identify the real need of the potential client and attack that one single point.

Why would a busy executive see and remember an ad? Does it somehow stand out because it awakens a responsive chord, "Yes, that's what I need," "Yes, this contractor understands our problems," "Yes, here is the approach for which we've been hunting."

Some large engineer-contractors advertise in institutional magazines, such as *Fortune, Forbes,* or *Business Week,* as a means of establishing credibility. Then when their sales representatives ("business development engineers") call, there is already a recognition of status and a favorable predisposition to the sales message itself.

Another approach can be aimed at the professional: the architect and engineer who specify or approve. While this approach is very important for marketing products, its primary use for engineers and constructors is in marketing specialized services and proprietary processes, which are similar to products from a marketing point of view. Raymond Concrete Pile has advertised for many years in key structural and civil engineering magazines. While this has reportedly been very effective, the question can be raised: Does not the architect also specify foundations? Is he not even more disposed to preselection than the engineer?

For this latter case, direct mail can also be very effective; for example, a periodic or intermittent newsletter can be sent to selected professionals, presenting news of new technical developments. It has proved most effective if such a newsletter is confined to one principal subject and if both conceptual and technical descriptions and data are furnished: data which can later be used in planning, estimating, scheduling, or designing. For the latter, great care must be used to avoid professional liability; this means not only a general disclaimer clause but also appropriate wording in the article itself, so as to prevent misunderstanding and misuse.

Advertising is seldom effective as a marketing tool by itself but must be part of a coordinated campaign. One international general contractor decided to enter a highly specialized and technical construction field. He advertised extensively and expensively but failed to follow up with personal calls on key decision makers. This is an extreme example of failure to learn how a market is sold, who makes decisions and how, and the need for an integrated campaign which in this case should place primary emphasis on personal contact.

If a specialist contractor or subcontractor is trying to reach the general contractors with numerous small sales, then he may find useful the advertisement in the daily or weekly newspapers or magazines which list jobs that are advertised for bid. Dewatering contractors have used this means successfully. Since many of their best jobs arise when a contractor unexpectedly encounters water problems and needs help in a hurry; he remembers the advertisement he has recently seen.

In summary, therefore, except for specialty contractors and product supplies, advertising should be used only as an introductory supplement to a marketing program. By itself, advertising by contractors is a waste of money, but properly used, it can create a favorable atmosphere and reception for a more concentrated approach such as the personal call.

11.9. *Community Involvement*

Everyone has read that he should get involved with the community. Insurance salespersons and stockbrokers often find that community involvement is a good method for salesmanship and advertising, for their market is individuals. However, engineers should be careful about such involvement, insofar as their professional career is concerned. The basic problem is that community involvement often does not bring the engineer in direct contact or involvement with potential clients. Generally speaking, community involvement can take too much time compared to the marketing benefits derived. However, there are some exceptions.

Certain companies feel that they are a major factor and participant in their community and that their clients are major businesses which are also involved in the local community. Therefore, they will work hard in a community program like the United Appeal. They will furnish one of their principals as a figurehead and assign one of their nonengineering personnel (to whom they give a title such as "assistant to the president") to work full time for the program. They feel that community involvement in programs like the United Appeal is not only their community responsibility but also pays off from a marketing point of view.

From an individual's point of view, you should do community work on your own time if that is your avocation and dedication in its own right. However, from a business point of view, community involvement may require excessive time in relation to its potential productivity.

As is discussed in Chapter 14, serving on a committee that represents your profession or your client's industry can be a very effective marketing tool.

The best salesman is he who gives satisfaction.
Forbes Epigrams

12

Product Development

Product development is one of the key elements of an overall marketing program. "Products" in this case can denote innovations of many types, ranging from a new method of construction to a new engineering concept and from a new management technique to a new contractual package. Seen in this broad light, products are developed in response to perceived needs. It is only if these new developments appeal to and better serve the clients that they will be accepted and used.

12.1. Innovation

Innovation is a currently favored subject by economists, who have just reawakened to the realization that economical growth and strength depend on ongoing innovation. Elsewhere we have repeatedly referred to the dynamic nature of marketing and the goal to attain dynamic noncompetitive advantage in a highly competitive world. Without innovation, competition focuses on doing old things ever more efficiently and for ever lower profit margins. While the increased efficiency is a highly valid economic gain, it quickly tends to a point of diminishing returns. Because the method (or contractual form) is no longer new, more and more competition emerges, and the inevitable law of supply and demand takes over.

Similarly, lower prices, with lower unit profit margins, may increase volume; however, in a professional and service industry, such as that characterized by much of the engineering and construction industry, the reduction in costs due to increased volume quickly disappears and eventually becomes negative.

Successful companies, therefore, are dynamic companies, searching for opportunities and encouraging innovation.

However, there are many institutional constraints acting to dampen creativity and change. These include building codes and standards, legally established bidding rules and rules regarding contractual changes and submission of alternatives, work rules of union agreements, opposition and negative selling by companies or industries that will be displaced, professional pride (the N.I.H. attitude, or "not invented here"), and, finally, opposition within your own organization for a wide variety of personal and business policy reasons. One of these latter arguments that is particularly discouraging is, "It will make our previous work look inadequate [or 'unsafe' or 'outdated']."

While all these constraints must objectively be addressed by the innovative entrepreneur, they often frustrate when they loom larger than the technical constraints.

Another and possibly more logical argument is that, judged on a cash flow and internal rate of return basis, many new innovations will never justify development. However, economists are beginning to rethink the validity of these methods of evaluation, since experience shows so convincingly that innovation and research development do pay off in the long run. The development of an innovation or invention to commercial status is an expensive and time-demanding affair, involving all aspects, from engineering to production to finance to, of course, marketing.

12.2. A Schedule for Product Development

The typical schedule for a new product seems to follow an inexorable pattern. The technical breakthrough is followed by a longer than anticipated and more costly than anticipated period of development. Eventually, the in-house concerns are overcome, and the new product is being employed on its first real use. Inevitably, difficulties arise. If it is new equipment, vibration develops, bearings wear, the unit proves to be underpowered. If it is a new contractual form, the clauses turn out to have different meanings to the two parties. A new estimating method turns out to have gaps. A new material develops unforeseen long-term corrosion or fatigue problems. Thus the first few such contracts are usually performed at a loss. Indeed, some sophisticated and experienced companies actually budget a loss on these early contracts.

Then if all one's resources and enthusiasm have not been used up in these early disappointing contracts, the real opportunity finally comes along and a major success is realized. For a while the market is almost literally knocking at the door.

Now the competition jumps in, if it has not already done so before. In the construction industry, it is generally rather easy to design around patents; and because of the fact that the construction project usually involves a great many conventional operations as well as the innovative one, the courts have generally not ruled in favor of construction patents. They are most supportive when it comes to protecting a new device, piece of equipment, or gadget, something that can be looked at separately.

So competition comes in. Often the competition sees a way to carry out your own idea more efficiently. The competitor is not stuck with the home-built equipment which you have used, so perhaps he can do it better or on a larger scale.

The sequence of innovation and product development is shown in Figure 12.1.

It will be seen that development costs are far greater than research costs. Also, it will be appreciated that the profits are much further down the line than costs; hence the discounted cash flow appraisal of innovations is not always favorable. In fact, from a strict point of internal rate of return on investment, one can often show that innovation does not pay; the risks are not worth the potential gain. Nothing can be further from the truth. Despite the pessimistic indications, companies which innovate survive and grow strong. Those which do not innovate tend over the years to lose their relative position.

12.3. What Does Happen in Reality?

The new product opens up a whole area or scope of work which may not directly involve the innovation but which depended on the innovation as an entry. For example, Ben C. Gerwick, Inc., was busy developing prestressed concrete piling. It was bogged down at point B. The opportunity came to participate in a very large project in the Mideast, which depended on the use

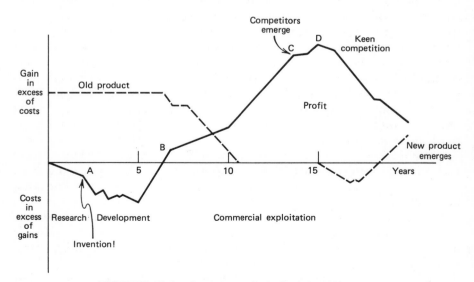

FIGURE 12.1. Sequence of product development.

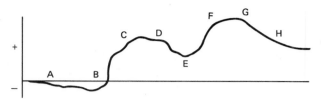

FIGURE 12.2. Extended sequence of product development costs and income.

of prestressed concrete piles. The returns then were not just those from the piles but those from the overall project as well.

The technique developed can often be applied to a whole range of related products. In the case cited, the techniques of prestressed pile design and construction could also be applied to bridge and building slabs. Your personnel develops a tremendous job satisfaction and esprit de corps out of having developed a new product. They address a creative attitude to all their work. The volume turns out to be much greater than anticipated. A better mousetrap attracts jobs as well as mice.

If the chart in Figure 12.1 is extended over a longer term, it will often be seen to have a second hump, EFG, which occurs after the product is well established, as shown in Figure 12.2. In this case, the competitors, who emerged at C and tried to usurp the market between D and E, fell out of the picture at point E and were not available when opportunity F arose.

To have the greatest chance of success, innovation should follow these rules:

1. It should be directed toward a substantial need of your clients.
2. An effort should be made to involve a client-user in the development phase, somewhere prior to point B.
3. A realistic appraisal should be made to ensure that adequate resources are available to exploit the first hump, starting at C.

12.4. Forecasting

Plotting life cycles of technology is currently a favored tool of business consultants. They emphasize that a technology does not exist in a vacuum, that actually there are a whole family of overlapping curves, as shown in Figure 12.3.

Examination of these curves emphasizes the points that the time to start R & D on new product B_1B_2 is just when the greatest success is being achieved from past development A_1A_2 and that B_1B_2 will then be in a position to supplant product A_1A_2 at the time of its obsolescence or decline.

Life-cycle technology curves may be modeled in a number of ways. For example, a straight-line forecast would work quite well for Figure 12.1 between years 10 to 15. However, between points B and C in Figure 12.1, an

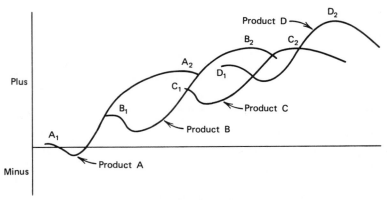

FIGURE 12.3. Products development.

exponential curve forecast would most clearly project the market trend. Looking at the overall picture in Figure 12.1 (between points A and D), a Gompertz curve trend analysis would provide a good forecast of the market trend. The Gompertz curve is a general growth S-curve that follows an introductory phase of slow growth, then a phase of fast growth, and, finally, a maturity phase of slow growth.

It is probable that the best use of these curves is pictorial, rather than quantitative, since the exact pattern and shape can be influenced by so many external and random factors.

12.5. Some Case Histories

Ben C. Gerwick, Inc., also developed the SPTC underground wall system. The term SPTC was denoted by one of its clients to symbolize "soldier pile tremie concrete." The period AB of development dragged out longer than expected due to both related and unrelated problems. Instead of one losing job, it experienced three. So when the really big opportunity was literally offered on a plate, Gerwick turned it down. Its psychological resources had been exhausted; that is, it did not have enough faith to make a commitment. Gerwick's arch competitor, using Gerwick's product and methods, made a very substantial profit.

Fortunately, Gerwick did stay in the picture long enough to realize a success in the second hump.

An interesting case of a market unfolding that was much larger than anticipated was with a tieback system for underground walls. It was developed for a specific project. When adopted, the equipment cost twice as much as budgeted. However, it worked twice as well as anticipated. This led to a demand for use on projects of which the inventor had never heard.

One winter while taking the train up to the mountains to ski, an executive of a friendly construction firm recognized the need to develop precast concrete snowsheds. A snowshed is the structure that protects the railroad tracks from snow and avalanches in the mountains. His own company was not interested, so he suggested it to his associate who was in the precast concrete business. Based on his suggestion, and discussions with the engineering department of the railroad, a system of precast concrete elements was developed which could be erected and bolted together by an ordinary track crew. A small model of this system was made. This model was certainly helpful in selling the idea to the railroad management. It liked the model so well that it ordered several miles of snowsheds on a series of successful contracts. The manufacturer had found a new long-term customer.

Foolishly, the manufacturer than raised his prices beyond the point of continued acceptance. The railroad looked elsewhere for its needs and finally found a manufacturer in Texas who could produce a similar system for significantly less. Since the in-house charges by the railroad for freight of the concrete segments from Texas to California were negligible, the original developer could no longer compete and eventually lost his innovation at point D.

One of the types of product development that may produce an early return is a new contractual approach, tailored to the needs of the client. For example, a way may be found to utilize the client's expertise and resources in carrying out certain aspects of a project. By contractually limiting the scope of your services, you may improve your batting average and still maintain your profit margin per job.

Certainly the target estimate form of contract and the construction management and engineering of contract were innovations of a high order, with minimum development costs and high returns.

Engineering-only firms can innovate by incorporation of CM services or by taking an equity position in the new endeavor. New computer techniques or adoption of aerospace system approaches, such as failure mode effect analyses and fault-tree analyses, can enhance market capabilities enormously.

12.6. How Can an Innovative Idea Be Protected from Competitors?

One conventional approach is to obtain a patent. In most European countries, all new construction ideas are patented since a major marketing tool in Europe is to offer an "exclusive patented system." In Europe an inventor simply applies for a patent, and it is granted; however, this patent is not worth anything until it is proved in court.

In the United States, the patent process is much more expensive and time consuming. Before a patent is granted, it has to be proved to be an invention to the satisfaction of the U.S. Patent Office. This means that a U.S. patent will require a great deal of detailed work on the part of the inventor and considerable legal expense. Also, after initial submittal, the Patent Office's examiner will typically keep coming back with rejections of each claim. This is not just arbitrary opposition on his part but a deliberate effort to interpose all probable challenges. By answering them, you are helping to establish the validity of your patent before it is challenged in court.

Unfortunately, the process often restricts the final patent allowance to a relatively narrow coverage. Thus the usefulness of a patent, particularly a construction patent, is often marginal. For this reason, in the United States, a "patent applied for" may be more valuable than a "patent issued": the competitor has no way to judge the extent of the claims which will be allowed.

What else can be done? Given the objective of not letting anyone else steal the idea, you can publish it in a journal with worldwide circulation. This will stop anyone else from getting the patent and will give you and your firm credit for the idea.

Another method that will help you hold rights to the idea is to keep a file on the invention with dated notes, sketches, and drawings. A contractor we will call Southern Construction was at one time involved in a patent suit where a competitor was infringing on its patent rights and was even demanding a royalty from Southern on its own invention. Southern's file on the early innovation was critical to defending its patent rights. Fortunately, early memoranda and sketches were found that proved the date of the invention.

One of the more successful developments by a precast concrete manufacturer was the pretensioned railway tie (sleeper). Posttensioned ties had been used in Europe but were judged uneconomical for U.S. application, as well as having some potential technical problems. On the other hand, pretensioned methods, to be successful, required development of mechanisms and procedures that would satisfy the technological requirements.

Although the development of the pretensioned tie by the Ben C. Gerwick, Inc., was original, it was discovered that other companies in Europe had previously tried the pretensioning system. Therefore, the tie itself could not be protected by patents. However, many specific steps in the manufacture could be individually protected. By patenting these, a whole package of constraints could be wrapped around the one innovation.

Meanwhile, years of effort had to be expended in engineering, market development, and trial applications. Fortunately, several railroads were found who were willing to install trial stretches, first in yards, then in their main lines. But precast ties could not be sold in quantity to the railroads because, on a tie-for-tie basis, the concrete was more expensive than the railroad's accounting costs for a treated timber tie. The railroads, in some cases, already owned both the forest and the timber-treating plant, both of which had been capitalized and written off. Further, they would charge little or no freight to themselves. So their in-house charge for a timber tie, representing labor and creosote only, was difficult to match.

Eventually the pretensioned ties were adopted for several Metro systems, which gave the needed spur to continued efforts.

However, the development story was far from over. Potential markets emerged in heavy-duty ore railroads, and several large contracts were obtained. Troubles developed in some of the tie-fastening details due to the extremely heavy axle loads in high speeds on curves. Further engineering was required and, fortunately, solutions were found.

Then Amtrak came into being, and their extensive testing clearly showed the advantage of the heavy precast concrete tie for anchoring down all-welded rail. The real savings turned out to be, not in the ties themselves, but in the reduced maintenance of roadbed and rolling stock. And so, 17 years after innovation, the company finally obtained a contract for the new ties between Washington and Boston, a $40 million contract that justified the 17 years of development.

Raymond Concrete Pile Company was very successful in maintaining a quasi-proprietary position for its cast-in-place concrete pile over a long period of years, far more than the 17 years of patent protection, by continually improving the product and patenting the improvements to the pile and to the installation equipment. Once again, it was the package of innovations that proved successful.

12.7. Creativity

Almost all construction and engineering personnel, both professional and nonprofessional, have an urge to create. All they need is liberation, guidance, and support. The new product is not the true product at all; the true product is the proven ability to respond dynamically to need on a continuing basis. Many companies have achieved a single great success, then affixed themselves to that one product unitl it and they finally ran out. The time to innovate, once again, is when you reach point B with one product.

It is generally recognized that the researcher is not normally the best developer, the developer is not necessarily the best exploiter. However, when transferring an innovation from the creator to the developer and exploiter, it is very important to let all feel a sense of identification with the ultimate success. A company must find a mechanism by which the innovator plays an advisory role, with adequate recognition, so that he does not feel that his idea has been taken from him and exploited. Ensuring that he is involved in the whole process will give him a realistic understanding of costs and time requirements, as well as the satisfaction of identification with his creation. Manufacturing companies give a great deal of attention to this, but do contractors?

12.8. Franchise Extension

In consumer marketing, there exists the established practice of introducing a new product by letting the new name hitchhike on the name of an existing product. Consider some well-known examples: Sunkist orange soda, Kodak copiers, and Easy-Off window cleaner. According to Edward Tarben, a University of Southern California marketing professor, this practice of " 'franchise extension [is] a method for a company to enter a new business through the leverage of its most valuable asset—the consumer awareness, goodwill and impression conveyed by its brand name.' "*

Although franchise extension has been used in the construction industry— for example, Hilti Anchor and Fastener products, Armstrong Roofing, and the Owens-Corning Construction Division—the idea is still somewhat of an untapped resource that could help a product development program. For example, several West Coast building contractors, who have been changing their operations from that of a lump-sum general contractor to that of a provider of construction management services, have been using the technique successfully.

In applying franchise extension, three rules are given by Richard Tongberg, marketing research manager at Miles Laboratories. First, "[look] at the strengths, weaknesses and image of the brand." Second, "assess the boundaries of the brand's franchise. What is its ability to be stretched to different product categories?" Third, "identify creative ways of connecting the present brand's image to a new category but in a way that is relevant."†

A similar philosophy has been adopted by the prestressed concrete industry, which has individually and collectively introduced new concrete products and applications by hitchhiking on the glamour and acceptance of the name "prestressed concrete." Today, less than half of the prestressed concrete industry's products are actually prestressed.

Succeeding comes only by trying.

Forbes Epigrams

---

* *Wall Street Journal,* Jan. 22, 1981, p. 25.
† Ibid.

13
Pricing

"Pricing" is the procedure by which the price is set for the construction or engineering services which you offer. It is fundamental to marketing in two ways: first, it is the tangible end point of all your marketing efforts and second, it can be used as a very effective and incisive marketing tool.

Pricing is a very subjective art, requiring far more skill and attention than just a flat percentage markup over costs. It is usually treated properly by most manufacturing businesses. Contractors and engineers, on the other hand, need to devote more consideration to it.

13.1. Proper Valuation of Construction and Engineering Services in Relation to the Market

Pricing has been very misunderstood and misapplied in the construction industry in the United States. In general, the construction industry prices its services much too low, resulting in inadequate returns to the enterprise. This, in turn, leads to inadequate incentive for capital investment, cripples internal growth, and curtails needed research and development.

This is not true in all countries: in many of these, a few large, well-entrenched firms dominate the industry or are organized in a cartel-like association. Their returns are generally much more in line with what would be demanded by a prudent investor in other enterprises.

There is, however, one benefit to the American pricing policy; it makes the contractors lean, hungry, and highly cost conscious, as compared with their more prosperous counterparts elsewhere. Perhaps, therefore, some rational approach can be developed that will lie in between the two extremes and will meet all objectives.

To return to pricing, if you ask a man in the street how much he wants for something he owns, he will likely go through several successive thought patterns. First, he will try to determine how much you will pay for it. Second, he will consider what it cost him. Third, he will try to ascertain how much that "something" is selling for in the market today.

He will not sell his something unless he can get more for it than its cost to him. He knows he is constrained by the market price. He will try to get as much as the prospective buyer is willing to pay.

Compare this to the construction and engineering practice where pricing has been traditionally and commonly based solely on costs plus a percentage markup. For example, the contractor will estimate his costs to perform the work and then add a relatively low percent of these costs as a markup which he often erroneously terms "profit," this representing his assignment for contingency and risk, general overhead and administration, financing, and profit. This procedure tends to relate his net income solely to volume rather than to the unique services he is offering. Competition tends to be focused solely on the fee rather than on the total service.

Pricing should preferably be based primarily on the value of the construction or engineering service, both in relation to the market and on the value received by the client. These should receive the initial focus and attention. Then an estimate can be made to see if the available markup is adequate to cover the risks, overhead, financing, and still leave an acceptable profit.

In run-of-the mill jobs where the market price is well established, the results may well be the same as with present practice. However, in unique, unusual, complex, and risky jobs, the markup will usually emerge at a much higher figure than is now the case.

If the available price does not yield an adequate profit, then you should not go into that market or should get out of the market if you are already in it.

13.2. Proper Pricing

Some areas of the construction industry do price their services at a healthy profit. One example from past years was the large-diameter concrete pipe and manufacturing business. Traditionally, concrete pipe has been priced on two factors: first, on the cost of manufacturing concrete pipe by the competition and second, on the pricing of steel pipe. Only then did the manufacturer consider his own costs of production. This strategy is termed "value pricing," or pricing according to what the market will bear.

Another example in construction where services have been priced correctly is the dredging industry. A dredging contractor often owns a single piece of very large equipment that represents a sizable investment. Perhaps the dredge will sit idle for a year at a time. Before bidding, a dredging contractor will usually prepare three estimates. The first estimate is an estimate of what the job historically will bring. This is often similar to the owner's estimate and represents the potential market price. The second estimate is an estimate

13.2. Proper Pricing

of what the competition can bid. The dredging contractor pinpoints where the other dredges are located in the area and how much work they have. (There are seldom many dredges in one area.) The third estimate is an estimate of what the contractor's costs will be. If the contractor cannot get a good profit on top of his costs, he will not bid. (Note: The clever contractor might, also, be able to find out what the owner's budget is through inside information, or it may even in some cases be published.) Assume the dredging contractor was given the following estimates:

1. Historical price, extrapolated $0.90/cu yd
2. Competition's cost   1.10/cu yd
3. Contractor's own cost   0.73/cu yd

Now the typical engineering construction contractor faced with this situation would not have evaluated (1) and (2), so he might bid $0.73/cu yd plus 10 percent; or 80 cents per cu yd. A typical dredging contractor, however, would assume that the owner might be willing to pay, say, 10 percent above the historical price, brought up to date, or $0.99. So he then decides to bid $0.97/cu yd. On this job the dredging contractor would get a 33 percent markup. Table 13.1 lists some pricing strategies.

TABLE 13.1. Pricing Strategy

Term	Description
Discount	Price subject to discounts on a predetermined basis. Example: Owner given discount for early payment.
Diversionary	Low basic price on selected services to develop image of low price; but profit made on other services. Example given in section "2.5. Proper Contract," when building contract has provision that contractor will perform all subsequent interior work at a relatively high fee.
Guarantee	Price is based on guaranteed results. Example given in section "4.1. Types and Scopes of Contracts," when contractor guarantees to accept all subsurface risks.
Loss leader	Intentional low pricing used for market penetration. Example given in section "13.5. Market Penetration," for prestressed concrete.
Offset	Low basic price to buy entry to a more profitable second market. Example given in section "14.1. Feasibility Study," where low-price feasibility study buys design contract.
Price lining	Price is competitive, but scope of services is adjusted to reflect changes in cost. Example: Owner contributes something of value, such as use of a crane or finishing materials.

13.3. Purchasing Decisions on Factors Other than Price Alone

People make purchasing decisions on factors other than price alone. Consider the traditional example of choosing a doctor when a person is severely ill. Factors such as reputation, health care quality, personal friendship, and schedule can be more important to the potential patient than price. If the patient needs an immediate appendectomy, a doctor with a low price that cannot take the patient for two weeks would not even be considered. The patient needs a doctor now.

The construction and engineering businesses need to find alternative methods to price their services. The foundation of these alternatives should be based on value to the client and not just the cost to the contractor or the engineer.

In the pricing of services, you also need to determine what the client will pay. Consider the case of Red Adair and Associates—the oil well fire fighter specialists. When there is an emergency or an oil well fire, Red Adair reportedly will not even give a price until the fire is out; then it submits a single lump-sum bill, with no breakdown. The oil company can pay the entire bill or nothing. If it chooses not to pay the bill, Red Adair will wipe off the bill and not charge it a penny; however, Red Adair will not work for that company again.

13.4. Tailoring Services Offered to Prices Available

The client will pay different rates for different services, as he perceives them. For example, a large supermarket chain might pay a very adequate price for a set of standard plans and specifications for its stores, a standard that will be used all over the country. That same large supermarket will pay a very low fee for the modification of the master plans and specifications required for each individual building location (the building permit drawings). It wants to get this work done as cheaply as possible. It is recognizing the difference in skill required and the long-term value to its company.

If the price available is not enough to cover your costs, then you may want to reduce the scope of your services and still get a fair market price. Many construction and engineering firms have used this method as a means of remaining competitive in a market that has developed a low-price syndrome.

Similarly, you can propose to the client that he provide certain services which will have a disproportionately favorable effect on your costs and, hence, justify a reduction in your price.
Examples are:

1. Client to provide advance funds upon which you can draw, so as to reduce your own financing costs.
2. Client to furnish certain materials or equipment or services.

3. Client to separately arrange certain services or insurances which may benefit both of you.

13.5. Market Penetration

Intentionally low pricing can be used for market penetration; a contractor may even price his services at no profit or at a slight loss. This is done for the long-term benefit of gaining access to the market or for building volume in a market. Low pricing for market penetration has two main disadvantages: the obvious one is that you do not get a proper price for your services during the break-in period and the second one is that most start-ups do not have the economy of an established operation. Many manufacturing companies will set up a cash reserve to be used during the period of market penetration. Unfortunately, the cost of going into a new market is usually underestimated. An example of this was found with one California company that wanted to go into the prestressed concrete market. This company initially priced its products very low to gain market volume. After a large volume developed, the company needed to raise prices substantially. Unfortunately, it had built its sales entirely on price. As a result, there was so much consumer resistance to price increases that the market dried up and the company finally went out of business. Market penetration is a tricky subject.

An enterprising firm started a plant on a Caribbean resort island and produced architectural wall panels. These were beautiful panels, sculptured in form, textured, and often colored. It produced a high-quality product.

The local architects wanted this product but said they would only specify it if the manufacturer agreed to meet the price level of the current alternative of plain, reinforced, cast-in-place concrete walls. Whenever the architectural manufacturer tried to raise its prices, the architects threatened to cut off their orders. The manufacturer was caught in a disastrous squeeze. Perhaps it should have had more courage and arbitrarily raised prices, or perhaps it should have undertaken a carefully planned market campaign to justify higher prices.

Conversely, a similar situation in western Canada was addressed more intelligently. Initial pricing was low, to gain market penetration, but a strenuous sales campaign was simultaneously launched to convince architects, engineers, and owners of the added value and to let them know in advance that there would be gradual price rises. By moving the price in small steps, it successfully raised it to an acceptable level in two years.

13.6. Unintentional Underpricing

Underpricing your services downgrades you in the eyes of your client. The client normally knows what a certain service is worth, and if your price is too low, he will feel you are either a fool or you are not going to give him the full service he requires. There is an exception with mass-produced work. A low price for a mass-produced item, for example, a standardized prefabricated steel building, can imply that the client is getting a mass-produced service of adequate quality.

A few years ago, a Texas land developer hired a San Francisco lawyer to advise him on a major project. In four hours the lawyer arranged for the developer to meet the people he needed to meet, showed him that his proposed approach would not work in San Francisco, and provided an alternate approach. He put him in touch with the proper architects and engineers. The lawyer submitted a bill of $50 per hour for four hours, or $200. The Texas land developer refused to pay. The Texan later told the architect that the lawyer's services were worth $5000 and that is what the bill should have been. He was not going to pay the bill, just to teach the lawyer not to be so foolish!

In billing a client, never be petty. Some lawyers will submit a large bill for several hours a week at $150 per hour and then add $1.25 for postage. This practice has a tendency to irritate many clients. A recent consulting engineer's bill was prepared by an accountant, listing many thousands of dollars for consulting services, expenses for a trip to Japan, and a $11.40 charge for a telex. Of course, the charge for the telex was deleted. Do not let your subordinates make you look petty.

It is easy to become trapped in a bad market. A West Coast contractor had started business as a pile-driving contractor on small foundations; he had developed an expertise and a reputation. He knew and liked his clients; he fought with but respected his competition. Unfortunately, the small pile-driving business had also become a low-priced market. Out of every 10 jobs, 2 were losses, 2 were good, and 6 broke about even. Other areas in the construction business were much more profitable; however, for too many years, he held onto the small pile-driving business, continuing to pour effort into a market with little return. He had allowed himself to become emotionally attached to the past. His parent company eventually forced him to pull out of this market. It was the right decision, even though it was a hard one to make.

Much larger contractors have made the same mistake, pursuing markets that were once profitable but no longer are, due to excessive low pricing.

13.7. Pricing to Anticipate Additional, Changed, or Future Work

In unit-price contracts, pricing can be established to anticipate additional, changed, or future work. In other words, if it is anticipated that specific items of work will be added or that certain items will be eliminated, pricing is appropriately set.

Sometimes, if carried to excess, this idea can backfire. On a multimillion dollar district contract, one small bid item was "5 cubic yards of mass concrete" for replacing unsuitable soil in the foundation. The contractor reasoned that the amount could increase, perhaps, 500 to 1000 cu yd, whereas if the item was eliminated entirely, it could only decrease by 5 cu yd. Therefore, the contractor priced the concrete work at several hundred percent above his estimated cost. It turned out that the contractor's evaluation was correct: the amount of concrete needed for the repair was almost 500 cu yd. However, faced with the excessive cost of concrete, the district redesigned the foundation and used piling instead of concrete. The contractor had priced himself out of the concrete business, and the district got a good deal on the new item of piling priced according to the contract as an extra work item, at cost plus a small percentage. The wise course, in hindsight, would have been to price the item high but not so high that it led to redesign.

13.8. Protective Pricing

Certain items should be priced protectively, such as covering a subcontractor's bid in a general contractor's bid. If a subcontractor prices reinforcing steel at $0.40/lb and the general contractor then bids the same item of reinforcing steel at $0.60/lb, the general contractor may have problems in negotiating his final contract with the subcontractor. The subcontractor may insist that the general contractor handle his materials, provide special access, and so forth. Or the subcontractor may later file a claim for being delayed. The subcontractor will feel there is so much markup on this bid that he should get more money. Even though the general contractor tries to explain the idea of bid unbalancing to the sub to justify the $0.60/lb price, the sub (who also understands unbalancing) will just keep reiterating to the contractor that his bid was $0.40/lb and the general contractor is getting $0.60/lb. So it is normally much better to price the work only slightly above the subcontractor's bid. Some general contractors even price the subbid item below the subcontractor's bid. Then they just sit and wait. The subcontractor, who, of course, knows the published bid price, thinks maybe the general contractor did get a special sub price from someone else and voluntarily reduces his price in a desperate effort to get the job. While the ethics of such a procedure are certainly questionable, there may be times when this can be ethically used to break a collusive situation among subcontractors.

13.9. Long-range Effects of Pricing Too High

There can, of course, be negative long-range effects of pricing too high: high pricing brings in competitors. A manufacturer of lightweight aggregate (for lightweight concrete) was pricing his material very high. The potential competition watched this practice and calculated its own costs to manufacture lightweight aggregate. After the manufacturer had built up the lightweight aggregate market, the competition entered the market and the price dropped close to cost. The manufacturer might have been wise to have progressively lowered his price while he had a monopoly and thus keep some of the competition out.

Excessively high pricing also brings in competitors in construction. On the first lock and dam contract on a major river development, there were only three bids. The engineer's estimate was $27 million and the low bidder was $32 million. The job was awarded, nevertheless, and publicly announced. On the second contract the low bid was below the engineer's estimate, and there were 30 bidders. In this example, the high price brought in the competition. The low bidder on the first contract may have been wise to bid high and take the money and run; however, he never succeeded in getting any future work on that river development. High prices, even if you are in a temporary position of monopoly, decrease your market. This is the traditional result of fundamental economic theory. In construction, consistent excessively high prices lead the client industry to find alternative solutions.

The above illustration confirms the generalized statement that the initial large job of a series usually gets the best price. On subsequent jobs, the other contractors have your published bid prices as a guide. They have an opportunity to see how the competition feels. This gives each estimator and each bidder more confidence. He may even have had the opportunity to observe the start of the initial contractor's operations. Thus each successive job tends to be priced a few percent lower than the previous one, down to the point where it becomes ridiculous. Then a new opportunity may arise, when most of the competition have given up that development as unprofitable and moved to other markets.

On the California Water Project, there were a series of very large pumping plants, of similar design. On the first job there were eight bids, at what was reportedly a satisfactory price. On the second, there were 16 bids, with prices 8 percent lower; the third had 14 bids, with prices 6 percent below the second. This latter job turned out to have been bid below cost!

13.10. Markup

For the Salinas River Bridge project, the Downriver Construction Company calculated its percentage of markup over costs as follows. First, it made an analysis of its annual business, based on the previous year's data.

Net worth of business	$1,800,000
Annual volume	20,000,000
Total costs (including depreciation)	18,000,000

13.10. Markup

Gross operating gain	2,000,000
Administration cost	600,000
Financing	400,000
Contingencies	500,000
Profit before taxes	500,000
Taxes	167,000
Profit after taxes	$ 333,000

Its markup for the Salinas River Bridge project follows. This is based on a total project cost of $3,020,000.

Administration	3%
Financing	1.9%
Risk analysis—contingency	3%
Return on investment—profit	2%
Total	9.9%

$$\text{Markup} = \text{total cost} \times 9.9\%$$
$$= \$3,020,000 \times 9.9\%$$
$$= \$299,000$$

The method of calculating the four components of Downriver's markup follows.

Administration

Downriver has a policy of adding 3 percent of the total cost (the direct cost plus the indirect cost) for the cost of administration. This figure (3 percent) is calculated yearly by using the average annual volume of business for the last two years and the actual average administration cost for the past two years.

Year	Annual Volume, in $	Administration Cost, in $
1974	18,000,000	550,000
1975	22,000,000	650,000
Average	20,000,000	600,000

$$\frac{\text{Administration cost}}{\text{Annual volume}} = \frac{600\ 000}{20,000,000}$$
$$= 0.03$$
$$= 3\%$$

230 Pricing

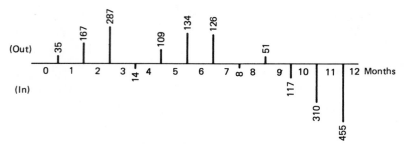

FIGURE 13.1. Schedule of costs and receipts.

Financing

The cost of financing for the Salinas River Bridge project was calculated by using a cash flow analysis. The total cash flow is shown in Figure 13.1. Given the total cash flow, a present worth (PW) analysis was made and the cost of financing was calculated.

$$\begin{aligned}
\text{PW} =\ & \overset{P/F\,1\%,\,1}{35^K(0.9901)} + \overset{P/F\,1\%,\,2}{167^K(0.9803)} + \overset{P/F\,1\%,\,3}{282^K(0.9706)} - \overset{P/F\,1\%,\,4}{14^K(0.9610)} \\
& + \overset{P/F\,1\%,\,5}{109^K(0.9515)} + \overset{P/F\,1\%,\,6}{134^K(0.9421)} + \overset{P/F\,1\%,\,7}{120^K(0.9327)} + \overset{P/F\,1\%,\,8}{8^K(0.9235)} \\
& + \overset{P/F\,1\%,\,9}{51^K(0.9143)} - \overset{P/F\,1\%,\,10}{117^K(0.9053)} - \overset{P/F\,1\%,\,11}{310^K(0.8963)} - \overset{P/F\,1\%,\,12}{455^K(0.8875)} \\
=\ & 57.7^K \approx \$58{,}000
\end{aligned}$$

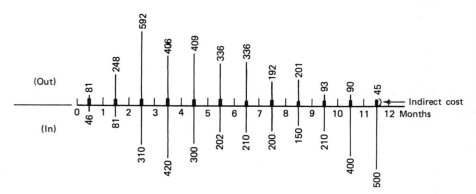

FIGURE 13.2. Cash flow.

Markup— Risk Analysis, Salinas River Bridge

Risk or Opportunity	Magnitude	Action to Minimize Risk or Maximize Opportunity	Revised Magnitude	Probability	Effect (+)	Effect (−)	
Labor problem	Work stoppage for 15 working days cost $3000/day	Try to predict when, so to minimize exposure	45000 0	P(WS) = 0.25 P(NWS) = 0.75		11250	
Excellent productivity	Labor in the area is extremely productive at present time.	Schedule work to try to save 10% of labor cost.	66000 33000 0	P(66,000) = 0.33 P(33,000) = 0.33 P(0) = 0.33	32670		
Flood	A major flood occurs once in four years. Most probable maximum damage is $300,000.	Builders risk is carried to reduce maximum loss to $200,000.	200000 100000 0	P(200,000) = 0.13 P(100,000) = 0.13 P(0) = 0.74		80000	E(X) = 38000 Utility = 80000
Major earthquake	1/50 chance; due to damage state would not cover.	Builders risk is carried to reduce maximum loss to $200,000.	200000 0	P(200,000) = 0.02 P(0) = 0.98		10000	E(X) = 4000 Utility = 10000
Minor environmental problem	Fine of up to $50,000.	Try to meet all environmental requirements.	50000 10000 0	P(50,000) = 0.05 P(10,000) = 0.10 P(0) = 0.85		3500	
Subcontractor has problem	Work Stoppage— loss 10 working days cost $3000/day and $5000 in legal expenses	Get competent subcontractors	35000 0	P(35,000) = 0.25 P(0) = 0.75		8750	
General contingency of ½% of total cost to cover unpredictable events. For example, falsework failure, oil shortage, etc.; certain problems can't be forecasted, but that you need some contingency for.						15000	

Total (+) 32670 (−) 128500 = 95830

$95800 Add for risk

$$\text{Risk} = \frac{95800}{3020000} = 0.03 = 3\%$$

Note:
Builders risk insurance is carried, and the more dangerous items
(the piling and cofferdams) are subcontracted.

FIGURE 13.3. Risk analysis.

232   *Pricing*

$$\frac{PW}{\text{Total cost}} = \frac{58{,}000}{3{,}020{,}000}$$

$$= 0.0192$$

$$= 1.9\% \text{ Financing}$$

Note: The total cash flow in Figure 13.1 was based on the (in-out) cash flow shown in Figure 13.2.

Risk Analysis minus Contingency

The analysis of risk analysis minus contingency is shown in Figure 13.3.

Return on Investment

Given the financial analysis of the Downriver Construction Company, the "profit" factor for return on investment was calculated as follows:

$$\frac{333{,}000}{1{,}800{,}000} = 19\% \text{ on investment}$$

The percentage on volume is calculated once a year and becomes company policy.

$$\frac{333{,}000}{20{,}000{,}000} = 2\% \text{ on volume}$$

14

Selling a Professional Service

Large professional engineering and architectural firms are having an increasing problem of getting enough business to keep continuity in their work and make an adequate profit. Some of the factors contributing to this trend appear to be high overhead and operating expenses, the need to have a large, highly technical, skilled staff, and the costs of carrying out equal employment opportunity and apprentice training programs. The costs of professional liability insurance and of legal advice to prevent incurring liabilities continue to rise dramatically. There exists an urgent need for effective marketing of professional services.

14.1. Feasibility Study

One effective method to get a good contract is to get an early or preceding contract for a feasibility study. (Feasibility studies can have different names, for example, a predesign study.) The feasibility study is performed to determine whether the project is technically, practically, and financially feasible and to study and compare various alternatives. Many feasibility study contracts are bid at or below cost. In effect, the low bidder is buying a possible entry for a later contract. By performing the early work, the firm can establish the proper relationship with the client and learn what is most important to the client—for example, financing, schedule, and so forth. By performing the feasibility study, the firm is in an excellent position to make a submittal for the final design. Various firms have used this very effectively on World Bank financed projects. Once the firm has established itself with the country's project, it has a high chance of being able to negotiate the final design contract and, thus, make a profit on its investment. This practice is good marketing; however, it has been abused. In the feasibility study, there exists a tremendous incentive for the engineer-architect to determine that the project is feasible. If the project is found not to be feasible, the firm loses its initial investment in the low-cost feasibility study. Therefore, the World Bank, as an example, has instituted review procedures to limit this practice.

Other governmental agencies have similarly provided for a review procedure on their feasibility studies.

A well-prepared and thorough feasibility study remains an excellent means of conveying to the owner your firm's competence, ingenuity, interest, and dedication. You establish high-level contacts: you become, in effect, part of the management team. Thus it is an excellent and highly successful marketing tool. It does have its costs, however, for it requires proper implementation input from experienced personnel and first-class engineering judgment.

14.2. Personal Contacts

Having the proper contacts can be an important aspect of selling a professional service. The client naturally likes to know the person he is dealing with. If the client has confidence in a person's capabilities, a relationship of trust will develop.

Personal contacts can be made by joining a variety of organizations—country clubs, dinner clubs, Rotary, and so forth. The key to what to join is based on the organizations to which the prospective clients belong. Some organizations, like the ASCE, contain only engineers. While belonging to such an organization may help develop technical competence, it does not put you in contact with clients, only with competitors.

Although personal contacts are very important in selling a professional service, frequently the engineer, architect, or contractor simply does not have a personal contact. Even without a personal contact, the professional can still be extremely effective in marketing his services by emphasizing his professional capabilities to the prospective client. In fact, the professional can take advantage of a sort of reverse phenomenon that exists because he does not have the personal contact. Many people become reluctant, after a while, to deal with their close friends. These people not only know their friends' abilities but also know the limits of their capabilities and the various mistakes they have made in the past. Therefore, almost as many times as a personal contact helps in getting a job, the professional finds that the absence of a personal contact and a long standing friendship is not an insurmountable obstacle.

Your services will not command a premium if your word has to be discounted.

Forbes Epigrams

14.4. Special Products or Services

At some point in the marketing effort, the professional without a previous personal contact develops some personal relationship with the prospective client through which he can assure the client of his technical, professional, and business capabilities. In this case, the professional should not attempt to duplicate the personal relationship developed by a long-standing friendship. He should keep his marketing effort on a very businesslike basis.

There is a growing tendency for people that represent companies or corporations to try to justify their decision of choosing a particular professional service. These people not only like to back up their decisions with written data but also be able to state that the selection decision was based entirely on professional capabilities.

Therefore, even if one has been fortunate enough to establish a fine personal contact with a prospective client, it is important to do a first-class job of preparation when submitting a proposal on a specific job: one that can stand on its own feet to justify the award to you.

14.3. Enthusiasm

Enthusiasm is the essential element in any selling, and this is especially true in selling a professional service. Enthusiasm is an essential part of creativity, and a typical client for a professional service is looking for a creative approach. A man who is enthusiastic tends to kindle enthusiasm in the client. If the professional believes in his ability, then the client can also believe in his ability.

Engineers that are subcontractors to architects frequently find that they are engaged to provide technical services only. The architect wants to be the creator, and he wants other people around him merely to provide technical backup. T. Y. Lin, the famous prestressed concrete engineer-designer, has overcome this difficulty, largely with enthusiasm. He is so enthusiastic about what he is doing that he sells the architect, as well as the owner, on what he can do. For example, he responds to the architect's concept with, "Yes, that is a daring design, but working together, we can do it with prestressed concrete!"

14.4. Special Products or Services

One way to sell a professional service is to cater to the known problems in the industry by addressing them head on. For example, the delays in procuring the soil samples and performing the laboratory tests are something that have been frustrating to many structural and foundation engineers. First, identify what are the key points that bother other professionals. Then find methods for exploiting these key points. In the soils case, one might set up his field operations and laboratory so that they can get the results out in two weeks instead of one month. This may be a tremendous selling point. Or, one may find that the solution is to provide a preliminary report even though it is not requested. This report could be issued before all the data are gathered and analyzed. This way the client may be able to start planning or working on his project. The problem of delays between professions is very critical in the industry. The architect delays in giving the engineer his basic layout, and the engineer delays in giving back the critical dimensions. Being aware of this and other problems can be an asset in selling professional services.

14.5. Maintaining a Proper Client Relationship

A professional service is always looking for repeat business. Maintaining a close relationship with the client is the key to maintaining ongoing work. That is why the professional should continue selling not only when he gets the job but all the way through to the end. The client should always remain as his present and future client.

Several engineering firms that have been taken over by large national firms have tried to set up a specialized group known as engineering sales. These people are those who go out and develop business. They may be called "business development managers." The company policy is for the engineering salespersons to get the job, and then the job management is turned over to other people. This of course is a self-destructive policy. Most clients resent having to negotiate with one professional before the job is awarded and then deal with another professional from then on. It is more effective for the company to require that the engineering salesperson stay with the contract right on through to completion. This also establishes a communication link with the people performing the work and the owner. As an example, a certain Bay Area client had engaged the same firm three times. During the third contract, he finally made up his mind never to deal with that firm again. Each time he had a project to let out for construction, he would be called on by a very impressive sales engineer from Company X, who would take him out to lunch and explain all the innovative things his company was doing to save costs. In fact, each time the client would find himself believing in this person and his company. However, as soon as he gave him the job, that was the last time he would see him until the next job came along. The client felt that he needed someone to whom he could talk and explain his problems during the job. He was looking for a person with authority who would stay with the job from inception until conclusion.

A professional needs to learn to listen to his client. Professionals tend to think they know best what the client needs; so they do not pay sufficient attention when he addresses them. One of the authors recently sat in during an interview of an engineering firm when the owner asked about the scheduling of the engineering effort for a large and complex multiphase project. The sales representative responded with a long, previously prepared presentation about the technical approach to one critical element of the project. That was not what the owner asked.

The engineer-salesperson presenting his company's case during a final interview needs to have a proposed plan of action: "If awarded this design contract, we have tentatively planned to proceed as follows. We would like your input if you have any suggestions as to how we might better fit your needs and schedule."

14.6. Public and Professional Relations

A professional needs to maintain his standing within his profession. That is why it is important to serve on technical committees and gain recognition in the professional field. However, these committees can become a full-time career in themselves. There are so many technical committees to serve on that there is no way an engineer can serve on all those in which he is interested. There are two basic motives for serving on committees. The first motive, of course, is to establish his own competency and prestige in the profession. Also, by serving on a committee and helping guide it, he is discharging his professional responsibility to that profession. Therefore, it is always wise to be on at least one professional committee, working with members of your own profession. The second motive—which is more important from a marketing point of view—is to be on a committee that represents his profession in his client's industry. For example, the American Mining Congress has a committee involved with the safety of underground construction. This committee is basically made up of mining engineers; however, one structural engineer volunteered to serve on that committee. This was an invaluable opportunity for that structural engineer to develop a close relationship, working with potential clients on mutual tasks. This is a marketing man's dream—to be able to demonstrate his abilities and establish contacts with the people who will be his clients in the future. Therefore, it is far more important for a structural engineer to serve on an American Petroleum Institute committee than to be a structural engineer serving with other structural engineers on the Structural Engineers of Northern California committees. Of course, unless one has the professional expertise which can be gained from committee membership within his profession, he will not be able to contribute to and gain from his client's committee. Therefore, a balanced approach of committee membership in one's own profession and in his client's profession is recommended.

Let your work be your best advertisement.

Forbes Epigrams

It is always important to identify a multiheaded client, as discussed in an earlier chapter. Consider the case of marketing the engineering of temporary foundation bracing. The contract will be awarded by the general contractor. However, the structural engineer may be the one who recommends you to the general contractor and also approves the subcontract award. For the foundation engineer there is also a third source of potential business—the architect. In most cases, the architect will know about the job before the structural engineer. It can be more difficult to establish a contact with an architect than with a structural engineer. Many of the leading architects do not participate in local chapters of A.I.A., although they belong to the national organization. Therefore, there is not always an easy way to reach a large number of architects. However, every once in a while the opportunity does present itself. For example, the architects may have a miniconvention or conference to discuss the problems of high-rise buildings. The foundation engineer involved in marketing should attend such a meeting. If possible, he should be one of the speakers. Also, becoming a member of one of their committees is a great opportunity.

A San Francisco structural engineering firm has been successful in being recognized by the general public and the engineering and architectural fraternity for its brilliant concepts in structural engineering. Much of this recognition has been due to its efforts in maintaining a public image for its work. As a result of this recognition, the firm can often sell its concepts as a major contribution to obtaining the project. Leading architects use this firm to help sell their architectural services to the client. Instead of just working for an architect, the firm becomes a partner with the architect and makes a major contribution to the marketing efforts of the venture. The firm has also succeeded in convincing the architects to increase its percentage fee well above the going rate.

14.7. Presentations

In selling a professional service, the engineer needs to clearly present concepts and details. Consider the case when a structural engineer or a soils engineer is making a presentation to an architect. He has to be careful to present just that portion of the material which is of interest to the architect. For example, he might use just a small picture of the complete structure and some nicely reduced and simplified pictures of the details. He should demonstrate his technical expertise without overwhelming or boring the architect. Therefore, in making the presentation, he should take the architect's point of view instead of his own. The architect may have little interest in structural details. He simply takes them for granted; the structural engineer, on the other hand, may have great interest in details and would personally like to deliver a half-hour sales pitch on them. However, it may not hurt if the engineer gives a brief statement concerning his technical abilities in design of connections to impress and ensure the architect of his abilities.

Sometimes a single sentence can be most effective. In the example where the structural engineer wishes to discuss a connection, a simple statement, such as "a typical connection for this framed structure will be highly stressed and require a thorough engineering analysis for both static and dynamic loads," may be sufficient and effective.

If the structural engineer's analysis is shown on a lengthy computer printout, it is best that only the final results be shown to the architect. The printout displayed on the table will adequately impress the architect of the engineer's abilities. The day has passed when one can impress and amaze a client with 30 pages of computer printout.

In presentations, it is always more effective if you can get the potential client involved in developing an idea with you.

For example, a structural engineer was being interviewed for a potential project involving a series of pipeway bridges in an existing refinery. The engineer had previously developed a new concept involving small-diameter pipe piles, with concrete sleeves, beams, and girders, one that had never been used before. In this conference, instead of presenting his preliminary drawings and calculations, he started to think aloud and sketch his concepts freehand, as though this was a new idea he was just conceiving. He got the refinery's engineer to participate with him: in fact, he took the sketch away and completed it himself.

The structural engineer not only got this job but a series of subsequent projects in the refinery.

In making a presentation, nontechnical ideas can often be beneficially discussed. More and more often, the delays and problems tend to be less constrained technically than institutionally. Of what use is the perfect design if it cannot be approved by the regulatory agencies and the governmental bodies having jurisdiction? One of the professional firm's greatest assets today is the demonstrated knowledge of these complex institutional steps and its ability to obtain approvals in a timely fashion.

In making your presentation, a frank discussion of the permit and approval problems may be an effective way of convincing your potential client that you will not get caught by surprise in the increasingly complex maze of bureaucratic regulations. Frequently, you can justifiably ask for increased compensation for efforts spent in facilitating and obtaining approvals.

14.8. Proposals for Professional Services

Sometimes a client will ask a professional to submit a bid for his services. This case often arises on government contracts. The latest criteria for awarding a contract usually states that price is not the only factor to be considered but will be a consideration in choosing the contractor for a professional service. The facts of life are, however, that price is often the major factor. The key to being successful in this type of situation seems to be the careful definition of the services and the exact scope of the work. Because of the very nature of professional services, it is often very difficult to define scope and details. That is why professionals do not like to do it; they want the freedom to work as the particular situation evolves and dictates what should be done. For example, a doctor is unlikely to give you a price for making you well. He will give you a price for performing a regular medical test or routine operation. If during the course of the treatment, the doctor determines the patient requires additional services, such as an additional operation, he will require that the patient pay for the additional treatment.

Those professionals that are successful in defining the scope of their work will specify very carefully such items as phases, the degree of involvement, and the number of alternatives that will be examined. This way they can offer a low fee for their services, which tends to make them competitive. In the bid they may also include a provision which recommends, for example, the examination of another alternative. If this recommendation is accepted and followed, then they can negotiate an extra fee with the client. Possibly, the professional will not even quote the extra fee at the time that the original recommendation is made.

The government, in particular, is known to be a hard and tough negotiator. It often does not seem to be bound by any particular ethical code other than trying to get the lowest price possible for the professional service, justifying this on the grounds of national interests. It will often try to get the professional to examine additional alternatives without additional cost. The professional may not opt to suggest too many alternatives. He may already know the proper solution, because of prior similar work. He might address this by adding a clause which states, "as the work progresses, it may prove desirable or necessary to examine other alternatives, in which case an additional fee will need to be negotiated with the client if the client decides to have the new alternative examined."

As stated before, the government is known for driving a hard bargain when negotiating for professional services. Some departments also use the practice of calling in the two or three finalists and asking them to go back, review their bids, and reoffer their final best price. This is nothing more than bid shopping. Unfortunately, the government often instructs its employees to use it. Some successful engineering firms have met this by coming back with a lower price; however, at the same time they also reduce the scope of their services. They recognize that the government is under severe budgetary constraints. Thus their profit margins remain the same.

Many engineering contractors that work for the government are well aware that government contracts are frequently expanded in scope. Therefore, they will knowingly accept a low initial fee for their services, recognizing that quite likely the scope of the work will be increased, and they will be asked to work on additional studies. A professional has to be extremely careful and shrewd in pursuing this particular approach.

14.9. Competition between Professional Services

The established professional societies have prepared some fine-sounding phrases about ethical and appropriate conduct for a professional and about competition. Actually the marketing of professional services is highly competitive and, on occasion, degenerates into areas of questionable ethics.

The architect is trying to get the entire job from the owner; the engineer is trying to get his job directly from the owner; the specialist engineer, for example, a geotechnical engineer, is also trying to get a piece of the job directly from the owner. He knows that if it goes through the architect, his fee will be lower, his contractual terms less liberal. Once the decision is made, each tries to get a slice of the pie from each other. In general, it is a highly competitive field, horizontally and vertically.

If a professional is going to stand out and obtain a direct contract, then he needs to find some way of persuading the client that he has something special to offer. For example, "dynamic analysis" is a specialty that, if available in-house, would tend to set an engineer's firm apart from the run-of-the-mill design office. As another example, a firm that specializes in design of underground projects may have a special competency in some of the new underpinning, bulkheading, tiebacks, and slurry trench techniques. Expertise in this area makes them stand out from the average foundation engineer.

The proper relationship between competitors has been discussed in other areas of this book. To further elaborate, there is a need to treat one's competitor as the professional which he is. If a fellow professional is not outstanding and is having a problem on a particular job, one should still be able to find something good to say about him. For example, if a competitor is being criticized, one could say, "He has done a very creditable job on all projects up to now; I am afraid that I do not know the particular circumstances concerning this present project." One can at least refuse to join in the criticism of a fellow professional. This just tends to downgrade oneself.

In the case where one professional is selling a professional service to another professional, the other professional is usually well aware of the problems and inadequacies of the subcontractor. The professional is usually in competition with his own subcontractor to some degree. Therefore, the professional may be tougher on the other professional than a nonprofessional would be on a professional.

Contrary to general assumptions by the public, the larger firms do not have a competitive edge on all projects. Being small can be an asset when competing with a large company on a highly specialized or technical job.

The small company can give the prompt, personal service that a larger company many not be able to offer. Recently, an owner was faced with choosing a process-engineering consultant. The choice was between a highly competent but large firm and a smaller but also competent consultant. With the large firm, there would be no way of determining exactly who would perform the work. It would be anyone from a recently hired man to one of the firm's best people. With the small consultant, the owner knew exactly who would do the work. The owner chose the small consultant.

15

Product and Professional Liability

Professional liability has become a serious problem in the United States. The current trend is for owners to be more aggressive ("litigious," meaning prone to sue) in their relationship with design engineers and architects, contractors, and turnkey constructors. So far, third-party suits against these entities have not grown as rapidly.

However, major constructors and the insurance companies are increasingly worried about the specter of class-action claims. These include such things as noxious fumes from a plant for which the designer conceivably might be sued on the grounds of negligence. Another specter is that of catastrophe, such as an explosion or a fire, with excessively large loss of life or property damage.

Internationally, the most serious situation exists in the United States. In Britain and Canada, the problem is growing, but the potential liabilities are more limited due to their legal procedures. On the European continent, where Napoleonic Code dominates, the problems are even less. So far, there has not been any problem in the Mideast or Southeast Asia of which the authors are aware.

15.1. Consulting Contracts

For design or consulting contracts, what is emerging as a solution is that the engineer or architect include a disclaimer clause in his contract document. A typical disclaimer clause that is meant as an example, not to be used without consulting one's own lawyer, is as follows:

Consultant agrees that he will perform his consulting services with that standard of care, skill, and diligence normally provided by a professional in the performance of such consulting services on work similar to that hereunder. Client shall be entitled to rely on the accuracy, competence, and completeness of consultant's services hereunder, but consultant shall not be regarded as a guarantor with respect to estimates, systems, plans, or any work product performed hereunder. Consultant's professional liability hereunder to the client, to all contractors and subcontractors on the project, and to any third parties due to consultant's negligent acts, errors, or omissions, shall not exceed in total aggregate the amount of $50,000 and shall be limited to a period ending one year after termination of this contract. Client agrees that if consultant is named as a defendant in any legal action arising out of the project which is the subject of this contract, client will hold consultant harmless, indemnify, and defend him against any claims which may be asserted against consultant, unless such claims are attributable to proven negligence on the part of the consultant.

By putting in a limitation on liability, the professional can limit his liability with the owner. Having a limitation on liabilities of, for example, $50,000 can partially insulate the professional from suit. What lawyer is going to take on a suit if he knows the most he is going to collect is $50,000?

Statistics show that most lawsuits for professional liability are between the owner and his engineer. The suits typically cover a wide range of legal matters from errors in design (for example, the plate-glass windows that fall out of the mullions of high-rise buildings) to the case when the engineer's cost estimate was exceeded by all contract bids and the owner proves he has been seriously damaged by this grossly erroneous estimate.

Responsible owners are usually willing to sign a contract that contains an equitable limitation on liability clauses. If the owner will not sign a contract with such a clause, your professional liability insurance may still cover you, but the premium will usually be higher. In such a case, the professional can offer to remove the objectionable clause if the owner will agree to a price increase to cover the extra cost of the liability insurance. Faced with this possibility, many owners will then agree to the limitation on liability clause. Sometimes, however, the owner will insist that the limitation on liability clause be removed and will agree to pay for the professional liability insurance. For the professional, there is a problem with this. Once a professional is involved in a suit on professional liability, his insurance company is likely to cancel him out. He may be protected on the first suit, but professional liability insurance may no longer be available to him.

Some consultants will also include a clause to try to protect themselves from lawsuits arising out of the actions of third parties with which they may be associated. A typical disclaimer clause that is meant as an example and not to be used without consulting one's own lawyer is as follows:

In the event the consultant is requested or authorized by client to retain or associate a third party as an expert or subconsultant or otherwise, consultant shall not be responsible for the acts or omissions of such third party for performance of that party's duties. No negligence on the part of the third party shall be imputed to consultant.

While clauses such as this and the preceding may help the consultant, the client may quite properly seek to modify them so as to place more responsibility onto the engineer.

The prudent engineer and professional keeps the owner well informed on all decisions. This matter of "keeping the owner informed" deserves some elaboration. The design professional and the engineer-constructor make many decisions on the basis of judgment. They must learn to record these decisions, note the reasons why, and send a letter to the owner, so that he has the opportunity to intervene. While an owner may be more than willing to take a calculated risk during design, if the event actually occurs during construction, he will be aggressively looking for a culprit.

Extensive records should be kept, and they should be sufficiently complete and accurate to stand up to future outside scrutiny. What the litigious atmosphere means for the professional engineer is that he must be more methodical and must document his internal decisions more carefully.

15.2. Construction Contract

Where a company is the general contractor only and performs only the construction (no design nor equity involvement), then the general public liability insurance policy of the contractor usually contains a "completed projects" endorsement to cover liability for injuries or damage caused by defective workmanship or materials. The contractor has a direct contractual responsibility to the owner for repair of such defects. He can, however, escape liability to the owner for contingent damage, such as loss of rent, by a suitable clause in the contract.

The emergence of "maintenance clauses" requiring the general contractor to perform all maintenance and repairs to the structure for a period of five years "regardless of cause," appears particularly onerous to the contractor. Does this include design errors by the design engineer? Does it include actions or lack of actions by the owner? Some courts have said that it does. Similarly, performance type specifications may impose upon the contractor the responsibility for selection of a product or design details.

It is legally possible for the contractor to write a contractual clause to the effect that the owner will indemnify him and hold him harmless for all claim of third parties for damages except those due to the sole negligence of the contractor. The words "sole negligence" are extremely important here. Contractual clauses limiting the liability of the constructor to the owner are quite often used by sophisticated engineer-constructors. The liability can be limited to specific amounts of money or, even better, by clauses limiting the liability for defective design to a requirement that the constructor re-design the work correctly and by clauses limiting the liability for defective construction to a requirement that the contractor reconstruct correctly. In other cases a clause may be inserted limiting liability of the contractor to the amount of insurance carried, this amount being spelled out in the contract.

It is very important to exclude contingent liabilities: loss of use, loss of production, loss of rent, defective products, lower output, and so forth. Such a contract should preferably also exclude the liability in regard to procurement of process equipment and specialty items. A typical clause states that the constructor will submit his procurement specifications and procurement contracts to the owner for his approval and that upon approval, there shall be no further liability on the part of the constructor for deficiencies in the equipment furnished.

The development of adequate insurance, as to amounts and coverage, requires the services of a large and highly competent insurance brokerage firm, with experience worldwide and especially in the American market. To reduce costs of these large policies, the major constructors (Kaiser, Santa Fe, Morrison-Knudsen, and so forth) are reportedly obtaining policies with very large deductibles (up to $1 million on the bottom; that is, they self-insure for the first $1 million of loss). Bechtel, who has very large overseas reserves, reportedly self-insures throughout.

There is generally no contractual way to adequately limit your liability to third parties. Most states of the United States have laws that expressly forbid the owner from indemnifying the contractor for the contractor's acts of negligence. Sometimes the contractor and owner do insert a clause by which the owner agrees to defend and hold harmless the contractor except where claims are attributable to proven negligence on the part of the contractor (as quoted earlier). The purpose of such a contractual clause is to keep the contractor out of litigation in the cases where the third party sues the owner and everyone else connected with the project for damages which are not directly attributable to negligence on the part of the contractor.

That still leaves a major area for third-party claims. For example, a large window blows out, injuring tenants. They then sue the owner and the engineer, alleging defective (negligent) design on the part of the engineer. The engineer's comprehensive public liability insurance does not protect him because it always excludes "design deficiency." While this is a serious risk in buildings, it is even more serious in process plants, where allegations of long-term pollution or poisoning conceivably could be made, including the so-called class-action suits.

Insurance is available to cover these third-party cases, but it is not easily purchased. It can usually only be obtained by a contractor who is undertaking a long-term involvement in many projects. The insurance pool will probably demand that it gets the insurance on all your projects, not just one, and that the contractor place all his policies, including public liability, with it, so that it can generate large total premiums.

The best protection against liability suits is to do your work well and keep the owner informed. Major decisions should be covered in writing: this can be done without offense if they are politely, factually, and accurately written.

15.3. Design and Construct Contracts

Professional liability is a special problem on design and construction contracts. Consider a case where an owner's roof leaks. With seperate contracts, the general contractor can blame the leak on the roofing subcontractor. The roofer can blame the leak on the architect or engineers who designed the roof. If this type of problem were to end up in a lawsuit, everyone involved would probably share some of the blame. However, on a design and construct contract where one party is responsible for everything, the owner knows for sure who is responsible for the leak. This is a whimsical example suggested to illustrate the point that design and construct contracts can lead to increased professional liability exposure.

On the other hand, knowing he is responsible, the design and construct contractor can step in promptly, correct the defect, and thus mitigate the loss.

When the contractor undertakes a design role as well as construction, the matter becomes extremely difficult. First, there is the psychological effect on all parties that a single large contract by a very large company has: injured parties are more prone to seek large damages and juries are more inclined to award large damages.

As far as relations between designer-constructor and owner are concerned, it is generally possible to limit the liability to the owner by insertion of a specific clause in the contract. U.S. law will not recognize a complete exclusion (a so-called exculpatory clause which says that the owner waives all rights against the contractor), since that is considered to be against public policy. However, it is generally acceptable to limit the liability of the contractor for defective design or materials or workmanship to a reasonable amount; a sum equivalent to a substantial portion of the total fee is usually acceptable as "reasonable under the law."

Where the contractor works for a public body or public utility, the owner may be forced to sue even when he does not want to. In a current case, on a nuclear power plant, the designer-constructor apparently made the decision some 10 years ago not to design the walls of the control room to withstand an extreme earthquake. The Nuclear Regulatory Commission subsequently demanded full seismic design for this control room. The owner had asked the designer-constructor to correct the deficiency and, originally, was apparently willing to pay for it. The plant was shut down while the work was going on, and the owner had to buy power from other utilities. Now, the state public utility commission, which controls the rates charged by the utility, has reportedly said that it will not allow the power company to include the costs of the purchased power in its rate base and has directed the owner to collect the contingent liability from the designer-constructor, suing if necessary.

Many large turnkey constructors are obtaining two insurance policies. One covers their general liability (public utility and property damage) as a contractor. The usual endorsement is called "completed operations" which is in effect a "product liability" coverage, covering damage to third parties as a result of constructor's operations. This will normally exclude damages due to defective design. The second is a professional or design liability policy, covering damages to the owner and third parties due to defective design.

To ensure that the policies mesh and that there are no gaps, especially where the damage could be partly due to design error and partly due to contractor error, there are two solutions. One is to buy a joint policy for both coverages from one company. The other is to get advice from a legal and insurance broker to ensure that the policies mesh. This is more complex but still fully practicable. These policies are expensive; sometimes the cost can be passed on to the owner. The cost of the premium can be reduced if the owner limits his rights against the contractor.

One possible way to pursue the insurance problem is for the parent company to set up two corporate subsidiaries: one for design, the other for construction, the way M-K Company has done with International Engineering Company. Then the construction company protects itself by buying a comprehensive public liability policy with completed products endorsement and by contractual limitations of liability with the owner, if possible. The engineering subsidiary protects itself by design professional liability insurance, which is relatively available, and by limiting clauses in the contract, wherever possible.

15.4. Case Histories

In a recent case on the Pacific Coast, a facility was to be constructed in a mountainous area notorious for slippages and landslides. In order to secure financing and as part of the investigation of the engineering feasibility of the project, the group of investors backing the project engaged a local soils engineering firm to provide a geotechnical report which included the design of the access roads and graded areas.

It asked that the engineer specifically address stability of side slopes and predicate his design on a conservatively stable slope. He accepted this contract without any qualifying or limiting clause.

During construction, one of the road cuts suffered a major slip-out. The owner directed the contractor to repair it. The owner took the position that it was either faulty construction or faulty design. Since there was no evidence that the contractor did anything wrong, the soils engineer had to bear the thrust of a major claim, more than 10 times his fee for the job.

Obviously, in hindsight, the geotechnical engineer should have insisted on a clause in his contract (1) limiting his liability and (2) noting that in a bad area such as this, one can reduce probability of failure but not eliminate it at reasonable cost. Further, he should have involved the owner, in writing, in the decisions as to what slopes to specify and how much drainage to install.

16

Negotiations

An engineer or an architect involved in marketing needs to possess or develop skills in the art of negotiating. Negotiating skill is obviously required in negotiation contracts. Equally important, negotiating skill is used in negotiating changes, change orders, and claims. If neglected, changes, change orders, and claims can cost the contractor his just profit or, even worse, cause a loss on a job that should have been a good job.

There is an excellent book on negotiations, *The Art of Negotiating*, by G. Nierenberg.* In reading this book, it should be kept in mind that negotiations are amoral, insofar as techniques are concerned. One needs to know the techniques, both those that he will use (within his ethical framework) and those that may be used by his less ethical opponent. Therefore, Nierenberg has given many examples and case studies, some of which may be shocking; certainly, they are informative.

16.1. Recognizing the Right Opportunity to Negotiate

Contract negotiations between the owner and the contractor are somewhat like pheasant hunting. In hunting pheasant, you can walk all day long, and the pheasant will just run on the ground. But if you stop and wait, the pheasant will get so nervous that it will jump. Contractors are the same way. As long as they are actively negotiating, they act rationally and negotiate with skill and patience. However, given a dead period (of about a week), one will usually jump. Someone gets so nervous he cannot stand it; he cuts the price by $10,000 or agrees to accept untenable risks; he "buys" the job and a loss.

---

* G. Nierenberg, *The Art of Negotiating* (New York: Cornerstone Library, 1968).

An important element in marketing is recognizing the right opportunity to negotiate. For example, consider a recent case of a lump-sum contract that has just been awarded. The fictitious bids are shown below:

$41,000,000	Engineer's estimate
$40,000,000	Bid A
$38,000,000	Bid B
$25,000,000	Winning bid

The winning bid was a correct bid; it was based on another method of performing the work. As an initial comment, if the low bidder had known or could reasonably have recognized that the competition did not have his method of performing the work, he should have bid higher. But this is a comment in hindsight. What can be done now? Perhaps the contract price can be negotiated upward. The owner is probably both happy and concerned about getting such a low bid and may be willing to negotiate an increase in price. However, on public works, this may not be legally permissible. At the least, however, the owner should take a positive attitude in negotiating legitimate change orders. The contractor should try to get such a commitment from the owner. The low bidder could take another tack; he can negotiate a favorable construction plan from his favorable price position. For example, he may get an agreement from the owner as to his sequence and methods of excavating and backfill. He may insist that the owner provide adequate survey and inspection teams so that there will be no delay in the operations. He may demand prompt payment in exact accordance with the contract (with which owners seldom comply). He may request an advance mobilization payment.

In some cases, since the owner has $16 million left in his budget, he may negotiate an extension to the contract for additional work, under a change order. The contractor is in a favorable position, since the owner undoubtedly feels that if he does not act reasonably, the contractor may claim an error and back out of the contract. A clever negotiator can exploit the worry of the owner, by slow and slightly hesitant responses, and by asking detailed questions "to ensure that he fully understands the requirements and mutual responsibilities."

The contractor should use this opportunity to clarify all ambiguous clauses in the specifications and to resolve all permissive clauses in his favor.

In this real-life case, the low bidder performed the contract at $25 million, but he did no negotiating. Unfortunately, he had trouble with the owner's engineers on trivial matters and almost lost money on this job, even though his method was brilliant. The inspection was slow, laborious, meticulous, and even petty. When the job was underway, the job-site representative of the owner took the position that "he is not going to let the contractor make up any money on his mistaken bid!" The contractor obviously missed a grand opportunity to develop a sympathetic understanding with the owner. After the job was completed, the contractor was seen to be correct in his method; it did work out. In the end, he even made a small profit, half that which he hoped for at bid time, but what an opportunity he missed!

The above example is a true-to-life recent case. In another recent real-life case, the winning bid of almost $200 million was 30 percent below the engineer's estimate and 12 percent below the second bid. One might think that the owner and his engineers would do everything possible to facilitate the contractor's work. Once, however, the contract was signed, the owner, a public agency, fell right back into its archaic habit of petty but costly and annoying delays. For example, a minor change in the prescribed detour to facilitate the contractor's equipment access was held up for two months and a credit (deductive change order) demanded. Shop drawings for temporary facilities were delayed in approval for two months. Not a very promising start for a contract in which the owner had already received a "bonus" in the form of a low bid.

Recognizing this persistent pattern of public agencies, a contractor with a low bid should examine all such points in the postbid, preaward conference, so as to obtain as many legitimate agreements as possible.

Consider the case where all bids have been rejected for exceeding a private owner's budget. Then there exists an excellent opportunity to negotiate a contract. In many of these cases, the engineering design and/or the architectural specifications were too costly; therefore, the owner will probably now be receptive to suggestions for modifications of the design or specifications by the contractor. The architect is now under tremendous pressure, for his client is unhappy; and if the project does not go ahead, he may have difficulty collecting his fee. Even though he may have previously rejected an alternative proposal, he now will be in an entirely different situation and will be predisposed to acceptance. For example, before the bids came in, the owner just had to have marble facing panels from Vermont and could not be talked into any other facade for his building. Now, the lightweight concrete panels (at one-fourth the cost) proposed by the contractor sound very interesting. This is the right time to sell to the owner a negotiated contract that he can afford.

In negotiating, timing can be all important.

16.2. Establishing a Negotiation Strategy

Before sitting down at the bargaining table, you need to establish a negotiation strategy in which you define your position. You need to define what objectives can be compromised—along with limits as to what can be compromised in exchange for other concessions. You should try to anticipate the position of the other side. Now that you have done this advance planning, it is important to remember that your negotiation strategy must remain flexible. Unexpected developments often occur at the bargaining table. If need be, you can always ask for a short recess in order to regroup your side and establish a new strategy.

One principle that can be used in negotiating is to use an unusual approach, to adopt a gambit. One developer-contractor would often follow this principle. One time he was trying to acquire a 10-acre parcel for part of a large subdivision development. Naturally the developer's objective was to acquire the parcel at the lowest possible cost. Upon inspection of the old estate's rather large house, he noticed a very nice Steinway concert grand worth about $18,000. Although he had no interest in the piano, he used the following gambit: he made a very low offer of $175,000 for the estate and the piano together. This was a low offer; the estate alone was worth about $250,000. Somewhat to the developer's surprise, the owner countered with an offer that he would sell the estate for $175,000, but his piano was not for sale. The developer had achieved his objective by changing the negotiation format. The owner did not negotiate as he probably had planned to.

16.3. Setting the Stage

Although the negotiation of a new contract is usually held at the owner's office or some central location, the negotiation of a contract modification should usually be held at the job site whenever possible. Then when questions arise concerning the contract modification, they often can be quickly resolved by a brief site inspection, which causes little delay to the negotiation. You can also bring in the superintendent or foreman most familiar with the event under discussion. This can often have a very favorable effect on the contractor's behalf. The very environment of the job site—the noise, the action—sets a favorable stage for the contractor and conveys to the owner far more effectively than words the need for practical and prompt decisions. Unfortunately, most negotiations are carried out in a conference room or office.

The physical environment of the room can have a great effect on the negotiations. Personal discomfort caused by a room that is, for example, too hot or too cramped, can turn the negotiation session into an unreasonable debating session where nothing is accomplished. A proper negotiating room should:

1. Be comfortably spacious
2. Have proper lighting and acoustics
3. Be at a comfortable temperature
4. Have comfortable chairs and a comfortable table
5. Be equipped with the necessary visual aids and supplies

16.3. Setting the Stage

Also, each side should have access to a private room in which to confer privately during recesses.

The most effective way to promote a productive negotiation session is to have an agenda. The agenda not only establishes the points to be negotiated but can put both sides in a frame of mind to make concessions. If you can, make up the agenda: this helps you lead and control the negotiation process.

As far as being on time, there are only three approaches that may be taken: promptness, punctuality, and not being late. If anyone is going to be late, let it be the other side who must apologize and perhaps make up for its late arrival.

At the start of the initial negotiation session, if you have not already done so, you should ascertain whether or not the other party has the power to bind its side to the agreement. There is a frequently encountered negotiation strategy, based on the use of a limited authority negotiator. Even if such a negotiator actually has a great deal of authority, this is never revealed to the other side. This results in lowering the expectation level of the other side. The other side tends to accept that a negotiator with limited authority may be able to make some concessions but cannot "give away the store." Further, the limited authority negotiator can exploit this limitation by, for example, stating that he will never be able to convince his principal to make such a large concession or that he will have to confer with his principal before proceeding further. The other side, anxious to reach a conclusion, may temper its demands to those that can be settled at the meeting. This is an advantage that many owners, because of their corporate size and power or because they are a public agency, often have over a contractor. The negotiators for the owner can often only promise to recommend the agreed action to their management, whereas the contractor is usually expected to live up to his commitment 100 percent. This unfortunately gives the negotiators on the owner's side the opportunity to agree "reluctantly" to a compromise and later come back with one more condition.

One clever Chinese contractor, to counter what he perceived as an unfair advantage by the owner's side, created a fictitious "chairman of the operating committee" to whom he promised to submit the compromise agreement with an equally "reluctant" recommendation for approval.

More seriously, you, as the contractor, can use this authority of yours to make decisions as a positive weapon in the negotiating deliberations.

Finally, just as an uncomfortable physical environment can have an adverse effect on the negotiation process, so can interruptions—which should be avoided or at least kept to a minimum. Therefore, you should clear your calendar of all business (including personal business) on negotiation days. If the other side is hurried by other commitments, this may be to your advantage. It may be willing to make some last-minute concessions just to bring the negotiations to a close.

One very tough administrator for a governmental agency used to take advantage of the typical contractor's impatience to "get back to work" on the

254   Negotiations

day-to-day affairs that dominate the average contractor's workday. He would purposefully pace the negotiations slowly. If he had learned by an "innocent" question at the start of the meeting that the contractor had to catch a 5 P.M. plane, he would drag out the principal matter until the end of the afternoon, placing intense pressure on the contractor to concede just in order to get a settlement. One contractor finally learned to use this in reverse: government administrators normally do not like to work late in the evenings!

16.4. Destroying the "Opponent" Concept and Replacing It with the "Mutual Endeavor" Concept

Negotiations should be conducted in an atmosphere of cordiality and mutual trust. This type of atmosphere makes both sides feel that they are working together to find a mutually acceptable agreement.

One way to develop the "mutual endeavor" concept is to get the client involved with the project too. Working together tends to tie two sides together. During the final negotiations for the contract, there is a tendency for you to want to jump right on the project, make decisions, and start moving dirt. Be patient. Get the client involved. Typical of the things you might ask the client to do as you start negotiations would be to obtain permits, to prepare and submit the environmental impact report, to look into design alternatives, to provide access or utility connections, to furnish job office space, to provide surveys, or to obtain additional borings.

The owner may use a similar tactic too. At the start of the negotiations, he can review some of the more attractive nonmonetary features; for example, his ongoing programs for capital expansion, the prestige attached to the project, or the opportunity for the contractor to develop new construction methods and gain experience for future work.

As noted earlier, at the time of award of a contract, whether private or public, there is a short period that is highly favorable for negotiations. For both sides, this represents the conclusion of a long and tiring endeavor: both are glad the decision has been made, and there is a natural letdown.

This is the time for the contractor to request a pre-start-up conference. Here he can meet at the same table with both the awarding authority or executive and the resident engineer. The executive will probably take a broader and more reasonable attitude than the resident engineer.

Policies can be established, covering approvals, time-extension for delays by the owner, inspection, payments, and a host of other matters.

If you intend to submit a value-engineering proposal, now is the time to arrange for its consideration. Procedures for submittal and response can be agreed upon timetables established, and principles for establishing the cost saving agreed upon.

Not until the contract has been completed and all claims have been settled will the contractor again have such a favorable situation for negotiation as he has in the few days following award.

One effective way to start a negotiation is to first establish the points of agreement and then the points of disagreement. The fact that both parties are sitting at the negotiating table indicates an agreement on the need to negotiate. In negotiating a claim, for example, both parties might be able to agree on the fact that the contract has been finished satisfactorily and the finished project is working as intended. Next, the contractor might wish to establish that both parties would like to settle the claim. This can destroy the "opponent" concept and replace it by a concept of mutual endeavor. Both parties want to settle the claim. At this point, the areas of disagreement can be brought into focus.

16.5. Various Appeals to Reason

At various points in the negotiation process, there will be appeals to reason. Initially, you are convinced that all the logic and all the points are on your side. However, the other party is not accepting them. The other party may be very logical, too, and may have some valid points that dominate its thinking. In this case, if both parties are logical, they will probably be able to find some middle ground. Sometimes, however, the other party may be illogical and emotional. In this case, negotiations are more difficult. The other party may have adopted this illogical or emotional viewpoint because its representative is afraid of being fired by his company for overrunning the budget on the job. He may be a contractor who is going bankrupt; this is his last chance. When the other party is being illogical, there exists a gambit that sometimes works. However, this gambit is risky. It is simply to appear to give up; then leave the negotiation.

A claim against the state highway department was once successfully negotiated in this way. The department had a policy of never recognizing claims, just or unjust. After two days of negotiation, the contractor was getting nowhere. Then he informed the other party he was giving up and said that this was it. He informed the other party he would not file any more claims and would not take any legal action. However, he said to the opponent that he was going to leave the books and job records showing the extra costs, just in case the other party wanted to see what actually happened. A few days later the contractor received the full amount of the claim. The tactic should be used only when you know the other party and, of course, only when your books and position are completely valid. It will not work if the other party is unethical. It will not work if the other party has no money with which to pay.

16.6. Avoid Arguments

The fastest and most effective way to reach agreement with the other party is to avoid an argument with it. The trick is to agree with the other party whenever possible while still guiding the discussion toward your own position. When you do have to disagree, it is very important to do so objectively. For example, instead of directly disputing the other party's statement, use a response such as, "Yes, you do have a point, but isn't it also important to consider. . . ."

When there are many points to be negotiated, you should avoid long, drawn-out discussions on any single point. Sometimes one party will intentionally lead the other party into a prolonged argument over a relatively minor point. This tactic might be employed to divert attention from other points that are unsupportable. Or this tactic might be employed to set up a point that appears on the surface to be significant but can easily be conceded as the negotiations progress.

Rather than try to argue a point head-on, it is better to explore other courses of action that are naturally acceptable. The number of possible alternatives is only limited by your resourcefulness and flexibility. For example, if you are far apart on price, you might suggest a reduction in the scope of work. Or you might ask the other side to suggest a solution that will meet its own requirements while still meeting yours.

When a serious argument does arise, a recess often has a positive, calming effect. During the recess, both parties have the opportunity to appraise their respective position with a cool head, following which both sides will often make concessions that will eliminate the basis of the argument. Also, a recess after specific proposals and counteroffers have been made can give both sides the opportunity to evaluate their positions in a low-pressure environment.

When you do reach a position that cannot be negotiated or argued further, you need to be tactful but firm. You do not want to give the impression of being arbitrary, capricious, or unreasonable.

16.7. Self-Discipline, Self-Control, and the Proper Use of Temper

The negotiation process completely breaks down when both sides lose their self-control. When only one side loses its self-control, the other side gains an advantage. For example, a contractor might become angry with the other side's line of questioning if he is unable, or unprepared, to defend the point in question. By retaining self-control, the other side will often succeed in winning the point.

Some negotiators who are good actors can effectively use apparent temper during the negotiations. One old-time contractor was very good at this. He was an extremely pleasant and personable man, and he could sit patiently for hours at the negotiation table. Then at a certain stage in the negotiations, he would lose his temper (he was only acting). His face would turn red, and his fist would hit the table; and he would make some statement like, "Damn it all, we are not getting anywhere! This is just ridiculous!" Then he would storm out of the room. He never really lost his temper, and he would not make any threats or say anything that would prevent future reconciliation. Usually as he was leaving, the other side would beg him to come back to the table, try to settle him down, and then ask him to restate his position. He would reply that he just had to have his last figure; he had lowered his price all day and could not stand any more of a loss. This tactic often worked—he was a real master.

There is one time it can be to your advantage to have the negotiation process break down, the time when the negotiations are just not going in a direction you can live with. In this case, it is best to start all over. You can make an unreasonable offer and let the other side call off the negotiations. When your side does resume the negotiations, you can always consider changing the negotiator or the format and try to find a new basis for settlement.

While the above may have to be used occasionally, it does represent a breakdown in the negotiation procedure. Is your side financially able to live with a long delay in settlement? You must carefully weigh the alternative course of persistence, and patience, hoping to wear the other side down, against the abrupt termination of the negotiations. In most cases, contractors simply do not have the strength to prevail over a large public agency, for example, after a breakdown in negotiations.

One large, successful, and famous contractor once stated:

Let's face facts: the statutes and procedures are heavily weighted against the contractor. Not only do we lose interest on the money at stake, but it inhibits our capacity to expand our operations. Legal fees are unbelievably high, but even worse is the time and attention our senior staff has to give to the case. Those are the very men who should be out running our jobs and making money. So we'll negotiate hard and long and tough on claims, and we'll always keep the ultimate option of litigation open, if for no other reason, to compel the owner to continue talking. But our policy will be to settle in negotiations, not in the courts.

16.8. Proper Use of Legal Talent

It is important to use legal talent properly during controversial negotiations. If the legal threat is introduced by one party, the other party will undoubtedly bring in its lawyer. The negotiations may then become a legal battle of wits, and in most cases only the lawyers will win. Each lawyer will be trying to display his skill. The legal threat can be introduced openly by one side bringing its lawyer to the negotiation or more subtly by making a general or indirect reference to a lawyer (perhaps a letter showing a copy going to a law firm). Some negotiators will try other subtle threats. For example, a negotiator may make such a threat by stating, "We should really try to settle this ourselves instead of turning the matter over to 'third parties' that do not know as much about the matter as we do." While this is tactful and subdued, it still is a threat and has in some cases aroused anger. So use it with extreme care, for he may reply, "Well, perhaps it is time to get our lawyers in." When you threaten a person, in general you get the exact opposite response you are after. Rather than buckling under to a threat, most people will stand up to the threat and fight even harder. With any threats, overt or subtle, the percentages for getting what you want are against you. As stated before, legal threats should be used only in extreme cases, as further negotiations may be much more difficult thereafter.

However, the mature contractor does make extensive use of the legal profession in preparing for negotiations. He will have reviewed the matter with his lawyer to find out just what his strong and weak legal points are. Remember that lawyers have their own jargon that gives away the fact that a lawyer is involved. So it is best to rewrite the lawyer's letter or ideas in your own language so as to hide the source. Then let your lawyer check your rewrite. If he tells you the ideas are the same but it is not as good as what he wrote, you are doing fine.

An overzealous contractor had a claim against a major public utility. In exasperation, he openly made a legal threat, "We'll have to let third parties work this out!" and was delighted to receive payment in full for the claim the next day. Later, he learned he had been removed from the client bid list and bid lists of some of the client's associates. A successful 25-year relationship with that client had ended abruptly due to his refusal to patiently negotiate the claim.

16.9. Involvement of Subordinates

The involvement of subordinates or team members in negotiations can be quite effective. This approach may be effective in presenting information from many sources—field personnel, accountants, or procurement personnel.

Subordinates are generally used to analyze data, investigate, and write reports, in preparing for the negotiations. During this process they often are in contact with their counterparts from the other side. When professionals are involved (engineers or accountants) they often talk openly among themselves; they feel it is simply one professional discussing the "problem" with another. (Accountants tend not to be as much at fault as engineers when it comes to talking too much.) If your engineer or accountant is listening instead of talking, he can often gain valuable information to be used during the later negotiations. Of course, it can work the other way around, too, if your subordinate is not well-trained.

Once, during a negotiation over alleged changed conditions, Ben C. Gerwick, Inc., had described the materials as "rock," and the soils engineer for the Corps of Engineers described the material as "mud." Gerwick brought in the dredging foreman, who stated that "mud was something that squeezes through your toes when you walk on it with your bare feet. What he was working in certainly would not squeeze through his toes!" This not only led to laughter but a settlement of the claim to Gerwick's satisfaction.

In using subordinates or a team approach in a negotiation, it is imperative that the negotiator maintain control over the team. Should a subordinate forget his role and enter into an independent or unrestrained discussion with the other side, the negotiator must stop him, perhaps by changing the subject or, if necessary, by asking for a recess. The subordinate or team member should only speak when the negotiator signals him to do so or at prearranged times when particular points are to be discussed.

16.10. Trades—But Not Too Soon: The Value of Taking Time

In a negotiation it is important to take adequate time: people do not want to settle too fast on a difficult matter that will cost them money. Lawyers are often purposely slow in negotiation in order to give the other party time to adjust its thinking to the lawyer's point of view. Some skilled negotiators use the waiting game; they let their opponent talk on and on, learning his entire argument and noting points for later attack. You should take sufficient time when responding to the other side, even though the answer may be at the tip of your tongue. A rapid-fire answer to an objection will usually just bait the other side into raising a new objection, one which may be more difficult to answer. Try building up to an answer, possibly by first restating the matter or objection and then giving a measured answer.

When you try to force a person to make a fast decision, his answer is usually "no."

Offering trades on big items is very difficult; and if offered, should not be offered too soon. However, both parties might very well want to settle the small items quickly so that they can move on to the big, important matters. For example, if you are asking $120,000 for one item and the other party is offering $100,000, you might be able to settle very quickly on $110,000 if there is a $1,000,000 claim item still left to be settled. Now let us examine the $1,000,000 claim. Suppose that you want $1,000,000, and the other party is offering $100,000. If you offer quickly to compromise at $500,000, then the other party will use $500,000 as your top asking price and want to negotiate a figure between $500,000 and $100,000. You can insist that you are still asking $1,000,000, but it will not do you any good. If you do offer a compromise, sometimes you can do so in the third person. For example, state that your boss or your client is reasonable and you think he might be willing to accept something slightly less: you state that you really do not know how much. Then you try to get the other party to make the first offer. Conversely, you may try the tactic that you would be willing to compromise but your boss is adamant. Can the other side suggest a solution which you can sell to your boss?

Later on during the negotiations, if a negotiator can predict agreement on the end-point difference between his last offer and the other side's last offer, he may wish to "split the difference." Splitting the difference does not have to be on a fifty-fifty basis; it can be a 90-10 split in your favor. He should offer to split the difference only if he is prepared to state and take a firm position on this offer. He should not let this offer serve as a basis for further negotiation: this should be the final offer.

Some negotiators will use time advantageously by keeping one unreasonable point unsettled until the end. The other party is sometimes willing to make an extra compromise on a major point at the end of negotiations just to get rid of the unreasonable and obnoxious item. Labor union negotiators often use this tactic.

Tied into the gambit of taking time is the gambit of using deadlines advantageously. If the other party is required to settle by a certain date, this puts pressure on it, and you can use this pressure to your advantage. For example, suppose the agreement to provide financing for a project will run out on a certain date. If the contractor can determine this, then by taking time he can make the owner feel forced to negotiate faster, and the contractor may get a more favorable contract. Labor unions use contract deadlines very effectively. The 45-minute coffee break that the union negotiator takes right before the strike deadline puts heavy pressure onto management.

16.11. Concluding a Deal

In negotiations, once you make your point and win, shut up! This can be very hard to do; you always have some more good points that could be made. There may be a report that you have worked on for three weeks and that you would like to show to the other party—but do not. Quit while you are ahead. Gerwick, Sr., and Gerwick, Jr., were once negotiating with the government on a claim they hoped to get $400,000 for. The government officer-in-charge came in, announced that he agreed on the basis of the documents submitted that they were entitled to something, but it was not one penny over $500,000. Both Gerwick, Sr., and Gerwick, Jr., were surprised. Gerwick, Jr., had a long report, an analysis, and other documents on the claim which he wanted to show the navy officer; but Gerwick, Sr., kicking him under the table, stated that the ultimatum put them in a hard position and asked for a half-hour recess to think about the offer. Both Gerwicks went to a private room and talked about other jobs and read the newspaper. In half an hour they reconvened. Gerwick, Sr., said, "You have put us in a tough position, but if you are adamant in your position, then we will have to accept the government's figure of $500,000." They had just made $100,000. If Gerwick, Jr., had persisted in presenting the report, all the subordinate representatives would have had the chance to attack it, and they might never have reached a settlement or, at least, not one half as favorable. Conversely, if they had accepted the offer without the recess, the officer-in-charge would have realized he had been too generous and would probably have found some excuse to reopen the matter.

This matter of cutting off conversation once you have won your point comes up time after time in conferences and negotiations. Invariably, if you try to give "one more reason" to justify what you've already won, the other side thinks of more contrary arguments.

16.12. Finalizing It in Detail: Writing It Up

At the end of each negotiation session, the results should be written up. It is probably late and everyone is tired; but it is important to make notes on the negotiation. Often a negotiation that appears to be settled at 11 P.M. starts over again at 8 A.M. the next morning if you do not finalize it in detail and write it up. The two parties that negotiated should prepare the final draft on the negotiated settlement. Remember, do not bring in lawyers openly; this can start the negotiations all over again. In private you can have your lawyer check the settlement.

It is imperative that all the points are stated honestly and fairly. Obviously if the letter states that you agreed on a point when you did not, you will lose all your credibility. The principal point is to tie down all details so that the agreement has no loose ends.

It is important whenever possible to wrap up points of agreement as they occur, even if you are not able to finalize all matters. In a typical follow-up letter you might thank the other party for meeting with you, and document that you both agreed a claim was justified. However, since you could not agree on the amount of the claim, you will go over your books again. You may suggest that the other party may wish to consult again with its cost estimator before the next meeting.

The side who is able to write up an interim or final report on the negotiations has a definite advantage. As with any meeting, there is a natural tendency not to want to be secretary and have to write up those minutes. However, the side that writes up the results has the advantage of being able to write up all the points favorable to its side in concrete and specific detail. Furthermore, while reporting accurately on the points favorable to the other side, an overly finalizing type of language may not necessarily be used. Therefore the side who did not write up the minutes should read the minutes meticulously and amplify the details to ensure its case has been fully and correctly stated. There is sometimes a tendency for people not to read minutes or to just glance over them, assuming their accuracy and completeness.

In a recent negotiation with a governmental agency, the contractor's appeal was granted in full and the staff directed to write it up. The staff, which had been overruled by its own board, wrote up the minutes but added clauses that vitiated the whole agrement. Only a careful reading of the document by the contractor's lawyer uncovered these devastating clauses and led to a simple but accurate rewriting of the settlement.

Table 16.1 lists 10 precepts for successful negotiating.

TABLE 16.1. Ten Precepts for Successful Negotiating

Precept	Description
1. Offer satisfaction as opposed to concessions.	In a negotiation it is possible to offer satisfaction to the other party without making material concessions. The other party would like (1) to be respected, to feel important, and to be liked by the other party; (2) to look good in the eyes of its boss while doing as little work as possible; (3) to avoid conflict and insecurity; (4) to conclude the negotiation and avoid future problems.
2. Structure the format so that the other side can win too.	Make half-hearted demands on which you can later make concessions, thus giving yourself room to negotiate. For example, if you need to get $100,000 for one claim, ask for a second claim of $75,000 also, then let the other side negotiate you down. Use of this tactic lets "the other side" feel like a winner too.
3. Make concessions slowly and carefully.	There are several reasons why you should proceed slowly in an important negotiation. First, if you make concessions slowly, the other side will feel that it is getting the best possible deal. Consider the Mideastern proverb, "All good things take time." Second, if you accept an offer too soon, the other side is likely to take back or modify the offer.
4. Lower the expectation level of the other side.	Lower the other side's expectations during the negotiation. Develop a concession pattern that decreases. For example, first make a $50,000 concession, then an additional $20,000, and finally, a $5000 concession (never the other way around). If you take back a concession at the end, it signals to the other side that you have gone as far as you can go. Also, whenever you make a concession, even if it has no value to you, ask for something in exchange. This also lowers the other side's expectation level.
5. Discover all possible concessions and take the best.	Learn all the different concessions the other side will make and then take the best combination for you. This may not be the combination which the other side had in mind. For example, an owner is told that his project with a short completion schedule will cost a certain high price and that his project with a long completion schedule will cost a certain low price. The contractor then negotiates a contract with a combination of a moderately long (but adequate) completion schedule and an in-between price.

TABLE 16.1. (*continued*)

Precept	Description
6. Use a pseudo-offer.	A pseudo-offer can be used to learn what concessions the other side is prepared to make. A pseudo-offer works as follows: you claim to have limited or no authority to make the offer and preface it with a phrase like, "I have no authority to make a compromise, but if I were running the company, I wouldn't settle for less than...." Then, quite likely, the other side will make a counteroffer, and you have not been forced to make a solid offer. You can then start negotiating at the point of the counteroffer.
7. Ask for a better offer.	You can often gain a better offer by asking for one. Children seem to be conditioned by a "you can do better than that" demand from their parents, where the child will do better if asked. This conditioning carries over to adults who will often respond positively to a demand for a better offer. However, a good negotiator will respond to such a demand for a better offer with, "We can only do better if you do better."
8. Use deadlines to your advantage	Pareto's rule applied to negotiations is that 80 percent of the concessions are made in the last 20 percent of the time. Given a deadline, both sides will usually wait until almost the deadline to make most concessions. During deadline pressure, discipline yourself to act rationally. If possible, set a deadline for yourself that has a cushion. Of course, do not tell the other side about the cushion.
9. Call a concession a concession and declare the other side the winner.	A concession is anything that has value for the other side; hopefully, you can find concessions (to make) that have little or no value for you. When you make a concession, get the most out of it. Call a concession a concession; keep reminding the other party about the concessions you have made. Also, at the end of the negotiation, you can declare the other side the winner. For example, "Our company has never built a project for such a low fee before." Tell the other side that it is a good negotiator and that it got a good deal.
10. Conclude the settlement in a carefully defined and explicit document.	During the writing of an agreement, it will be discovered that there are many loose ends. Insist that each of these be tied down specifically so there can be no future misunderstandings. Be fair but firm; many of these small matters can prove essential to a satisfactory agreement.

17

Changes, Change Orders, and Claims

Why discuss changes, change orders, and claims in a book on marketing? These only arise after a proposal has been accepted, and the contract has been consummated.

Marketing, however, is a continuing function. The successful submittal and resolution of contract change order requests and claims is an extremely important aspect of marketing and may well determine whether the contract ends in a profit or a loss.

Changes poorly handled can erode the original profit and even lead to an undeserved loss. Conversely, prompt and effective submittal of requests for additional compensation can increase the return to the contractor and can transfer responsibility for developments beyond the contractor's control from the contractor to the owner. Such "marketing" of changes can even turn a losing job into a profitable one.

Therefore, the marketer who has landed the initial contract should follow the job through to the end, maintaining contact with the owner and participating in the preparation, submittal, and negotiation of the changes or claims. Too often, the small changes have been left to the project manager, and the large claims have been handled only after completion of the project by the executives of the contractor's firm and its lawyers.

The skilled marketer will try to resolve these changes as soon as they arise, keeping them in the change order category as far as possible, instead of their becoming claims, and avoiding the lengthy delays and adverse cash flow aspects that accompany most claims. (Table 17.1 highlights the distinction between changes, change orders, and claims.) He will particularly endeavor to keep the changes within the framework of the contract and avoid any steps that can categorize the claim as one for breach of contract, recognizing the long delays and high legal costs of such a course of action. He is aware of the present value concept; that is, the high value to a contractor of a

17.1. Claims Caused by Changed Conditions

TABLE 17.1. Changes, Change Orders, and Claims

Term	Description
Change	Given a construction contract, any additional, deleted, or differing work that is not explicitly or implicitly defined in that contract constitutes a "change."
Change order	If the parties involved agree that a change exists, then by mutual agreement a change order can be issued that establishes what additional work will be performed and the adjustments which will be made in contract price and time for completion. A "change order" is a modification to the existing contract that covers additional work required by a change.
Claim	Often the parties involved cannot agree that a change exists. For example, the contractor feels that there exists a changed subsurface condition and the owner feels that there is no changed subsurface condition and refuses to issue a change order. In this case, one party may file a claim. A "claim" may be within the contract, that is, alleging that a change order should be issued in accordance with the terms of the contract, or it may be a claim for breach of contract, which, of course, throws the case immediately into legal proceedings.

settlement and money in hand as compared to an unresolved claim for an indefinite higher amount.

In the negotiation of a change order, one argument that is usually not effective is the argument of equity or fairness. Contractors always seem to make the mistake, however, of using that argument. The typical reaction of the owner to this argument is that he is sorry about what has happened to the contractor, he agrees that it seems unfair that the contractor has encountered difficulties, but he has to follow the contract: "His hands are bound by the department's regulations or by the board of directors' policies." An owner is normally not in a position to award additional money to a contractor because he thinks he has tried hard and overcome difficulties, or because he thinks he did a good job. Even under the most adverse circumstances, the contractor should stay with the facts when negotiating and use the contract to substantiate his arguments.

17.1. Claims Caused by Changed Conditions

Claims often arise because of changed conditions. Contracts modeled on the U.S. Government Standard Form 23-A refer to two types of changed conditions. The first type is "conditions differing materially from those indicated in the contract." The second type is "unknown physical conditions differing from those ordinarily encountered and generally recognized as inhering in work of the character provided for." Definitions of changed conditions depend on the particular contract. Some examples of changed conditions follow.

W. W. Contractors had a subcontract that required the driving of sheet pile cells for a marginal wharf (quay wall) structure in a harbor. Early in the operation, the sheet piles hit something that would not move. As the contractor excavated down into the river sands, he discovered some large logs deposited by an ancient flood. Luckily this occurred on a government contract that contained a changed subsurface condition clause intended just for this type of problem. It was very easy for the governmental agency to handle this problem: the agency could pay for the additional work that was required without have to state that it had made a mistake in the contract preparation or administration.

Some contracts do not have such a clause: in this case the contractor may have a serious contractual difficulty. He can take the position that a changed subsurface conditions clause is implied in "this type" of construction contract and that adjustment in contract terms for such an event is universally recognized throughout the industry. There exists a danger that the other side will insist that when a contract, particularly a government contract, does not contain such a clause, it was purposefully left out of the contract.

The State of California Highway Department purposefully did not have a changed subsurface conditions clause and vigorously opposed all change requests or claims for changed subsurface conditions. Ben C. Gerwick, Inc., encountered this very difficulty on a bridge project for the California Highway Department, as subcontractor for furnishing and driving very heavy walled steel piling. The state-furnished soils data indicated that the pile length would be 100 ft, ending in end bearing or rock. Gerwick therefore based its cost estimate, equipment selection, and construction procedures on the installation of 100-ft piles. However, the rock structure at 100 ft was so badly fractured that 130- and 140-ft piles were required in order to develop proper bearing. Therefore, the contractor's equipment and procedures proved inadequate for the job, and all the piles had to be spliced and redriven. The California Highway Department paid only the bid unit cost for the extra pile length, whereas the contractor contended that he should also be paid for its costs of splicing and redriving. These costs were significant, and the total cost greatly exceeded the income. In this case the state successfully rejected the contention that there existed an implied changed subsurface conditions clause, and Gerwick was left with a well-performed but money-losing contract.

Not all contracts end as badly as that one. A few years later, Ben C. Gerwick, Inc., was awarded a dredging subcontract on a Bureau of Reclamation project in Alaska. At that time the Bureau of Reclamation contracts did not have a changed subsurface conditions clause—they do today. Wash borings had been taken by the bureau's engineers through the winter ice, and

17.1. Claims Caused by Changed Conditions

only mud and silt were recovered. Therefore, the bureau classified the material along the dredged trench as mud and silt. The contractor expected to encounter mud and silt, and its selection of dredging equipment and its bid prices were predicated on this assumption. However, its operation quickly ground to a halt as large cobbles were encountered—Gerwick was in glacial till. Initially, the bureau's position was that the specifications clearly stated that the boring logs were furnished for the bidder's information only and that the contractor was to make its own evaluation of the subsurface conditions. The bureau maintained that it was Gerwick's responsibility to move 200,000 to 300,000 cu yd of underwater glacial till. Gerwick obtained legal advice that recommended emphasizing the fact that glacial till was encountered at the very site of a particular boring that showed only silt. Following the principle enunciated in the previous chapter on negotiations, lawyers were not used openly during the meetings with the bureau. Gerwick was finally able to convince the bureau that since the soil at the boring itself was changed significantly, a changed condition did indeed exist. The bureau agreed and did pay for the extra work that was required.

In one epic case for the California Highway Department, a contractor encountered huge quantities of sandstone on a contract in the desert. The borings, which were correct, showed only sand. The contract contained no changed conditions clause; and since the borings were correct, the California Highway Department refused to negotiate any payment. The court, however, ruled that the amount of sandstone was so great that no contract existed because there was no meeting of the minds. However, this settlement took several years, sizable legal fees, and immense effort on the part of the contractor to achieve. Would he have made more money with the same expenditure of effort on new work?

In San Diego, a California-based contractor was awarded a contract for the construction of underwater caissons. The soils data supplied by the owner indicated coarsely graded sand with a 70 percent relative density. The contractor's procedures were planned on the assumption that the data were reasonably accurate. Actually, the soil conditions turned out to be extremely fine micaceous sand which had the potential for acting like quicksand under even minor differences of hydraulic head. This changed condition required that the construction method be significantly modified at major additional expense. The contractor repeatedly tried to negotiate a "changed condition" adjustment with the owner, but the owner kept turning him down, claiming that no changed condition existed because "sand is sand." Finally, the contractor's engineer personally talked to the soils engineer who supplied the data. (He had wondered why the soils engineer had not been present at the previous negotiations.) The soils engineer turned out to be a one-man firm. The soils engineer told the contractor's engineer that he was sorry about this situation. He had submitted a $6000 price to take undisturbed samples; however, the owner ran out of budgeted money and told him to do the best he could for $50. Therefore, the soils engineer had gone down to the beach on a weekend and had taken one sample with a water jet pipe. What he recovered was relatively coarse, and he so reported. Armed with this information, the contractor was successfully able to negotiate a change order with the owner. The contractor was initially so incensed that he thought about filing a claim for breach of contract, but wisely did not. There was obviously no "intent to deceive"—only gross negligence; hence he would have lost its case besides incurring heavy legal expense.

17.2. Claims Caused by Change Orders

Claims often arise out of change orders. For example, the owner's engineer orders additional reinforcing steel to be installed and is willing to pay for the cost of the additional reinforcing steel. However, he is not willing to pay for the extra costs to the contractor for removing and replacing the forms that had already been installed before the change was received.

A contractor we will call Nuclear Construction Company was in the process of sinking a 10,000-ton concrete caisson for a large power plant when the owner's representative ordered that all operations be halted while a minor modification was made to the caisson. He wanted to add two anchor bolts to the caisson's steel liner at a point one foot above the current waterline. The following day, the owner sent a man to the site by airplane with the bolts. It took one welder one hour to install the bolts. Until the welder finished, 40 men and all the equipment (a large crane, pumps, and so forth) remained idle. When it was time to start sinking the caisson again, it would not move. For the next four days the contractor tried jetting, digging, and vibration to get the caisson moving again. After four days, the caisson began to advance once again. If the caisson had not been stopped in the first place, it would presumably have been sunk at a steady rate of 4 to 6 ft per day, but following the suspension of operations, five days advance had been lost. All parties agreed that a change had occurred, but each party had a different idea of how much the change should cost. The owner wanted to pay for one welder for one hour—about $40. Nuclear Construction wanted to be paid for five days downtime and the extra costs incurred to get the caisson moving again—about $70,000. This is an example of how a claim developed out of a simple change. The claim took many months to settle and eventually the contractor had to accept a settlement at half what he believed due him.

Consider a similar but more usual case with concrete work. The owner's engineer issues a change order for a minor change that requires a small amount of extra concrete and reinforcing steel. The owner is quite willing to pay for the extra concrete and steel. However, the contractor wants to be paid also for the extra form work and scaffolding. Also, assume that the new work causes the contractor to wait three extra days before it can strip the formwork. The contractor will then want to be paid for the delay. Even a simple case like this may degenerate into a claim.

17.3. Claims Caused by Impact or Ripple

A new type of claim has emerged in the past 30 years—the ripple or impact claim, which is the overall effect caused by an excessive number of change orders on current and subsequent work. This is often the biggest legitimate element in claims. Contractors have found that when they have an excessive number of change orders, they usually lose money, even if the direct cost of each of these change orders is paid in full. This situation causes an impact or ripple effect that prevents the contractor from planning his work ahead of time. The workers become discouraged, productivity falls, materials have to be rehandled, and equipment is not utilized efficiently.

Suppose that the owner sends out an average of three change orders per day and he is willing to pay for all the direct costs involved, plus a nominal percentage for overhead and profit. However, the excessive number of change orders forces the contractor to continuously reschedule his work and realign his operation. There are so many change orders that the contractor cannot work effectively on either the new work or the old, unchanged, work. The excessive change orders are causing an impact on his efficiency.

Some of the first examples of ripple/impact claims occurred on military contracts for missile silos in the 1950s. On one single contract there were over 3000 change orders. The contractor was unable to perform the work in a reasonable manner. In this case the contractor filed a ripple/impact claim and eventually did collect extra payment from the owner.

Impact claims can also be caused by acceleration. Claims for acceleration, which are caused when an owner demands that work be completed sooner than the contract requires, are easy to pursue. However, a more difficult case arises when an owner issues a number of changes or when changed conditions are encountered and the owner still requires project completion by the originally scheduled date.

17.4. Claims Caused by Owner-Furnished Items

One problem that often leads to claims is the matter of owner-furnished material or services. For example, the contract document states that the owner will furnish the steel H piles. The owner is interested in procuring materials and services at the lowest possible price. The owner often is not nearly as aware as the contractor as to the importance of having the material delivered on a particular schedule. Also, the owner is not particularly interested in whether the items need to be re-sorted or reassembled by the contractor; in fact, this matter probably does not occur to him.

Consider an example where the owner is an oil company. It may very well buy all the piping and furnish it to the contractor. The most inexpensive way for the owner to purchase the piping may be in "40-ft double random lengths," which in practice means that the individual lengths may vary from

16 to 54 ft. On the other hand, the contractor may have his operation set up to handle and weld pipe lengths between 38 and 42 ft in length. Any other length will require special handling. Another matter arises. Has the owner's purchasing agent specified out-of-roundness tolerances? If not, the contractor may have great difficulty in fit up of the pipe sections for welding. Obviously, problems of this nature, with hundreds or thousands of units on the project, can lead to substantial claims.

Now as to the marketing considerations on the particular example.

The owner's purchasing department will likely fight such a claim because the lack of special specifications on length and out-of-roundness tolerances reflects on its ability. The owner's contract prople will have a negative attitude because it means greater expenditures on the project, expenditures that may exceed the budget. Who will be understanding? The owner's construction and project management personnel—who are interested in job completion on schedule and who are in the field and able to visually see the problems—are the people whom the contractor should enlist to help resolve these matters as a "change order for extra handling and fit up" rather than as a "claim for misspecified owner-furnished materials."

17.5. Claims Caused by Inspection

Inspection claims are very messy for both sides to handle. A typical inspection claim occurs when the inspector has demanded that the contractor do work beyond that which the contractor felt was reasonably required by the specifications. For example, the disagreement may be centered on the interpretation of the specifications with respect to tolerance or finishes, qualitative matters that are hard to express in writing.

A contractor we will call P. Construction Company was constructing a bridge with large quantities of cast-in-place structural concrete. He would spend all day aligning the forms within the ¼-in. tolerance specified. By the next morning, due to changes in temperature and humidity, the forms would be out of tolerance and the inspector would refuse to allow the concrete to be placed. Was this a reasonable interpretation?

From the contractor's point of view, he should not tolerate unreasonable inspection. An unreasonable inspector will never get any better as the job progresses. Unreasonable inspection is seldom if ever the policy of the people at the top. The top people are never concerned with whether the forms are ¼-in. out of tolerance; they want the job built. One thing that the contractor can do to stop an unreasonable inspector is to go above him to someone who will make a reasonable decision. The contractor may have a problem if he approaches the resident engineer. The resident engineer may feel that if he overrules the inspector, he will undermine the morale of his employees. Therefore, it might be wise for the contractor to go above the resident engineer. Even better may be to bring the matter up at a high-level conference when both the senior administrators and the resident engineer are present.

Unreasonable inspection is really a human-factor problem that needs to be straightened out. It is best if this problem is straightened out in a manner so that everyone can save face. If the contractor tries to solve this problem early enough, without anger, it can usually be solved: "I think this is unnecessary. It's going to delay the job unnecessarily. Let's go and look at it."

The contractor should not let unreasonable inspection go on, or the job can progressively turn into a loser.

From the marketing point of view it is important that the adverse inspection be corrected at an early stage. Experience shows that continued suffering, followed by a claim at the end of the job, receives little sympathy and is seldom successful. The public owner will take the position that this was what the specifications required and is what the contractor agreed to do. He's sorry if the contractor underestimated the costs of conforming to the specifications, but he, the owner, cannot legally pay any more.

The private owner will take the position that the contractor should have brought this matter to his attention earlier: that he was interested only in a completed project, on time, and of sufficient quality to serve his operating purposes. Unfortunately he has no money in his budget to cover the extra costs.

17.6. Claims Caused by Differences in the Interpretation of the Plans and Specifications

On the contract for the reconstruction of a major two-level tunnel, a West Coast contractor encountered a problem with the interpretation of the plans and specifications. These contained a note stating that concrete surfaces were to have a class 1 or class 3 finish, "as applicable." There is an extreme difference between these finishes. A class 1 finish requires the finisher to work and polish the concrete until it has an umblemished finish. This type of work is usually required for architectural concrete at eye height. A class 3 finish is designed for places that cannot be seen. The purpose of this finish is just to achieve the proper durability. After the West Coast contractor had finished the job and was awaiting final inspection, the resident engineer stated that the underneath surfaces of all the precast double Ts that formed the roof of the tunnel required a class 1 finish and the job would not be accepted until this finishing work was performed. To change the finish at this point was very expensive and difficult. The workers were required to perform the finish work overhead, while standing on scaffolding above heavy automobile traffic! On the plans, a class 3 finish was shown; however, the state took the position that the specifications governed over the plans and, in this case, the specifications required a class 1 finish on any surface that could be seen by the motorists on the road. The only way a motorist could see the underside of the double Ts would be to stop his car and look straight up. This appeal to reason did not sway the project engineer. As a final insult, after the finishing was done and the state had accepted the work, the owner had the surface sprayed with a heavy coat of reflecting epoxy under a separate contract, so the finish really did not matter. This was one change the contractor did not collect on, even though he fought for a long time. A contractor has to decide where he can most effectively address his efforts and not let his emotional reactions involve him in an endless and fruitless dispute. There comes a time when it is best to get on with new business.

Once again, in hindsight, if an alert contractor-marketer had taken the time and trouble to discuss tolerances, finishes, and other qualitative matters with the resident engineer early in the job, perhaps more reasonable interpretations could have been agreed on. If not, at least the extra finish work could have been scheduled for performance in the precasting plant instead of at the site.

Sometimes in a set of plans and specifications, some minor-detailed requirements will be intentionally omitted, assuming the contractor will naturally realize what is required for a completed job. Such a contract typically includes an "all incidental work thereto" clause. Under such a clause, the owner's agent or engineer can require the contractor to perform all customary work. In one famous building construction case, the plans and specifications did not contain any electrical work, so the contractor did not include any electrical work in his estimate. When the owner ordered the elecrical installation to be performed under an "incidental thereto" clause, the contractor asked for a change order. The owner refused and a claim arose. In this case, the contractor finally did receive extra payment for the electrical work, on the basis that it was "substantial," not "incidental."

"Incidental thereto" clauses can be abused. If the owner intentionally fails to disclose material data that would affect the bid, he may be guilty of fraud. Since this immediately throws the case to breach of contract, requiring long legal proceedings and proof of intention, the contractor should be extremely careful before going this route and proceed only on the basis of top legal advice. It will usually be far better to negotiate a settlement, even a compromise settlement, early in the job.

17.7. Claim Letters

If a claim is not documented in the manner required by the contract, the contractor may be held to have released his rights. With government contracts, if a claim is not made within time limits specified by the contract, the government entity may legally refuse to permit the contractor to pursue the claim. Then why are contractors so dilatory about writing claim letters until the job is finished? The reason is that they are usually afraid. If they write to the resident engineer, for example, stating that his inspection requirements on forms are unreasonable, he may get angry and intensify the inspection on the earthwork.

17.7. Claim Letters

Sockem Construction was in the process of driving piles through sand into rock on a state highway contract. During this period, it filed a claim with the resident engineer on a different matter entirely—a change in some concrete work. The resident engineer's reaction was, "You probably have something on this claim, but it is making me look bad. You can submit this, as I suppose you have a legal right to, but I have a clause that says jetting shall be performed whenever directed by the resident engineer! I'll make you jet every one of those piles in rock. You'll wish you had never submitted this claim."

A contractor should avoid getting into an adversary position with the resident engineer. A contractor needs a good working relationship with the resident engineer to resolve the many minor differences and problems that arise on any job. There are many ways in which the mature project manager accomplishes this, but all include these elements: recognition that the other has an important job to do; respect for the other's competence; and maintenance of communication on a free-flowing basis.

When claims do arise, experience indicates that it is the best policy to submit them in accordance with the procedures specified in the contract. The claim should be worded factually, positively, but courteously. A contractor should avoid creating problems for himself by using aggressive language; a politely worded claim is legally as effective as an argumentative assertion.

The contractor should pay particular attention to contract clauses defining time of submittal and method of modification for potential changed conditions. Almost all contracts contain clauses to the effect that the contractor will notify the owner or his agent in writing within a specified number of days if a change is encountered. Also, there is usually a clause that requires a follow-up letter within 15 days, further reporting on the change. If this procedure is not followed, a lawyer will note that the contractor is not in compliance with the contract, in which case the contractor now has two problems instead of one.

Therefore, if a contractor encounters a potential condition that might require a change order, he should immediately write the owner a letter to protect his legal position. This letter should state that the contractor has run into something that could develop into a change order; he should never use the word "claim" in his letter. In discussing this potential change, he should request that if the matter does develop into a significant problem, then an appropriate adjustment should be made in contract time and price. Here, again, additional time and money must be requested in a timely manner in order to protect the contractor's legal position; otherwise, the owner may take the extreme position that "you told me about the changed condition but didn't ask for a contract change, so I assumed you were doing it without extra pay."

In the initial letter a contractor should emphasize that he is attempting to stay on schedule and to limit additional costs. For example, he may have

SUBTERRANEAN CONSTRUCTION COMPANY

November 23, 1980

Mr. John Serdip, Manager of Construction
University of California
Building Department
California Hall
Berkeley, California

    Subject: Strawberry Creek Diversion Conduit
                    Contract 3478A

Gentlemen:

During the course of excavation for subject conduit, in the zone just west of Campus Drive, we have encountered an extensive obstruction which we have so far been unable to remove. It is not clear at this time whether this consists of a buried concrete structure or a bedrock outcrop.

In accordance with paragraph 17C of the specifications, we are notifying you of what appears to be a changed subsurface condition, in order that we may jointly determine its extent, its effect on the performance of the contract, and the extra costs and time involved.

We are making arrangements to procure samples by means of test borings.

In the meantime, we have shifted our operation to the section farther west so that minimum delay and extra costs will be incurred.

                      Very truly yours

                      Subterranean Construction Company
                      by A. Mole, Project Superintendent

cc: San Francisco office.

FIGURE 17.1.   Subterranean Construction Company—Letter 1.

encountered rock and is moving his work operation to another area where rock is not present in order to minimize additional expenses and time. In the initial letter he should also state that he is trying to determine the full extent of the possible changed condition.

Since the contractor seldom knows exactly how big the potential problem will grow to be, he is always wise to take adequate time before final settlement on price and time. Ben C. Gerwick, Inc., once made the mistake of settling too soon. In its initial letter it stated that its estimator calculated the additional cost of the changed condition to be $12,000. The owner wrote back immediately and issued a change order for $12,000 as the agreed extra price for removal of the rock. The final cost, however, turned out to be

$28,000. Unfortunately, Gerwick had committed itself to the $12,000 figure. Here then we have a paradox: in an earlier section, a prompt settlement was recommended; in this present section, the dangers of settling too soon are noted.

The proper course of action usually lies in between the extremes. One suggestion is to promptly settle the fact that changed conditions do exist, followed by an interim cost adjustment based on specific quantities, noting that "a further adjustment may be necessary when final quantities are determined." The lump-sum plus unit-price form of contract could well be applicable to this situation.

If the potential change does indeed turn out to be one that the contractor wants to pursue, he should send as many follow-up letters as necessary to document the change, for example, progressively document the results of additional investigations. In the case of the discovery of a possible rock outcropping, a follow-up letter could record that further investigation has determined that the obstruction is not an isolated boulder but is a bedrock outcrop. The follow-up letters not only keep the owner up to date and informed on the problem but also maintain the records needed for future negotiations and to document a claim. Once again, if there is no follow-up letter, the owner may take the negotiating stance that "he assumed the rock to be only an isolated boulder: that if he had known it was bedrock and extensive in area, he would have had the basement plans changed."

As an example, consider the case of the Subterranean Construction Company who needs to write several letters to protect its legal position concerning a possible claim due to a changed subsurface condition. Letter 1 (shown in Figure 17.1) is a proper letter that initially documents the possible changed subsurface condition and wisely attempts to gain mutual participation in the investigation.

Now let us assume that the university resident engineer, who received the letter, handled the case badly. He made a threatening telephone call to Subterranean saying that the changed subsurface condition was not the responsibility of the university. Then he flatly stated that there will be no consideration of changed conditions. Finally, he reminded Subterranean of the contract's heavy liquidated damages clause that will be implemented in case of late performance.

The university resident engineer's response to the first letter was so adverse that Subterranean wrote letter 2 (shown in Figure 17.2) to the engineer, with a copy to the university manager. This letter was required to maintain Subterranean's legal position. Also if Subterranean can get the owner involved in investigating this claim, it should be easier to negotiate an equitable settlement. Hopefully, the owner will recognize that there really were changed subsurface conditions and will be sufficiently acquainted with contract procedures, rights, and responsibilities to act in a fair and rational manner.

SUBTERRANEAN CONSTRUCTION COMPANY

Mr. John Serdip, Manager of Construction
University of California
Building Department
California Hall
Berkeley, California

        Subject: Strawberry Creek Diversion Conduit, Contract 3478A
        Re: Our letter of 23 November 1980

Gentlemen:

    On 23 November our project superintendent wrote to you notifying you of the encounter of an apparent obstruction in the zone just west of Campus Drive.
    We are indeed disappointed that you have not yet sent your representative to check this situation.
    Mr. Mole informs me that the only response to his letter was a telephone call from your Mr. Jones on 28 November 1980, denying any responsibility on the part of the university and threatening to assess penalties if we did not maintain schedule.
    We are sure that this call does not represent the university's true position in this matter and once again request your instruction.
    We have now taken borings at three locations and have determined that the obstruction we encountered is an old brick culvert or drain, crossing the new alignment on an oblique angle, over an interfering length of about 240 feet. The report of Williams-Jones Associates, Foundation Engineers of Oakland, California, is attached.
    While we believe the contract provisions must speak for themselves as to responsibility for payment, we believe it will be to our mutual advantage to have a clear record of the facts. We therefore suggest that daily records be kept of the actual removal operations and be agreed to by signature of both your inspector and our superintendent. This would be without any admission of liability on either party's part but will be of value to both parties in later discussions and settlements.
    If this is agreeable to you, will you please so instruct your inspector.
    Since we have now completed all other excavation for this project, we plan to move to the zone in question on Thursday, 4 December. If you have any directions concerning realignment, etc., we will, of course, need them one or two days in advance so we can properly lay out the lines and grades.
    I believe both of us recognize the need to complete this conduit at the earliest possible date as we are now well into the rainy weather and potential flood season.

                                                Very truly yours,

                                               Subterranean Construction Company
                                               by E. Beaver, Vice President

Enclosure: Report of Williams-Jones
Associates of 1 December, 1979.

FIGURE 17.2.   Subterranean Construction Company—Letter 2.

17.8. Documentation

A nationally known foundation contractor was very effective in getting changes where other contractors would have failed. The contractor was smart enough to know that big business and big government need a substantial file in order to justify a significant adjustment in contract price. If an owner's representative authorized a substantial payment based on a single sheet of paper, he may not look very good in either the eyes of his boss or the bureaucracy for whom he works. Conversely, if the contractor has submitted many pages of data, daily reports, letters, and a summary analysis, then the owner's representative can use this file to justify his authorization of the additional work and payment. Also, this mass of data has the psychological impact of showing the owner that the contractor means business. There is a wry observation that the value of a large claim is about $100,000 per inch of file.

When a serious potential claim arises, a special file should be set up and kept up to date. Of course, the home office is always notified. It is often advisable to request that the home office assign a person solely to monitor the potential claim. This individual will probably pay for his salary many times over. The project manager should not attempt to handle a serious claim all by himself. Inevitably, the file is not kept up to date; and if a job problem or another change order arises, it may keep the project manager from properly following the first claim. The subordinate monitoring the potential claim should keep a detailed diary and collect photographs (which are dated) of the workers and the equipment doing the actual work. In a serious case, a professional photographer should be engaged to take and date the photographs. This way the contractor will have a disinterested third party that can testify if litigation arises. Dated photographs are particularly valuable because, inevitably, there is disagreement between the contractor and the owner as to the additional time, equipment, and manpower that were actually employed on the changed work. From the owner's point of view, such photographs and daily records are an effective means of rebutting exorbitant claims.

It is in the best interest of both parties for the contractor and the owner to work together on the gathering of data for such change. It is clever and contractually proper for the contractor to formally request by letter that the contractor and the owner cooperate in the gathering of data. The contractor may request a daily meeting with the resident engineer to agree on the actual amount of work that has been performed. In making this request, the contractor should point out to the owner that the owner is being asked to verify the data "without obligation as to liability." In other words, the owner is not agreeing that a change exists nor that he is required to pay for the change. The owner is simply monitoring the amount of work that is actually being performed. When the resident engineer works with the contractor to collect data, he becomes more sympathetic to and aware of the contractor's problems. From the owner's point of view, the mutual collection of data provides a stopgap against overstated demands. The mutual investigation ensures fairness to both parties and helps ensure the accuracy of the data. If the owner refuses the contractor's request, this action will not look good either at the negotiating table or in arbitration proceedings. If the owner refuses to mutually collect the data concerning a claim, the contractor should still send daily reports of the investigation to the owner and ask him to notify the contractor of any disagreement with the facts.

How can a contractor build up and most effectively present a large claim that has impacted other aspects of the job? At job-site level. two cost bases must be established—the costs before the change (the costs of the unchanged work before the change) and the costs that resulted from the change (the increased costs of doing the unchanged work after the change, plus costs of the extra work itself).

Too many claims are filed using the following logic. The estimated cost for the building was $5 million and the work ultimately cost $6 million (including the contractor's standard fee). The contractor had encountered a changed condition—some extra excavation of rock. The changed cost must be $6 million minus $5 million, or $1 million. This logic just does not stand up; however, many contractors will attempt to use it as a basis. The fallacy is in the contractor's unproven assumptions that his initial estimate was correct and that his performance matched his estimate. One needs first to establish that the work would actually have cost $5 million without a changed condition. Therefore, it is necessary to take representative periods before and after a change and establish the actual costs in the two periods—for example, the cost of form work, the cost of placing concrete, and the cost of placing reinforcing steel, both before and after the change order. The cost of the changed work and the cost of the unchanged work need to be thoroughly documented in order to pursue a claim effectively and fairly.

Therefore, cost data are needed, as the project progresses, before a claim arises. For example, the contractor needs to know the cost of excavation prior to the time the water main—that did not show on the plans—burst. A contractor needs to document the costs of the work without changed conditions.

A clever, if ethically questionable, practice of many contractors is to maximize the estimate of the extra costs due to the changed conditions. First, the contractor uses his normal estimating procedure to calculate the costs. Then he breaks each item into five or six sub-items, each with its own estimate of performance as though it were an isolated item of work instead of a concurrent operation.

Consider the following example using structural steel. Assume that in his normal estimate, he used $8 per ton as a local delivery cost for structural steel on a big job. When submitting his claim, this item could be broken down into several items:

1. Loading at warehouse (crane, crew, overhead, and supervision)

17.9. Using the Owner's Records to the Contractor's Advantage 279

2. Hauling (trucks, special flag, cars, and drivers)
3. Unloading at job site (crane, crew, overhead, supervision, and maybe even preparing the site)

With such a detailed breakdown, an estimator might build the delivery cost up to $30 per ton, and an owner's representative might not be able to refute it! On the other side, some owner's representatives are very astute at challenging cost submittals if they have the detailed data with which to work. One notorious government resident engineer used to boast, "Once I can get a contractor to submit a full breakdown, I can beat him down below cost!"

At some point both parties will need to meet to establish the responsibility and the additional payment due, if any. It is assumed that the contractor has previously submitted this data along with direct and indirect costs, with a cover letter to the owner. In general, the side with the most complete and voluminous data has a definite edge in the negotiations.

When it is time to present the claim formally, imagination is invaluable. Charts and graphs are extremely effective. Having your draftsman use the proper scale to make your points stand out works well with these charts and graphs. By merely changing the ordinate scale, one can effectively dramatize the degree of change involved. Color can also be very effective, for example, red for cost overruns. A computer printout or voluminous data from which the charts and graphs were prepared should be taken to the negotiations. In effect, the contractor is stating that his chart or graph has to be right, since he has the data and computations to back it up. Many people are somewhat intimidated by computer printouts and will not challenge data that have been calculated and tabulated on the computer. On the other hand, engineers who work with computers are highly aware of the garbage in–garbage out (GIGO) rule and may not be as easily overwhelmed by a computer printout.

Time-lapse photography can be a wonderful tool for substantiating claims. A West Coast contractor used this approach successfully on a large coastal outfall contract. The owner had furnished the contractor with inadequate oceanographic data for the specific site. As the actual conditions turned out, the working conditions were severely affected by the high-sea states. The time-lapse photography of the construction work clearly demonstrated the problems the contractor was experiencing. The shots of the workers attempting to carry out work with waves breaking around them proved to be the most effective means of convincing the owner that the low productivity was not the result of the contractor's ineptitude.

17.9. Using the Owner's Records to the Contractor's Advantage

The owner's own records can often be used in the contractor's favor when pursuing a claim. Many owners will simply give copies of the records to the contractor if they are asked. If the owner refuses to do this, it will appear he is hiding something in the negotiations. If a claim goes to court, then, of course, a contractor can always subpoena the records.

Similarly, a subcontractor may be able to use the general contractor's records against him. In one particular case the subcontractor had been performing work in a very negligent manner. The general contractor was finally forced to terminate the contract, since the subcontractor had caused numerous delays and had performed some work that was totally unacceptable. The subcontractor then asked for arbitration of his claim against the general contractor. He formally requested and received the daily diaries kept by the general contractor's project manager. During the arbitration, the general contractor presented his case, which seemed to him to be factually conclusive. The subcontractor then presented his side of the argument, based entirely on some handwritten entries in the project manager's diaries. On an almost daily basis, the project manager had entered longhand notes in his diary castigating the subcontractor, notes such as: "The sub is making a hell of a mess of this job. I'll fix him later."; "Today was just like yesterday. The damn excavating sub fouled up our access. I'll back-charge him until he goes broke"; and "If our lousy excavator doesn't bankrupt himself on this job, I'll do it to him myself." This type of entry had been made day after day. The subcontractor's representative simply pointed out to the arbitrator that these entries clearly demonstrated the impossible conditions under which his client was forced to work. (These diaries were hard to refute since they had been made by the general contractor and not the subcontractor.) Upon almost this one bit of evidence alone the arbitrator quickly ruled in favor of the subcontractor.

A piling contractor successfully negotiated a claim on the first Arkansas River Lock and Dam contract. The contract contained a clause stating that no jetting of piles will be allowed except with the permission of the contracting officer. Once on the job, the contractor found that piles had extreme difficulty in penetrating the dense sand unless jetting was employed, even when using a very large hammer. The contractor repeatedly kept requesting permission for jetting, and the resident engineer repeatedly refused to allow jetting to be used. When the project was completed, the contractor had documented the many extra costs that had occurred because jetting could not be used. Fortunately, even with the difficulties and extra costs, the contractor had still made a profit on the job. Therefore, in the negotiations, he was not as desperate as most contractors usually are when involved in a large claim; hence he could pursue his claim rationally but with persistence. This is an unusual but tactically wonderful position for the contractor to be in.

The contractor asked for and received the daily records kept by the construction inspectors. (It is amazing how often contractors fail to ask for such records.) The records contained numerous daily entries stating that the pile driving would have been much easier if jetting had been allowed. The contractor also asked for and received the records from the second and third Arkansas River Lock and Dam contracts. These records documented that jetting had been allowed on these contracts which contained identical clauses to the first job.

The contractor's contractual position was that if the government wanted no jetting under any circumstances, it could have said so. By adding the phrase, "without the approval of the contracting officer," it must mean that reasonable jetting would be permitted.

During the negotiations, the owner's representatives recognized two things about the contractor's position. First, they could see that the contractor had an impressive file of records and data to substantiate his extra work and costs. Second, they knew that the contractor must mean business—he had obviously spent the time and money to prepare those records. It was obvious that he intended to persist in his claim. In this case, the contractor was successful in negotiating an acceptable change order.

17.10. Legal Assistance

Legal assistance is normally required when preparing a claim. It is practically impossible for an engineer to keep abreast of all new legal developments. Advice from a competent attorney is often best used covertly, at least initially. The contractor learns from the lawyer the strong and weak points of the case. At this point the contractor should leave it at that. For example, the contractor may have learned that recent court rulings would indicate that he could be paid for a claim based on the impact effect but could not be paid on a claim based on ripple effects. In this case he would use the right term—impact—when discussing the claim.

It is advisable for the contractor to have his lawyer check all written documents for legal soundness. It is normally unwise, however, to have the lawyer rewrite those documents in legal language. If so rewritten, this will usually force the other side to bring its lawyer into the negotiations. The legal approach is, therefore, normally used only for claims which have been rejected.

Some contractors worry about paying the costs of an engineer at $200 a day (total costs) to monitor and document a serious claim and then wind up paying a lawyer $1000 a day to help prosecute the claim that has gotten out of hand.

A national contractor of high repute once had a policy of pursuing each claim vigorously up to the point where he needed to get lawyers involved. At that point, he would take the best settlement he could get. In addition to avoiding the excessive legal fees, he also avoided having his best men involved for weeks or months in claim preparation and negotiation. His personal opinion was that his key people were more effective and productive on new work than working with a lawyer to substantiate a claim. The major cost of legal proceedings is often the time devoted by the management and key people rather than just the attorney's fees.

One California contractor stated that he seldom ran into a lawyer that did not agree that his claim was "airtight" and that if he could pursue the claim right on through, he would get 100 percent of that claim. Many lawyers seem to be overly optimistic as to their probability of winning a major claim. Conversely, experienced lawyers, specializing in work for contractors, take a more objective viewpoint of claims and will often advise a settlement as opposed to an extensive legal battle even though the latter may get a higher monetary return.

17.11. Advantages of Deductive Change Orders

Deductive changes are much easier to settle than additive changes. The owner is in a positive frame of mind because he knows he will be getting back some money. Therefore, the contractor is in a favorable position to negotiate. Obviously, knowledgeable parties should pay as much attention to a deductible change as to an additive change. It certainly warrants maximum attention by the contractor. The contractor can legitimately state that the owner is only entitled to get back the incremental savings in the deductive change. This is because the contractor is still required to have essentially the same equipment and men available, although the materials may underrun. He can argue that he is still turning over the same completed facility.

17.12. "Walking Off the Job"

When you do have a dispute, it seldom, if ever, pays to stop work and walk off the job. The typical reaction of the owner will be to stop all payments, file suit for breach of contract, and file an action under your bond. Instead of walking off your job, the proper action on a serious dispute is to write repeated letters requesting a change order for the work in dispute—the payment requested should include direct cost, the cost of delays, the cost for additional insurance, the cost of finance, and a proper amount for profit.

It is reported that on the first Caldecot Tunnel (near Oakland, California), the joint venture of six contractors stopped work and abandoned the job because they could not negotiate reasonable settlements on major change orders for additional tunnel sets and special shoring in squeezing ground. Eventually the court ruled that the joint venture had the financial resources to finish the job while pursuing their claims in accordance with the contract and that since completion was not "impossible," they had breached the contract by stopping work. As long as the contractor has the financial resources to finish the job and the job is not physically impossible, the law appears to require the contractor to finish the job.

Twenty-five years later, two major contractors with an outstanding record were forced to near bankruptcy by walking off the job. The catastrophic contract was for a dam for a northwestern public utility. As often happens with large dam projects, a fractured zone of rock was discovered which required major amounts of additional excavation and concrete work. The owner agreed to pay for the additional work but would not grant a time extension. The contract contained a $20,000 per day penalty for a late finish. Since the contractors were unable to negotiate what they felt was a reasonable time extension, they shut the job down. The owner then terminated the contract and brought another contractor in to finish the job on a cost plus fixed-fee basis. The new contractor did finish the job on time by working three shifts a day, seven days a week. Because of the change in contractors, and the accelerated schedule, there was a $3 million overrun in cost. The courts ruled that the original contractors were liable for the overrun. The court said that the original contractors had the obligation to stay with the job, as long as it was physically and financially possible: that their proper recourse was to file a claim.

Although it is dangerous to just walk off the job, on small cases S. J., area manager for a nationally known foundation contractor, would use a clever negotiation gambit whenever he ran into what he considered unreasonable treatment. He would stop work offering the typical excuse: "We really cannot figure out how to handle this particular problem." For example: "We have run into pile lengths that are 40 feet longer than expected. We are still working on trying to solve the problem every day; however, we cannot seem to move ahead on the job. We need time to figure this out. While we're trying to solve this problem, we'd better get these contractual matters settled." Through this tactic, the owner would become worried that he wasn't going to get this job done and would often accelerate settlement of the claims. This is a very dangerous approach, a "walking the tightrope." Public owners would normally not fall for this tactic. They would simply write back and demand that the contractor continue work immediately.

17.13. Vindictive Owners

Unfortunately, some owners will be vindictive and petty toward a contractor after that contractor has been successful on a claim. For example, on a large pipeline-aqueduct contract, the contractor was successful in pursuing a claim for changed subsurface conditions affecting the pile driving. Thereafter, the district unofficially instructed its engineers to demand meticulous compliance with every tolerance and detail and to find every little thing which could be deducted from the contractor's final payment. For example, in computing the quantity of structural concrete, they deducted the volume occupied by the reinforcing steel. In hindsight, the contractor probably should have filed a new claim. Timidity seldom pays off.

A California contractor was successful in negotiating a claim against the California Highway Department on a large bridge substructure contract. Immediately, the resident engineer recalculated the pay quantities, meticulously searching for deductions from the nominal volumes; for example, deducting the rounded corners of the footing excavations from the earthwork quantities and the volume of concrete occupied by embedments. This particular contract did not state in detail how these quantities were to be calculated. This recalculation had the effect of costing the contractor more than the value of the claim it had previously won.

17.14. Lost Causes

There is no point in pursuing a claim if there is no money available to collect. A major construction company was involved with a large claim on a dam project on a California mountain river. The bond issue that financed the project had an upper limit of $80 million for the project. Also, the owner had contractually limited its legal liability to the $80 million of the bonds. The contractor compiled a large claim for cost overruns; however, there was no additional money available to pay the claims. It appears in hindsight that a better course for the contractor would have been to try to get the city to reduce the scope of the work and stay within the $80 million limit. This problem of limited funding and liability can also arise on small contracts. It is wise for management to find out just how and when payments will be made and how overruns in budget will be handled.

Similar problems have arisen with New York City due to its financial difficulties of the 1970s.

17.15. How to Avoid Claims

One secret to the successful negotiation of claims is to avoid the claim in the first place. Everyone should be concerned with avoiding claims.

Claims should be for the rare situations in which a dispute occurs that cannot be settled by negotiations at the project level. If all parties concerned strictly adhere to the contract, there is less chance for a dispute. With a few exceptions in the "changed conditions" area, claims are a net loss for everyone.

How can claims be avoided? The general answer is by having adequate communications and discussions between the contractor and the owner. This is a very complex subject. A few general ideas follow. The owner can hold a prebid meeting where the contractors that will bid are strongly encouraged

17.15. How to Avoid Claims

to attend and have an open and frank discussion on the plans and specifications. During this meeting the owner can specifically ask if there are any ambiguities or inconsistencies. Once the job has been bid, the owner and contractor can hold pre-award and pre-construction meetings. These meetings should involve such principals as the chief engineer, the vice president, the project manager, and the resident engineer; all the parties that will be involved in a responsible position with the project. During this meeting, the interpretation of the contract documents, plans, and specifications is discussed. If the meeting is successful, many potential claims will be eliminated. Finally, after the project is underway, both parties should periodically sit down at regular conferences. At these, they may agree informally, for example, on how a particular contract clause will be interpreted. Or the contractor will notify the owner about a developing problem, so that early action can be taken to minimize its effect.

It is important that the project manager have a thorough knowledge of the contract documents and specifications. It has been stated that 30 percent of claims arise out of the fact that the contractor's representative on the job has not thoroughly read and really understood both the contract itself and the specifications. For example, one project manager did not realize that the specifications required a special finish. In another case, the project manager simply did not recognize the requirement to obtain written authorization from the owner before proceeding with a major additional item of work. Because of a lack of understanding between the contractor and the owner, a simple matter which is not properly administered often degenerates into a claim. There are too many cases in which project managers, who usually read the specifications in great detail, have not bothered to read the contract even once.

Subcontractor's claims against the owner present special difficulties for both the subcontractor and the general contractor. The general contractor has the legal duty to pass these on to the owner and not to do anything to reduce the subcontractor's ability to obtain proper settlement. For example, he cannot legally trade off a subcontractor's claim against one of his own.

At the same time, the subcontractor's claim may appear excessive or unwarranted. Aggressive pursuit of this claim may endanger friendly relations with the owner.

Once again, most such problems can be avoided by frequent communications, for example, inclusion of the principal subcontractors in weekly jobsite meetings with the resident engineer.

18
International Marketing

18.1. How to Sell to the International Market

The marketing of construction and engineering services on an international basis requires a deep and quick sensitivity to needs, resources, cultural constraints, international finance, and competitive pressures. Above all, it requires mature judgment to distinguish opportunities from dead-end promotional schemes, and sound business from adventurism.

There is a romance to marketing internationally that cannot be denied. The fraternity of entrepreneurs who can operate and survive in this jungle is small indeed, but it must be admitted that it has its allure, like all searches for the Golden Fleece from time immemorial. The sucessful marketer develops a canny sense that enables him to judge where to concentrate and where to strike, and to know when to parry the urgent invitations until a more opportune timing and event.

First and foremost is to decide if the client really can pay for the project. Who is the ultimate client? The host country? A multinational enterprise? The World Bank? Who will make the final decision for award and on what basis? Certainly price alone is rarely the dominant factor in determining award.

Second, what are the political and fiscal uncertainties surrounding the project? Will you be able to carry out the work on a straightforward basis, normal schedule, normal operations? Will you be able to bring home your investment and its profits, if realized? If the answers are negative or doubtful, are there ways of hedging your bets, providing for advance payments or insurances?

Perhaps carrying out the work on an expedited schedule may anticipate any undesired political event. Conversely, you may see an opportunity to pace the job at so slow a rate that it remains immune to change.

18.1. How to Sell to the International Market

Some highly successful contractors have seen a way to cover such uncertainties by the use of subcontracts, whereby the risk was diversified. Governmental guarantees may or may not be available to you as a prime contractor, but may be available to subcontractors of other nations.

If the client cannot pay in hard currency, can he pay in soft currency? You may be able to presell soft currency, at a discount, of course, in the international money markets.

Alternatively, consider barter. A great many international contracts in today's market, as of this writing, are going on a commodity barter basis, with payments to be made in future deliveries over a long period. Strange as this method of payment may seem to a contractor-engineer, it is attractive to import-export firms, who will buy such future contracts, for cash, but at a very large discount. For them, it represents a guarantee of supply and a hedge against inflation.

All of the above discussions relate to contingencies and risks other than those inherent in the work itself. Working internationally implies working in new areas, where soils, weather, water table, and labor may all vary greatly. These site-specific conditions must be covered by thorough and detailed site inspection. Often opportunities, such as availability of local aggregates or fill material will be discovered: conversely one would hopefully have found most of the adverse conditions prior to submitting his proposal.

Most international contracts incorporate a large number of conditional clauses which can provide partial protection for changed conditions, both physical and man-made. It is customary in such proposals to stipulate provisions concerning matters within the control of the client such as duties; import licenses; rental fees for land and royalty fees for aggregates; road-use permits; ability to unload directly at the site (as opposed to clearance through an established port); provisions for advance payments, mobilization payments, and interim payments; prompt approval of drawings, and so on.

Clauses giving protection for events beyond the control of the client are also included: changed conditions, delays and suspensions of work, escalation, hostilities, currency devaluation, and so on. These, however, are of value only up to a point. If the client has no more funds, he cannot pay for changed conditions. If his currency devalues, he is often powerless to rectify the matter.

Many of these situations can be handled by such devices as prepurchase; that is, purchase and pay for key items of materials at the start of the contract. In other cases, one may elect to reduce the scope of his own services by having the client purchase items directly, in order to exclude the risks involved with specific items that have a large potential exposure. If this approach is adopted, care must be taken to ensure that the specifications are complete and the schedule is firm.

As described above, marketing internationally can be visualized as very complex, very involved, and very risky. It is. However, on the other side of the coin is the opportunity to use ingenuity in covering these risks by guarantees, subcontracts, contractual terms, and prepayments. Thus marketing becomes a matter of trading.

We of the West must then recognize the shortcomings in our cultural heritage. We are not and cannot be as shrewd as traders from the East. We do have something on our side, however, and that is our openness, forthrightness, and sincerity. We are, or should be, in a position to deliver what we promise, and on schedule. These facts are respected and transcend cultural differences worldwide.

Our problems are greatly reduced when our client is a multinational enterprise. Then we and our client are able to work on similar terms and understandings.

There is one proviso. As the multinational client operates in the foreign country, it is constrained by the country's laws and regulations. It is also constrained, to some degree, by the laws and regulations of its home country. We, as marketers, have to determine what these constraints are and, as necessary, obtain guarantees from the client's parent company that the contractual terms will be honored.

Matters such as these may be stipulated in proposals. However, experience shows that it is best not to be too blunt. The client may have had the initial intent of passing the risks on to the contractor. Later, the client will probably find out that all contractors who are competent to carry out the job are also sophisticated enough to demand the same terms. So perhaps it may be wise to merely stipulate that you wish to discuss these matters and arrive at mutually satisfactory terms.

For many clients, the need is urgent and the opportunity is available, but there are no funds to pay. Earlier, barter agreements were mentioned. Another approach is to provide financing with your proposal for engineering and construction. There is a vast reservoir of capital available in the world, anxious to be employed, but only on favorable terms and rates of return. Sometimes the source of financing wants not only interest but a tie of the principal to an international currency index such as SDRs (special drawing rights) plus, sometimes, an equity position in the venture itself.

One mechanism for handling projects of high risk is to provide financing, but price the job at a level that will cover the costs of finance and the risk, that is, at a level that will attract the investment. This is very expensive for the client but he may have no other choice. Cases have occurred where the job has been awarded to the high bidder who provided financing, rather than the low bidder who provided none. Sophisticated international engineers and contractors devote major efforts to lining up financing for projects that would otherwise never materialize.

18.2. Marketing Techniques

In international contracts, one has to be extremely careful to ensure a commonality of understanding. Specifically, this applies to the units for pay, the way in which they will be measured, who accepts responsibility for overrun,

18.2. Marketing Techniques

and so on. Many international contracts contain "provisional sums" to cover expenditures which in U.S. domestic practice would be risks assigned to the contractor. Obviously, one must be familiar with the metric system of units and the currency of payment. Whenever practicable, use of the client's units and systems in your proposal will facilitate understanding and acceptance.

Negotiations themselves should be clear and specific. If English is used, speak slowly and distinctly, use common words and phrases, avoid colloquialisms, side remarks, extraneous comments, and jokes. Be modest and accurate, stress credibility. Behind all the negotiations, the host country is buying technology and managerial competence. The latest technology has a major appeal to developing countries, particularly if there is included training and development of the proficiency of host country technologists to operate and utilize the new facility. Each developing country feels an urgent need to possess the latest and the best.

If visual aids are used, they should be simple, clear, and logical. However, they should not "talk down" to the viewer; after all, the client's representatives are often highly educated and highly intelligent engineers, and they understandably resent any implications of inferiority. As with all potential clients, models may be more effective than slides.

At the same time, the submission of accompanying technical data in great detail and sophistication, but without any attempt to present it verbally, may lend credibility to your sales effort, while constituting a recognition of the recipient's engineering competency. This can also be best handed to the client's representatives as a "technical supplement" for them and their engineers to read.

Nationalism and local pride rank strongly with companies and agencies in many countries. The experienced international marketer recognizes this and words his proposal accordingly. This is especially important when it comes to such matters as the use of local subcontracting and material suppliers, the development of local industry, and the training of local labor. Even the spelling of words (for example, labour versus labor) may be important. Similar words and phrases may have grossly different connotations when translated.

In "design and construct" proposals or "design only" competitions, it may pay to incorporate a feature which can become a source of national pride. Thus one designer-constructor successfully obtained a series of bridges in developing countries by incorporating a stayed-girder span with tower piers, that is, a notable aesthetic structure, where more conventional low-level spans were felt by the other bidders to be both adequate and cheaper. In one such case, the submittal with the "feature" won the contract even though it was the high bid. The stayed-girder span was later featured on the country's currency and postage stamps. Appeal to pride is not only good marketing but a recognition of a developing country's legitimate need for a focal point of achievement.

It is an error to dwell at length on "offers to help," as though you were doing the client a favor. In the first place, such excessive altruism both sounds and is hypocritical and insincere. The client knows you are there to make a profit. In the second place, it degrades the client; he does not want a handout but rather wants to be sure he is getting a good project at the right price.

Each developing country wants to keep up with its neighbor; hence one sale of a specialized facility may well be used to establish credibility in a neighboring country.

Unfortunately for Western and especially U.S. contractors, prices are often if not usually determined by negotiation, both before and after. The age-old practice of haggling over price is a prolonged semi-ritualized procedure that is expected in many parts of the world. Fortunately, the growing trend toward firm pricing followed by the World Bank and its offshoot agencies has diminished, but not eliminated, the practice.

If haggling is to be done, changes in scope can always be negotiated to match changes in price. Careful and slow, but dignified, negotiation can often result in maintaining the desired profit margin.

Although in many cases the practice of representatives of developing countries during negotiations is not to take notes, these individuals have highly trained memories that recall literally almost every word said, to a degree that often astounds the Western negotiator.

Perhaps the most universal rule to suggest in international marketing is the use of a local partner or associate. He must be selected with great care to ensure that he can really contribute and participate, both before and after award.

In some areas, the "agents" are 5- or 10-percenters (maybe even 15-percenters), who offer their services on the basis of their inside relations with the awarding authority. Such agents may or may not be able to produce results. One clue to determining reliability is whether or not the agent can actually serve as a minority partner, that is, provide services and some construction performance so that he becomes a partner in fact as well as in name.

Amusing stories are told about "10 percent agents" who managed to gain representation for every bidder on a particular project. This is the kind to avoid.

A much more difficult situation occurs in those countries where custom has established a practice of payoff as a reward for the award of a contract. This is discussed further in Chapter 19.

Negotiations are a way of life in many Eastern countries. We in the West proceed too fast, are too forthright and blunt. "All good things take a long time" is a Mideastern proverb. Indirectness in approach is valued above directness. No one should ever be forced to directly give in or back down; he must be given a way in which to "save face." Negotiations should never be rushed, concessions should be made slowly, important points should be maintained with polite but firm persistence. Firmness is respected. Too many westerners are over-eager for an agreement. It is well to remember the objective for all marketing, which is to obtain new work on *favorable terms*.

International markets are generally more open and receptive to alternative proposals than are domestic markets, being less constrained by laws and regulations. Thus, despite clauses which state that "alternatives will not be considered," the fact is that alternatives often will be considered, particularly if the initial bids come in over the budget price and available financing. One successful international contractor makes a practice of ascertaining the maximum amount of financing available. If his estimates then indicate that all bids will probably exceed the available funding, he either gives an alternative design, reduces the scope of the project, or offers delayed payments and/or barter.

The opportunities are there. The world has enormous needs which have to be fulfilled. It is a question of matching services and capabilities with needs and resources. Marketing of international construction and engineering extends the range in which the competent marketer may bring to bear his full capabilities for ingenuity, experience, and skill.

18.3. Short-Term Business Development*

Short-term business development is quite often the means of entry into the international market for an engineer or constructor. It frequently affords the least costly approach and usually provides the experience necessary to determine the desirability and, more importantly, the potential profit for continued or expanded international work.

After all the window dressing has been removed, there is really only one motive remaining which prompts people to seek international work—profit. Your success is important to your government and its trade balance—it is important to the host country who needs the expertise you provide—it is important to your country's manufacturers who will furnish much of the equipment and materials you specify or design—it is important to your country's subcontractors and consultants who seek an opportunity to assist you—but its ultimate importance to you is the resulting bank balance. With this in mind, let us explore for a few moments short-term International Business Development.

The primary objective is a single project, or group of related projects, in which you have definite confidence of your ability to perform. The ideal situation hopefully will include minimum or no local competition, assurance of prompt payment in U.S. dollars, no involvement in arranging financing for your client, stable government in the host country, and conditions conducive to performing the major part of the engineering in your home office. Probably the last time all of these conditions prevailed was in

---

* This section is an excerpt from a 1971 presentation entitled "Marketing and Short-Term Business Development" by J. M. Lane. It was previously published in *Proceedings of Multi-National Engineering and Construction Conference on Finance, Planning, and Managing* in 1972 by Continuing Education in Engineering/University Extension and The College of Engineering, University of California, Berkeley.

Guam or some other United States territory immediately after World War II, but it does not do any harm to hope.

A short-term project usually results from factors favorable to your particular situation. One condition which improves your competitive stature, for example, is the requirement of knowledge, experience, and reputation in a field where you are a recognized authority. The project would thus fit the unique assets you possess. Another condition which will enhance your chances is where the project meets the immediate needs of a long-term client for whom you have successfully performed on similar projects. The job preferably should be one which can be completed without an unusual expansion of your staff or emergency financial arrangements beyond normal business practice. You should be on firm and familiar ground, and risks should be minimized as far as possible.

Your immediate goals for a short-term project are contract award under favorable conditions, preparation of an accurate estimate, ability and flexibility to attack the job in a workmanlike manner, on schedule performance, and realization of a reasonable profit. These are really no different from the goals of a domestic job, but international complications of geographical distance, slow communications, different business practices and language difficulties complicate the picture. If you have compensated for these circumstances in your estimate and realistically constructed your project schedule, you will have a good chance of success.

We have mentioned short-term projects from the ideal position, but we must also recognize that a short-term approach is valid under circumstances which are less than ideal. Perhaps your market analysis indicates that a certain overseas area is potentially attractive for long-term involvement. Establishing overseas sales and engineering offices is an expensive exercise, and, too often, years of developmental "spade work" are necessary before the dotted line is signed on your first contract. A more satisfactory alternative is to seek a short-term project in the area, and through it confirm the long-range prospects. When the work nears completion, you can determine the advisability of planting more permanent roots. In any event, you are already established, much of the spade work is finished, and you have a better appraisal of the long-term potential. You are truly in business if you want to be.

The short-term approach is also indicated in areas which are unstable financially or politically. Even though you can protect yourself with insurance against risks of instability, it is wise to approach involvement in such an area on a short-term basis. After the project is completed, you can decide more rationally whether the future potential is worth the risk.

The marketing efforts you exert for short-term development sometimes are expensive and, unfortunately, they also can often be totally unproductive. Many such efforts are totally developmental, but all are educational. You finish the exercise much wiser than when you started.

Efforts for short-term involvement are usually based at home because you do not have overseas bases from which to operate. A wealth of information is available in the United States from reliable sources both governmental and private. You can get a pretty good reading of the market without ever leaving your office, but, like all good things, this condition suddenly comes to an end.

Even if your client is from your own country, you will eventually realize that you must go to the project site. While there, you will investigate physical conditions which affect the work, and you must also acquaint yourself with local laws, regulations, and customs. Preliminary discussions will be indicated with an attorney and governmental officials, and, in some cases, it may be advisable to acquire a resident representative.

Your only hope of recovering the expense involved in this pre-proposal stage is to be awarded the contract. Under these circumstances, an unsuccessful offer is a bitter pill indeed. To minimize such risk, be very selective in your choice of client and project. You cannot afford the luxury of kidding yourself. Rationally analyze every factor, and logically plan every step you take. It will pay dividends. Marketing effort for short-term projects must be carefully and skillfully done.

Unfortunately there is no patent or foolproof approach which will guarantee instant success. There are, however, some basic principles which, if followed, can greatly improve your chances.

Be certain your capabilities establish you as a specialist or authority in the type of job under consideration. Foreign buyers usually do not exhibit pioneering instincts. They want a known quantity, and they almost never back a long shot. Point with pride to similar projects successfully completed by you, and have the staff who can do it again.

Offer your services to a long-term client or a friend of one. They know what you can do and they approve of the way you do it, and they can get further evaluation of your talents from others whose judgment they respect.

If the client is new to the area, offer him pre-project assistance in every way possible. Demonstrate knowledge of local conditions, determine what permits and licenses he needs and help him get them, assist him in answering questions from his banker, introduce him to important local people, and, in short, impress him with your competence at every opportunity.

Be certain your prequalification correctly establishes your ability, and then make it available to every person or agency who will be consulted concerning the contract. Cover all the bases. Do not assume that anyone, even one having only a small interest, is unimportant and can be ignored.

Grasp every opportunity to participate in pre-feasibility studies or actual feasibility reports. The more exposure you get, the greater the opportunity to prove your worth.

To make the short-term development a reality, pull out all of the stops, and make a maximum effort. Stress every advantage your selection will provide for your client and project.

Past experience in the same or similar work must be emphasized convincingly. Do not be reluctant to demonstrate your competence. False claims will be as evident as the proverbial sore thumb, so do not go overboard. Your past performance can be checked and evaluated. However, modesty has no redeeming qualities either. Be proud of your accomplishments, and, what is more important, be certain others know how well you can perform. It is capability and quality they are seeking.

If you have patents, processes, or other proprietary items which enhance your position, be certain the client is aware of it. This places you in a niche entitling you to special attention, and is greatly preferred to a position where you are one of several seeking the project on the same old terms. By possessing special expertise, you give the client reason to provide special consideration. All that remains is to be certain that he recognizes this fact.

If previous projects demonstrate that your staff produced innovations or designs representing advanced or improved concepts, this can also be a benefit. Clients usually appreciate the chance to avoid stereotyped facilities, provided the alternative offers advantages of lower cost, simplified operation, greater efficiency, or all three.

Of prime importance to any client is ready ability to serve at the designated time. Give assurance that your staff is available, and necessary equipment is ready to perform. Never give the impression that only after a contract is awarded, will you then see what you can do about it.

In summary, there are many circumstances where short-term, or single-shot development provides the best approach to international business. To be a successful participant, define the project realistically and convince yourself that you should be selected. Once this is accomplished, establish your goal, formulate your marketing effort, plan your promotional steps, and see that every advantage you offer is emphasized to the hilt. After you get the job, perform well, and bring home a reasonable profit. This, more than anything else, will get you the next one.

18.4. Long-Range Business Development*

Few foreign ventures have justification for being conceived except as part of permanent long-term business development. A short-term involvement in which a national contractor goes abroad to tender on a particular construction project, expecting that when that job is finished he will dispose of his plant, his remaining possessions and repatriate his key personnel, is not as common today as it was during the 15 or 20 years following World War II. The fast buck operation overseas is rapidly passing. Because of the competi-

---

* This section is from a presentation titled "Long Range Business Development" by Donovan Jacobs. It was previously published in *Proceedings of Multi-National Engineering and Construction Conference on Finance, Planning, and Managing* in 1972 by Continuing Education in Engineering/University Extension and the College of Engineering, University of California, Berkeley.

18.4. Long-Range Business Development

tion from Europe and elsewhere, it should suffice to say that there will be fewer and fewer opportunities for American firms to hop abroad in order to pick off the fat and juicy plum and hustle it home while the rest of the world is asleep.

The first prerequisite for obtaining business is a workable organization; because organizational structure and function is so important to eventual success in sales. In the field of engineering services there are two basic ways for Americans to participate in overseas design work. The first is by exportation of services performed in the home office located in the United States. This method will work best for projects located in underdeveloped nations. The second is to establish an office in the country where the project is being constructed. It involves determining what portion of the work performance personnel must be imported from America or elsewhere and what may be recruited locally. Somewhat the same consideration faces the American contractor who contemplates a foreign job. Will it be possible to staff the job with a minimal number of imported management people supplemented by local superintendents, foremen and other staff, or must all key personnel be brought over at great expense from the United States?

The U.S. engineer or construction superintendent on an overseas assignment is a very expensive individual indeed. The high salary plus living, travel, and family allowance usually paid to a U.S. technical employee overseas total several times the amount paid to his counterpart in the European or Asian firm. The high cost of American overseas personnel is the principal reason why the U.S. contractor has found it very difficult in the last several years to underbid European contractors for large foreign contracts, especially in the underdeveloped nations. Because present trends tend to work against the profitability of short-term foreign ventures, most construction and engineering firms today who have urges to travel are planning for long-term involvement.

Opening up a foreign activity must not be a matter of "let's have a go at it for a year or two and see what happens." One could cite a number of cases of failures of ventures begun under such a commitment which folded within 18 months for reasons such as lack of understanding of basic goals, disgruntlement of key staff, loss of interest by principals at home or basic lack of long-range planning. A new overseas venture should be planned for a minimum period of three years, preferably five. The planning includes work objectives, personnel, financing, sales effort and home office support. Before venturing aboard it is very important that a firm have a good reason for doing so, and it must be more than wanderlust, enthusiasm or the desire to invest money.

It does not make much sense for an American highway contractor to seek work in Japan or Italy because in such locales there is nothing the American can do that the local contractors cannot do better. If that same highway contractor turns toward Panama he will probably find better chances for success because of a dearth of skills in his particular specialty. In other words, before planning a sales campaign you must have a saleable commodity. The first and most essential commodity that an engineering or construction firm must be able to supply, in order to insure success for a multnational venture, is technical know-how in a specialized field. The technical ability must be in a field for which there is a definite need within the country of destination. In most foreign countries a defensiveness is building up against the invading expert from overseas. In engineering or construction, that defensiveness sometimes turns into outright hostility on the part of the local consultants and contractors.

One of the first long-range planning tasks that must be undertaken as soon as the firm makes the bold decision to venture abroad is the establishment of both appearance and fact of permanency of resident in the host country; a process of identification with the host country by the company becoming nationalized or domesticated in that country. Let us study some cases where multinationalism has been accomplished successfully by American industrialists. General Motors Holden in Australia is controlled by GM in Detroit. You will find few Australians who are not particularly proud of the fact that Holden is the one and only motor car which is completely made and manufactured in Australia. The success of the Holden over the years is the result not only of technical know-how but of excellent public relations. A similar situation exists in the Mount Isa mines in Australia which are owned by the American Smelting and Refining Company.

Caterpillar Tractor Company, a few years ago, eying the expanding industry in Japan and fearing the threatening competition from Komatsu in the tractor business, solved the problem by forming a partnership with Mitsubishi to manufacture Caterpillar machines in Japan. One effective way of establishing residence quickly is to do as Caterpillar did and affiliate with a local firm by forming a working partnership or merger. A more cautious procedure would be a working agreement stipulating a several year joint working venture to be followed by merger negotiations.

There is the matter of language ability. If the language of the host country is not English, it is very important that the manager or his principal assistant be able to speak the native tongue. Unfortunately, many Yanks are darn poor linguists. But before the first airline ticket is bought to ship a man abroad, especially if he is in a managerial position, he and his wife should be submitted to an intensified language course so that he arrives in the country with the ability to converse. At the beginning of an overseas operation it will be necessary for management to be in the hands of an American sent from stateside headquarters. Since we are planning long range and expect the new enterprise to remain in business for many years, the fact must be faced that an expatriate American is expected to remain away from the United States from three to five years. In some countries the tax situation is such that the American cannot afford to continue to earn income for longer than about four years—especially in the countries which have a treaty with the United States. There are other countries where an American can stay abroad for longer periods without paying income tax in the United States. For these

18.4. Long-Range Business Development

reasons and to help promote previously mentioned local citizen participation it is important that, very early in the organizational stage, local technical, administrative and supervisory employees be recruited and placed in positions of responsibility.

Continuing attention must be given to building up a pool of local technical talent. In all but the most undeveloped countries there will be local people who possess high degrees of technical skill. Some Americans are quite surprised when they discover for the first time that local engineers are better educated than some of those imported from the United States. Quite frequently the local man excels technically but is short on practical application and management capabilities; if that were not the case there would be little excuse for the American to be there in the first place. Personality traits as well as technical ability of local employees must be carefully considered for they are the ones on whom the firm will have to rely in the future.

In line with the establishment of a local image, the new firm should early in its existence investigate local associations and organizations and seek membership in those that promise to be of benefit. Likewise, employees should be encouraged to join local technical organizations and apply for local professional registration where it exists.

The successful United States contractor or engineer has learned at home long ago most of the facts of life about business procurement methods. Once he is alone in a foreign country, however, he will find that he has a whole new set of complex rules to learn including government, administration, local laws, customs and geography. Much more diligence in sales methods will be required in the new country, primarily because he is a stranger fighting for recognition which for years has been taken for granted back in the homeland. The basic mechanics of overseas business procurement are those which have been developed after years of successful application in the United States. Business development starts once a foriegn office has been properly organized and staffed; but technical ability alone will not do the job. To the technical ability must be added large degrees of integrity, dependability, sincerity, good public relations and good political acumen. This is all part of the important process of reputation-building which must be a prime activity during the first three to five years of the subsidiary's existence.

In most foreign countries the American will find that personal contacts require much more traveling time than at home. Internal communications, telephone or telegraph are almost never equal to U.S. standards. Air service is often disappointingly poor, roads bad and living accommodations sometimes substandard. On the whole, more time and money should be budgeted for personal travel for business promotion on overseas ventures than would be required at home.

Once the foreign operation has been established on a permanent basis, no opportunity should be passed to utilize to full advantage the ties with the home office in the U.S.A. It will be discovered that most public works officials in foreign countries live constantly in hope of being sent soon on another overseas trip, usually to the "States." No better opportunity exists for improving public relations than to assist such people make their travel arrangements, take them to visit interesting jobs and perform other friendly services during their stay in America. In the reverse direction, it behooves the U.S. office to seek and cultivate American industrialists with possible interest in the foreign country. Similar travel assistance and guidance services can be provided to those junketing Americans when they come to visit the host country.

19
Ethical Considerations in Marketing

19.1. Bribes and Payoffs—International Problems

Throughout the world payoffs and bribes have been used in the past and undoubtedly are being used today to get contracts. Lately, this problem has been in the news media, concerning primarily overseas contracts. In some countries payoffs and bribes are a way of life. This presents an ethical problem to contractors, architects, and engineers wishing to do business in such countries.

The following article discusses some of the problems that have occurred in the past.*

BRIBES AND BUSINESS, U.S. FIRMS SAY '77 BAN ON FOREIGN PAYOFFS HURTS OVERSEAS SALES

Harvey Trilli is fuming with frustration, and he doesn't care who knows it.

Mr. Trilli is president of a large Pittsburgh-based engineering and construction firm—Swindell-Dressler Co., a subsidiary of Pullman Inc. He contends that his company recently was beaten out of a number of big overseas construction jobs because of payoffs to foreign-government officials by European competitors. One of the projects was a $40 million brick plant in Iraq.

"We thought we had the thing all wrapped up," Mr. Trilli says. "We had a team of people there for five or six weeks. All the terms were agreed to, including the pricing, and we were told the contract would be signed in a month. Then out of the clear blue sky, a German firm got the contract."

The Pittsburgh executive names the West German firm involved and says he believes it got the contract "because they made a big payment to a high official in Iraq." Mr. Trilli concedes that he wouldn't be able to prove that a payoff occurred.

---

* Reprinted by permission of the *Wall Street Journal*, © Dow Jones & Company, Inc., 1979. All rights reserved.

But even if he could, it would be of little avail. Bribery by West German companies to obtain foreign contracts isn't illegal in West Germany, and the companies may even deduct the costs as a business expense for tax purposes.

$100 Million in Lost Sales

Chicago Bridge & Iron Co., an Oak Brook, Ill., fabricating and construction concern, likewise has lost "in excess of $100 million of sales over the last three years" in situations "where we think the element of bribery was at least present," according to William M. Freeman, senior vice president for finance of the company.

Increasingly, American corporate executives are complaining about what they contend is a continuation of under-the-table payoffs by foreign competitors whose governments either look the other way when overseas bribery occurs or surreptitiously encourage the practice in order to increase their exports and improve their balance of payments.

Midland-Ross Corp. of Cleveland cites two instances of suspected payoffs this year in overseas contract negotiations. Both involve paper-mill equipment for two West African nations, which the company declined to identify. In one case, although Midland-Ross was favored by the consulting engineer for the project, it lost out on a $4 million contract to an Italian company that came in "at the zero hour." In the second case, Midland-Ross dropped out of the competition on a $4 million to $6 million job when it learned that it would have to kick back 10% of the contract's value to the president of the country and another 3% to "the president's sidekick."

A Solitary Campaign

Paying off foreign officials to obtain lucrative contracts isn't unknown among American companies, of course. More than 300 of them have admitted to the Securities and Exchange Commission that they had made questionable payments or engaged in misleading accounting practices in their foreign operations. But since December 1977, U.S. companies have been operating under the provisions of the Foreign Corrupt Practices Act, which makes it a criminal offense to offer a payment to a foreign-government official to assist in obtaining or retaining foreign business.

The law provides for prison sentences up to five years for violators and fines of up to $1 million for their companies—the largest fines ever authorized for imposition on American business firms. Companies doing business overseas also are required to set up rigorous accounting safeguards to detect the existence of slush funds and other financial devices for making illegal payoffs.

So far, however, no other industrialized nation has seen fit to impose any such strictures on its businessmen, and an anti-commercial-bribery treaty, proposed by the United States, is languishing in an inactive committee of the United Nations Economic and Social Council with little prospect for early agreement even on a working draft.

Meanwhile, the unilateral American effort to upgrade the ethical standards of international business, critics of the 1977 law complain, is blocking off large chunks of the globe as unsafe areas for U.S. companies to solicit business. These "sensitive areas," businessmen say, include not only many developing nations in Africa, the Far East, and Central and South America but also the oil-rich kingdoms and sheikdoms of the Middle East.

19.1. Bribes and Payoffs—International Problems

Paying an Entry Fee

"The U.S. brand of morality hasn't been successfully sold to a lot of areas yet," says Robert F. Conley, vice president for international marketing of Lockheed Corp. "In many countries, rewarding the decision makers is still the way that things are done in the business community. It's pretty obvious that we now have less ability to get an audience or even the attention of the decision makers when they know we're restricted in paying fees."

How much business is being lost by American firms as a result is impossible to determine because of the private, if not clandestine, nature of most international contract arrangements. But to judge from the intensity of criticisms of the 1977 act, the hardest-hit firms are large international construction companies that deal mainly with foreign governments or with government-run industries. Some of these companies say that in certain countries, it is impossible even to get on the bidding lists without paying what amounts to an "entry fee" to a local agent who has good connections with the government in power.

Last week, the lid was lifted a little bit on how large these fees can sometimes be. A Belgian firm, Eurosystem Hospitalier, which had been working on a $1.2 billion hospital project in Saudi Arabia, was plunged into bankruptcy. The crash was attributed to "excessive" secret commissions, estimated at $282 million, paid to get the Saudi contract. The Belgian royal family is involved because Prince Albert, brother of King Baudouin, was a leading member of the Belgian business mission to Saudi Arabia that obtained the contract.

The United States, which in 1976 ranked first in its share of the overseas construction market, dropped to fifth place last year, trailing Japan, Korea, West Germany and Italy, according to the National Constructors Association, a Washington-based trade association of large construction companies.

"A good part of the reason for this," says an executive of a West Coast construction firm, "is the 1977 law." He adds, "The commissions we used to pay are still a social or business custom in some countries, but now they're crimes as far as we're concerned."

Granville Kester, an executive vice president of Michael Baker Corp., an engineering consulting firm based in Beaver, Pa., says his firm is no longer trying to do business in the Middle East because "we feel we would have to pay big commissions to agents in those countries that could be considered kickbacks by U.S. authorities."

The problem with having to rely on commission agents, he says, is that "it is difficult to determine whether their services are legitimate and yet we're criminally liable if it turns out their services aren't legitimate."

Mr. Trilli of Swindell-Dressler says his company doesn't even attempt to do work anymore in Nigeria or Libya, partly because "we've been led to believe that in order to get business there, you have to pay off somebody." A similar problem exists in Mexico, according to several companies. Says Eugene Myers, vice president for finance of Marley Co., a manufacturer of water-cooling towers that is based in Kansas City, "We know we can't sell to certain people down there." And Joy Manufacturing Co. of Pittsburgh suspects it lost a big contract to supply pollution-control equipment to Mexico's government-owned power company because a Swedish competitor made personal payments to company officials to obtain the contract. "I couldn't prove it, but I have a strong feeling that a payment was expected" in order to get the contract, says a Joy executive.

"A Rough Area"

South Korea is mentioned by one large chemical company. "It's a rough area, corrupt as hell," says a company spokesman, who adds that his company has withdrawn from the area because payoffs to government officials were required. "We decided their business just isn't worth it," he says.

Earlier this year, the State Department received a cable from the U.S. embassy in Kinshasa, Zaire, expressing concern that the U.S. might lose access to Zaire's markets and its vast natural resources because other governments aren't showing the same zeal as the U.S. in policing their businessmen's activities in Africa. "Strict federal regulations against payments to facilitate sales put U.S. salesmen at a distinct disadvantage in a system which would rather not work at all than work without oil (payments)," the cable stated.

Criticism of the 1977 law in U.S. corporate circles has tended to be low key and often off-the-record for obvious public relations reasons—any company that complains too loudly inevitably raises questions in the public's mind about its own overseas sales practices. Many companies, of course, profess to have no difficulties with the legislation; Du Pont Co., for example, with $3 billion in foreign sales last year, says its own corporate code of ethics is more rigorous and predated the passage of the U.S. law. And an official of a major aircraft company says the law has been a boon because "since it was passed we have been getting about a fifth as many requests for payoffs as we used to."

An Incident in Qatar

So far, the government hasn't brought criminal charges against any corporation for violating the law, but in April the Justice Department filed a civil suit against businessman Roy J. Carver and an associate. According to the complaint, Mr. Carver approached the American ambassador to the sheikdom of Qatar and asked, "Who do I see now?" to win approval of an oil concession for which the two men allegedly had paid $1.5 million in bribes during 1976. The two businessmen consented to a court order enjoining them from offering further payments.

Despite intermittent grumbling about the anti-bribery statute, Congress isn't likely to amend it, certainly not this year. However, a White House task force on export disincentives is considering ways to eliminate "ambiguities," especially a provision subjecting company executives to criminal prosecution for "having reason to know" that independent sales agents abroad were making payoffs. The law could be costing U.S. exporters $1 billion a year in lost business, a task force spokesman says, citing "very rough, preliminary figures."

The "sweeping language" of the law as it relates to a corporation's responsibility for the actions of its overseas agents is particularly disturbing to Dresser Industries, a spokesman says. And Westinghouse Electric Corp. concurs, declaring in a recent position paper on export policies that this provision has led some companies to adopt "an unnaturally conservative marketing approach overseas."

Cincinnati Milacron Inc., one of the nation's largest producers of machine tools and other industrial products, has lost business because of "restrictions on where we can pay legitimate commissions," says James A. D. Geier, president, "but we don't know how much." Mr. Geier adds, "However, there are other boys in the world who don't play by the same rules."

One place where the game is fast and loose is the international aviation market, according to some U.S. plane builders. At one time they themselves were active participants, paying out millions of dollars in commissions and "fees" to some of the most highly placed "sales agents" in the world, including a former prime minister of Japan, the husband of the queen of the Netherlands and the commanding general of Iran's air force. Now, under the provisions of the 1977 law, they are required to sit on the sidelines. One big competitor they are watching enviously—and suspiciously—is Airbus Industrie, a European consortium owned mainly by French and West German aircraft makers, including Societe Nationale Industrielle Aerospatiale, the French state-owned aerospace company. Airbus in recent years has had phenomenal success selling an airliner called the A300 to government-owned airlines.

Comments one U.S. plane producer: "The Aerospatiale folks are completely at liberty to take on anybody they wish as a consultant for a retainer or a commission. In the last 18 months, they've had success in selling the A300 to Singapore, Indonesia, Thailand, Malaysia and the Philippines. It may be a coincidence, or it may be their day to shine, but the sales happened all at once and at a time when U.S. companies were restrained from employing their past sales practices."

In Indonesia, for example, Airbus's agent used to be a high official of Pertamina, the government-owned oil company. He now is a minister in the government. Recently, Garuda, the Indonesian state airline, placed an order for six A300 jets, with an option for six more.

Indonesian Sugar Mills

Another Indonesian business opportunity currently is being watched closely by businessmen around the world. The Indonesian government has asked for bids for the construction of six sugar mills for about $40 million each. All the bidders are European except for two U.S. concerns—Arkel International Inc. and a consortium headed by Katy Industries Inc. Neither company will discuss the project, but a source involved in the competition says, "Whoever gets the contract there must pay for it, whether you call it a bribe, commission or consulting fee. That's a fact of life in Indonesia, has been and always will be." For an American company to get an Indonesian contract "without getting caught violating the U.S. law, some very sophisticated paper work has to be done," the source adds.

Businessmen and government officials abroad, meanwhile, are taking attitudes ranging from commiseration to amusement at the controls now imposed on their U.S. competitors. An informal poll of more than a dozen British and European trade officials indicated a nearly unanimous opinion that the U.S. has lost overseas business because of the restrictions of the 1977 law. But none could provide specific examples or make an overall estimate of the amount of business lost.

"The main problem," says Sir Frederick Catherwood, a former chairman of the British Overseas Trade Board, "is extortion, rather than corruption—that you can't do business (in many parts of the world) unless you pay the entry fee."

Sir Frederick says the question involves the extent to which a nation can impose its laws and regulations overseas. "Can you make illegal in your own country something which is only nominally illegal—but not enforced—in another country?" he asks. "The U.S. has said more or less, 'Yes we can,' and other countries have said, 'No, we can't.' "

Foreign-Trade Dependence

Another factor, Sir Frederick says, is a nation's degree of dependence on foreign trade—exports account for only 8.5% of gross national product in the U.S., against 30% for Britain. He leaves little doubt that legislation against illegal foreign payments has minimal public support in the United Kingdom. "We, who are so much more vulnerable, would be killed," he says.

This attitude is reflected in the lack of zeal shown by the British government in following up recent payoff scandals. When a subsidiary of British Petroleum, partly state-owned, was shown (through SEC filings) to have made payoffs totaling up to five million pounds to win a Saudi computer contract, the British government refused even to discuss the situation. Similarly, there hasn't been a public investigation of state-owned British Leyland's Mideast contracts, which have been widely publicized as bribe-aided.

Business payoffs are even less of an issue in West Germany and Japan, the two nations whose companies are most often suspected by U.S. competitors of making illicit payments. Neither country has ever taken any steps to investigate the business practices of its nationals. And neither country has any permanent agency similar to the Securities and Exchange Commission, the U.S. agency charged with enforcing the corrupt practices law. Gen. Douglas MacArthur, in reorganizing the Japanese government after World War II, set up a Japanese SEC, complete even to the name. But it was one of the few MacArthur innovations that didn't take; Japan disbanded the agency after a few years.

In Tokyo, in fact, the man-on-the-street's reaction to questions about payoffs is that they are an established way of transacting business. Even after former Prime Minister Kakuei Tanaka was arrested and jailed on charges of receiving a $1.7 million bribe from a Japanese agent of Lockheed Aircraft Corp., he was reelected as a representative to the lower house of parliament, receiving one of the largest votes cast in the election.

19.2. Ethical Solutions to International Problems

Bribes and payoffs are now prohibited by the law of the United States, applicable to all U.S. citizens and companies wherever they may operate. However, the practice continues, often disguised or cloaked in ambiguities, and thus one is faced with his competition's use of this marketing tool even if he himself refrains.

How can the ethical contractor meet this competitive challenge? Sometimes the demands are veiled, mere suggestions, whereas in other cases they are blatant. He is forced to find other ways to market his services. Actual solutions vary widely; there are no magic formulae. What methods have worked?

19.2. Ethical Solutions to International Problems

A first and reasonable rule is not to act self-righteously when a payoff is proposed. The custom is rooted over many centuries or even millenniums and has in the past served as an inverse means of compensation to the contracting party. Consider the case of the waiter in the United States, where a "tip" is a payment to the one who furnishes the service. Not very many years ago, waiters were paid no salary by their employer, hence the tip was their only means of compensation. Similarly, officials and executives in many countries are grossly underpaid, because it is assumed that they will receive fees from the contractors to whom they award contracts.

Therefore, one must treat such requests with respect and find alternative solutions. The laws of this country and of the host country must be obeyed. There are legal requirements, for violation of which the penalties may be extreme: cancellation of the contract, or even criminal proceedings against the contractor. One non-U.S. contractor who was convicted of offering a bribe to a government agent in Saudi Arabia was barred from bidding on any further projects in that country and his resident manager was sentenced to jail.

Further, a contractor who does break the law may find himself unable to properly negotiate claims or obtain final payment; the violation gives the other party a hold over him.

The reason the matter has recently reached such prominence and attention is that the amounts of payoff became unreasonable, actually ridiculous, far out of proportion to past practice. Therefore, many host countries have recently enacted laws restricting this practice. These laws aid the ethical contractor considerably.

In the Mideast, some clients have reportedly demanded a brown paper sack full of cash—"bakshish." American firms try to avoid such payoffs, but some Mideast clients are reportedly very blunt in their demands. They demand the cash immediately or else they will find another contractor. About 15 years ago a major oil company developed a policy of no more payoffs. The company was very polite in implementing their policy and continued to make calls on the clients to demonstrate friendship still existed. After about a month the demands stopped, and the company is still doing business in the Mideast. For others doing business in the Mideast, payoffs can still present a problem. It is the authors' opinion that this situation is improving.

One way previously practiced was to take a constructive middle ground by forming a joint venture with a foreign national and then inserting a clause in the joint venture contract that all partners agree to abide by the laws of their own and the host country. The foreign national is by inference not bound by U.S. law. Thus he may or may not make a payoff, but the U.S. company will not know about it. This has recently been ruled invalid; the U.S. company is required to maintain controls to ensure that the U.S. laws are obeyed. If a payoff is made by the foreign national partner, whether you officially know about it or not, you could be liable under U.S. law as acting against public policy. Present law requires you to institute adequate controls to prevent this. Hopefully, the foreign national joint-venture partner will have enough influence and proper legitimate connections in his country so that payoffs and bribes are not required. This way the entire issue may be avoided.

As has been pointed out in numerous recent articles in influential newspapers and magazines, the current extension of legal restrictions to overseas practices of U.S. businesses has seriously curtailed their ability to obtain contracts in the international arena. Time and again, a U.S. contractor or engineer has successfully marketed a major contract, all the way to the point of signing a contract, only to learn that the contract has been awarded to a firm from another country under conditions highly indicative of a payoff. For many of these other countries, there are no comparable restrictive laws or, where there are such statutes, they are not enforced. The effort by the U.S. Congress to impose the domestic morality on the rest of the world has had very limited effect except to tie a chain around American contractors and engineers.

However, as of this writing, amendments to the U.S. laws are being considered by Congress, amendments which will probably exempt U.S. businessmen from criminal responsibility for the payment of such fees by joint venture partners or associates in the host country.

Regardless of the laws, however, the problem of unethical practices remains and constructive alternatives must be developed. The contractor or engineer is forced to emphasize all his other marketing capabilities and make maximum use of his reputation for getting the job done efficiently and economically. The worldwide reputation of American contractors for performance is perhaps their strongest marketing asset.

The existence of the U.S. law is widely known. This helps the contractor when a demand is made. In most cases it is subtly phrased, but the intent is clear. A polite but positive refusal, citing the U.S. law and company policy, accompanied by emphasis on the advantages that will accrue to the country and to the official by the award of the contract, is often effective. The successful completion of the project may lead to recognition and promotion of the official making the award. Properly presented, this approach has worked surprisingly well.

One contractor tried the alternative of extensive entertainment in lieu of payoff, a means that had worked in the past. In this case, after a week of entertainment, the official said bluntly, "Let's cut out this foolishness. I'm talking about money, and I want cash!" Even without regard to the law, a contract with such a person would probably be a bad one; additional demands would undoubtedly be made before progress payments and final payments would be released.

Dignified entertainment, at a standard appropriate to the country and customs, has helped to soften the effect of a refusal to pay. A proposal of a trip and visit to your office in your home country and a visit to your projects elsewhere may often be a legitimate sales approach as well as a means of accomplishing needed communication and understanding.

In one case, it finally developed that repair of a local road that gave access to the project, although not required as part of the contract, would resolve a major political problem for the official concerned and, at the same time, legitimately facilitate conduct of the contract work.

A contractor may decide to form a joint venture with a local company and emphasize national participation and national pride. In one case, Company Z, a large international company, formed a joint venture with a good and honest, but small, local contractor. He then put the local name first on the joint venture's letterhead. He got the job.

As in all forms of marketing, but even more so in construction marketing, personal relations are of great value. A man's word is valued more highly than a legal document in many parts of the Mideast.

Difficult and frustrating as these situations may be, it is far from hopeless for the ethical and competent contractor.

Restrictions work temporary havoc for U.S. contractors in some areas when competing with other internationals who are free of restriction, but some are surviving. They have to be keen, they have to offer a better package, they have to perform better. The message eventually gets through; and after the inevitable scandals involving others, the honest contractor may well get preference at last.

19.3. Bribes and Payoffs—Domestic Problems

One time the Ethical Construction Company (ECC) was asked to negotiate a contract for a large marine facility. The proposed contract was for a design and construction project of the type that ECC specialized in. Just when they were about to sign the contract, which had been negotiated at a price favorable to ECC, the client's representative, Mr. X., pointed out one final problem. He would like a new home on Lake Tahoe and already had the lot. He also just happened to have the plans. He stated, "You are a building contractor with spare materials and labor, and it will really not cost you anything." Mr. Smith, the contractor's executive, politely but very firmly told him that due to company policy, they could not and would not build him a house. Mr. X returned the contract to his briefcase, unsigned. However, due to Mr. Smith's tact and restraint, they parted on friendly terms. A little later Mr. Smith invited him to dinner to demonstrate that they were still friends and that Mr. Smith held no contempt toward him because of the demand, which was never mentioned again. Two weeks later ECC was awarded the original contract. Apparently Mr. X felt that ECC's high ethical standards would assure his company of getting a fair and just treatment on the contract.

Another time ECC was in the process of negotiating some change orders with a city commissioner. In private the commissioner asked ECC whether it could buy him a new car on ECC's fleet policy so that he could save $500. It appeared that what the commissioner was really saying was that he wanted a new car, that he wanted ECC to buy the car and then tear up the bill. ECC tactfully told him it would need to investigate how their fleet policy worked. Calling him back the next day, ECC said it was sorry; the terms of its fleet automobile policy did not permit such a transaction. The commissioner never repeated his demand. Strangely enough the relationship from that point on remained excellent. In more cases than most people think, requests of payoffs can be tactfully declined and the proper relationships with the other party still maintained.

However, unfortunately, this is not universally true.

In the early 60s, W Construction had a problem with payoff demands on a project in Louisiana. It flatly refused to make payoffs; and as a result of the policy, it lost money on the job because of hostile inspection and labor problems. It can be difficult to do business in an area with which you are not familiar. This indicates an alternative which often seems to be effective; namely, use as many local suppliers and subcontractors as possible. Take the time to establish personal relations. Coupled with a polite but positive refusal to pay off, this often works.

Some of the older employees at a public utility district recall petty unethical practices that were common many years ago. Each Friday the general contractor working for the district would put a fifth of whiskey under the seat of the district inspector's pickup truck. In effect, the inspector was selling his honor for the price of a bottle of whiskey.

Small payoffs can easily get out of hand. The ABC Construction Company was awarded a large government contract in Alaska. At the end of the first week, a government inspector asked for a case of whiskey. ABC provided it. The next week and following weeks, the inspector asked for and received a small cash payoff. Then one week the inspector asked for a significant payoff. ABC refused and discontinued giving any type of payoff. For the rest of the job the inspector made ABC miserable. When the contractor complained, the inspector threatened to go to the FBI and tell them of the small bribes. He himself had little to lose, and he knew it, but the contractor was in an untenable position. However, ABC held fast. As a result of excessively strict inspection, it lost $200,000 on the job because of the hostile inspector. The only proper course of action would have been to refuse the first request; and, if repeated, to report the payoff request to the government.

19.4. Legitimate Fees for Services

There can be a misunderstanding as to what is a payoff or a bribe and what is a normal expediting fee. A normal expediting fee is completely ethical and proper: lately, U.S. Law has been changed to allow normal expediting fees. Consider the case of getting a telephone installed in Thailand. To get the phone installed, the telephone installer demands a $50 payment. Since the installer in Thailand is not paid by the telephone company, or inadequately paid, this is a reasonable legitimate fee for services. (It is similar to giving a waiter a tip in the United States.) Under U.S. law this payment is now legal for an American company. Another example of a normal expediting fee is the fee a foreign customs agent may charge for clearing goods through customs.

Now, let us examine a gray area. Consider an expediting fee demanded to process a progress payment on an overseas job. According to the country, the fee might be entirely ethical and proper. However, if the fee is to process a progress payment in conjunction with a change order, then the fee might be considered to be a simple bribe in return for the change order. Therefore, this question can be quite involved.

19.5. Ethical Considerations in the Negotiation Process

Ethical considerations become a part of the negotiation process. You should never, of course, make untrue statements. On the other hand, it is not necessary to volunteer arguments against yourself. For example, assume the other party has asked the contractor making a claim for changed conditions if he put any money in the bid for contingencies. A contractor could answer truthfully and say "yes" and thus weaken his negotiating position. The contractor could lie and say "no." A third approach would be to sidestep the question by reminding the owner that the question at hand is whether a changed condition exists. Does the contract state that extra money will be paid for a changed condition? To be effective, a negotiator needs to establish a reputation for frankness and honesty.

19.6. Boycotts

When two countries are at war, or technically at war, they normally boycott each other's goods and services. It can be treason to trade with the enemy. For example, during World War II, the United States boycotted German goods and services. In international construction marketing, there always seems to be a case where one country is boycotting another. For example, some Arab nations currently boycott goods and services from Israel. It is currently illegal to sign an agreement stating that you will not do business with firms doing business in Israel. One way a U.S. contractor can comply with the boycott at this point in time and still not violate the law is to ask for and receive a list of acceptable subcontractors from the client and then use only these subcontractors.

The Japanese contractors handle the problem of doing business with both China (P.R.C.) and Taiwan by setting up a holding company with two subsidiaries: Construction Company A and Construction Company B. Company A does business with China only, and Company B does business with Taiwan only. Hilton Hotels appears to have effectively used a similar arrangement in Arabian countries and Israel.

19.7. Legal Standards versus Ethical Standards

Ideally the two standards would coincide, but in the real world, they often diverge.

The contractor-corporation is a legal entity responsible to the directors, the stockholders, and ultimately to the government which has legalized its existence. It performs its work under contracts—legal documents. Therefore, it must comply with the laws applicable to them, whether they refer to payoffs or taxes or affirmative action plans. A contractor who frets and becomes emotional about the laws affecting his work is wasting his time and energies that can be better used to get and carry out jobs.

Ethics is the loosely defined code of conduct that becomes a set of rules for the game of construction contracting. There are, however, few cases which are wholly black and white. Most are gray. Yet everyone in the business recognizes the ethical contractor and everyone knows the unethical contractor.

It is interesting to note that through the years, the ethical contractors manage to stay in business whereas most unethical contractors slowly disappear or, in not a few cases, become gradually more mature and more ethical in their practices.

Business can be as ethical as religion.
Forbes Epigrams

20
Business Development in Your Career

20.1. A Changing Profession

A high rate of change in societal demands, needs, and goods is affecting today's engineers and architects. The changes in technology and the professions are extremely dynamic if not revolutionary. This offers challenge and the opportunity for advancement.

When one of the authors graduated from the University of California, Berkeley, in 1940, he had received approximately the same education that his father, who had graduated from Ohio State University in 1906, had received. There was not much of a change in engineering education. Some of the textbooks were actually the same. Both referred to the same "Wellington's Economic Principles" in relation to engineering decisions. Imagine how a prewar engineer, who had not kept abreast of his profession, would feel if he walked into today's classroom to be confronted with finite element analyses or a sophisticated computer program. Or imagine how that same engineer might feel when confronted with today's environmental or public policy engineering classes. Many elements of today's engineering education were not even suggested in 1940. Changes are certainly occurring and the rate of change is accelerating. These changes affect the career of today's engineer and architect.

20.2. Professional Responsibility

Civil engineers are being challenged by today's society to develop social awareness. The civil engineer historically was the military engineer who saw the opportunity to serve society. Because of this awareness, he recognized the needs of society as opposed to the need to destroy the castle on top of the hill. The civil engineer is the one who translates societal goals for clean water and a bridge across a bay into real structures and facilities. Therefore, the civil engineer is basically socially aware in the practice of his profession.

The professional civil engineer will be more socially effective if he concentrates on being a good engineer and not a half-baked sociologist or environmentalist—society does not want that. Society wants an engineer who responds to its needs by translating its goals and objectives into structures and facilities that satisfy its needs. This recommendation applies only to professional practice. What an engineer does as an individual outside his profession is his own business. Outside of the professional practice is the place to be an environmental activist, if you so wish.

The civil engineer is a sort of a liaison between knowledge and action. He integrates science and the practical, tangible world. Remember the Hindu philosophy: right thought, right knowledge, right action. The civil engineer is a bridge between the two aspects of man: thinking and doing. His responsibility to society is to practice his profession well and to carry out his engineering functions in a way so as to maximize the benefits to society as seen by society and not necessarily as seen by himself.

At the same time a civil engineer also has a responsibility to himself. What is his responsibility to himself? As in Hamlet, "This above all else, to thine own self be true—and it must follow as the night the day, thou canst not thus be false to any man." The concept of truth as an ethical standard has emerged through many societies. The Pharaoh Akhenaten's creed was "living in truth." Cyrus the Great's motto was "Shoot straight and tell the truth." Consider the Christian philosophy: "You shalt know the truth and the truth shall make you free." The concept of the importance of truth has been with humanity for a long time and is with us today. An engineer must be true to his profession. Being true to his profession is different from the ethics that professional societies often publish. Although they have been improved in recent years, they are still somewhat self-serving. They are almost like the old guild idea of supporting one's fellow engineer. Being true to the profession is a meaningful commitment to design and build according to the needs of the project with full consideration for the environment, safety, efficiency, and economy. Engineers do have an obligation to safety and the environment. However, engineers also have an obligation to efficiency and economy. Economy means dollars that can be directly translated into man-hours—a portion of life. A saving of approximately $1 million represents a saving of one person's life, a life that will be able to do other meaningful work.

20.3. Education and Experience

The president of a large engineering and construction company once stated that he will not hire graduates with an advanced degree anymore. His basic complaint with them was that they were willing to start as project manager, "provided that in about a year they jumped to vice president." Other large contractors will want to start people with graduate degrees in construction at the bottom and let them work their way up. Evidently, one university pro-

20.3. Education and Experience 313

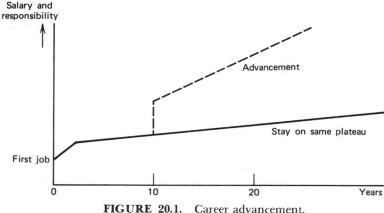

FIGURE 20.1. Career advancement.

gram teaches their students to settle for nothing less than the top. However, nothing one can be taught in school can replace on-the-job experience. If one does not know construction, he is not going to be useful. One needs to be involved with field operations to learn the physical problems of working. Therefore, early in one's career, he needs to get out on the job. Hands-on training breeds confidence. Until one has completed an "impossible job," he does not realize his own potential.

One construction company once had a man who was very good at administration but who had never been out on the job. This man was too closely associated with office work, estimating, procurement, and planning. He was extremely good at what he did. When it came time for the vice president rank, he never made it. He did not have the ability to meet with a client and talk about construction with confidence and credibility. He did not have the ability to direct workers in the field with confidence. He always needed to go back and check with someone else.

Although experience cannot be taught, education makes experience more meaningful. One does not have to make each mistake twice in order to recognize it. With advanced education, one still makes mistakes but hopefully recognizes them the first time—one becomes more observant. One learns to recognize the mistakes made by others and add them to his own storehouse of experience. Additional education is most meaningful as one's career advances. With a B.S., the road to project manager might take about 12 to 15 years. With an M.S. or an M.E., the road to project manager may take only 6 to 8 years. Most important, there is no limit to upward advancement into management ranks after 15 to 20 years. An advanced degree can help one move into a management or executive position.

Since an advanced degree produces only a slightly better starting salary, what is its main advantage? Advanced education can help one head into a position of responsibility. Typical engineering or construction careers follow the graph shown in Figure 20.1. At some point in one's career, usually after about 10 to 12 years, an engineer reaches a plateau, one which slopes upward only gradually. Many engineers remain on that plateau for the rest of their professional lives. Others make a jump into a position of responsibility—management, marketing, or creative engineering. Society pays for responsibility, and society rewards an individual who can produce in tangible terms. It is at this point in one's career that advanced education is most helpful. It has prepared him to make a jump.

20.4. Planning Your Career

Given all the broad statements, at this point, the individual's career should be discussed. The discussion covers not just the initial career (the first job) but also the future career. Figure 20.2, a flowchart, illustrates the process that is described below.

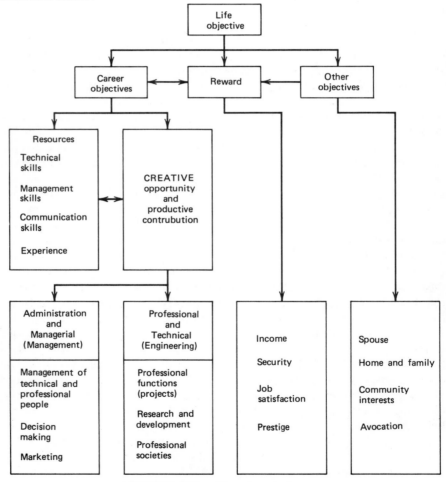

FIGURE 20.2. Planning your career.

20.4. Planning Your Career

To start with, a person needs to determine specifically what he wants to do. Most people have an intangible, cloudy idea as to what they want to make of their life. They need to set a career objective or objectives and then look at the opportunities. One's objectives do need to be examined with respect to constraints that can override one's decisions. In planning a career one needs to assure that he or she will make a productive contribution to society; in the long run, unless one is making a productive contribution to society, he will not be paid. Society may tolerate nonproductive people for a few years or even a generation but not for the long term.

There exist three or four paths of work one can choose. The most direct is the practice of one's profession. Other paths are research and development in the university or in the industry or university teaching. Research in the field of construction is on the increase. A last path of work, although it is usually not paid work, is service in the professional societies. Since one is not getting paid for this service, it is usually peripheral. After a few years, one can start moving into administration and management. A middle-management position can eventually lead to an executive position.

In a management position one is putting together and organizing the work of professional people. In management decision making, one often makes very broad decisions where there may not be a completely right or wrong choice, whereas in professional decision making, one usually knows whether he made the right or wrong decision.

Once the opportunities have been determined, then the individual's personal needs have to be determined—income and security. Most people like to get up in the morning and know they have a place to hang a hat, just in case hats come back into fashion. People like the praise of their peers: although it may not seem a legitimate reward, prestige can be a need. Job satisfaction is a need. If a person does not like his career type of job, he should change jobs right now. It is not in one's best interest to stay with a career that he does not like. This suggestion is not applicable to the temporary frustration that affects all of us—for example, a supervisor that one does not like. This suggestion is made for a job that is not satisfying when everything is going well. Also, in the evaluation of the opportunities, other job objectives and constraints need to be determined. Needs or objectives, like family, personal life, and interest in the community, should also influence the decision process. Keep your family informed, and make them participants in the decisions which affect your career and which therefore affect them, too, directly and indirectly.

How should all the information be evaluated? For the past 20 years the engineering profession has provided many opportunities and choices of varied employment. Therefore, most engineers are always deciding between several alternatives. Figure 20.3 provides a flowchart mechanism that can be used for the evaluation process. This is a method to evaluate alternatives in light of satisfying the points above. Once a final decision has been made, it should be checked once more to make sure it fits into one's objectives. In

316 *Business Development in Your Career*

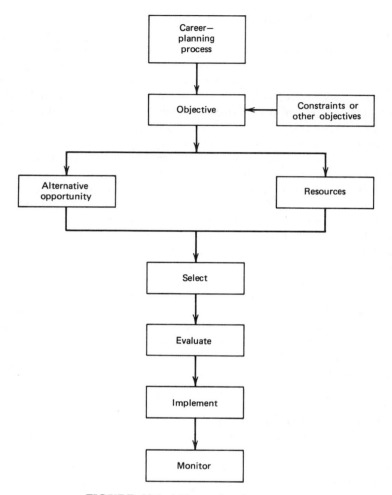

FIGURE 20.3. The evaluation process.

one's career, just as in engineering, the decision needs to be implemented. One needs to work out those details, and then do it! After the decision has been made and implemented, it should be periodically monitored to see whether it is really meeting one's objectives. Now with respect to the word "monitor," in the last 20 years, there have been so many opportunities for engineers that there has been an overconcern with monitoring and reevaluating—the question of "Did I make the right decision?" One can get to the place where he is constantly monitoring and not implementing. Therefore, it is suggested that one reevaluates only on a three- to five-year basis. For example, take a week during a vacation, and reexamine the situation—make sure the right decision was made. Having done this, do not examine the alternatives for another three to five years. Because, otherwise, one is spoiling his performance, fun, and enjoyment. This type of approach seems to work out well for many people in career planning.

20.5. Finding the Right Job

Once the decision has been made on the type of job to pursue, then the job seeker should take the initiative in seeking the proper employer. It is true that employers are seeking people to fill positions; however, it is in the job seeker's best interests not just to respond to the first opportunity that opens but to actually find the job that he wants. Then when one finds an opportunity, he is in a position to take it. He knows what he wants; he does not have to shop around only to find that the first job is not there anymore because it has "just been filled." It is similar to looking for a parking place: most people pass up the first ones available, hoping they will find a better one in front of the store.

One of the first steps to getting a job is getting an interview. Many students and some alumni will use a university employment service as a way to get an interview. This idea is regarded by the authors in a negative fashion. There are several problems with the university interviewing system. First, consider whom the construction company will most likely send to do the interview. It will seldom send a key executive—he is too busy. Therefore, it will go down the line a bit. The person sent is usually from the personnel department, probably has the ability to talk well, likes people, but probably has not yet found himself technically and professionally. He is someone the company can afford to send. So, the interviewee does not necessarily meet the top people. Once in a while the company sends an older man who is semi-retired (and they are the most effective, by the way), but he is not in the mainstream. Second, a person doing interviews will go to, perhaps, 10 campuses and interview 20 people at each campus—that is 200 interviews. One recent interviewer at Berkeley had been all around the country and had actually interviewed over 1100 students. After all those interviews, even the most astute personnel officer can become jaded. How can he possibly recognize the person he wants?

After a person is hired by this process, he shows up in the personnel department, fills out the standard forms, and starts at the standard wage. Now, he is sent to the operating department. The operations person informs the personnel person, "Hell, I did not want this kind of man." The personnel person informs the operating person, "This is the best I could get. I think he has a lot of potential." And the operating person reluctantly accepts the new hire. This is a traditional way for a young engineer to get a job—one in which he may not be happy.

Now reverse the situation to what the authors recommend. The job seeker gets to know the firm and what he can do for the firm. Then he approaches an individual in operations. Any company from Kaiser or Fluor to a one-person office in San Francisco is looking for a top-notch employee to work with it. People in operations are actively looking for good individuals. If the job seeker can get in and talk to this type of person, the operating manager, he is getting it right off the firing line. This type of person may not be smooth and polished, but he is interested in the job seeker as a potential employee to work for him. If he decides he wants the job seeker and the job seeker decides to work for him, he calls the personnel department and asks them to hire the job seeker. He also might add that he wants them to give him everything that the job seeker wants, for "that is the man I need." Now, the personnel department is not going to be so eager to cut down the job seeker's salary request. It has to satisfy its boss in operations. Also, the person in operations has hired just the worker he wanted. The job seeker and the employer have a good relationship to start with. Personnel departments do exist, and one does have to go through them: they have the necessary forms, data, and red tape. But from the point of view of getting a job, one should try to go through the operations or management route, if possible.

Once you are offered a job, one approach that can be taken is to accept the job when offered—with enthusiasm. You may be able to do this only if you are interviewing at the very company for which you want to work. This, of course, requires that you do your homework of learning about the company before the interview(s). When you do enthusiastically accept a job this way, it creates a very favorable impression on that person in operations, your future boss. He knows you really want to work for his company and subconsciously will feel you will do your best and do a good job. The job seeker that responds to a job offer with an answer of "I want to see what Engineering Company X has to offer before I make a decision," may be starting his job on the wrong foot.

To get a job, one needs a resumé. In preparing a resumé, remember many personnel departments get hundreds of resumés per week, and these are filed but not really read. Therefore, try to make the resumé specific and oriented to that company. Once in a while a resumé does attract attention because it is specifically oriented to a company's needs. When one of the authors was involved in the foundation business and when a statement on a resumé, telling of a thesis on deep cofferdams or slurry trench walls jumped out at him, he would retain the resumé. Otherwise, the resumés would end up in the personnel department's file. Resumés are written in order to open doors and so that people will remember one as an individual.

When sending a resumé, it should be accompanied with a cover letter. The cover letter need not be long. Resumés without cover letters will seldom be read and possibly will not even be filed. Most companies feel that anybody who does not care enough to write a cover letter really does not care about a job in the company.

Once Ben C. Gerwick Inc., received a letter from an engineer who was interested in cofferdam construction. The engineer stated that he had heard that "Gerwick did the best work in cofferdams of all the cofferdam builders in the country, so he wanted to work for them." Gerwick was naturally flattered to receive such a letter and sent him an airplane ticket for an interview. At the end of the interview, the applicant had a job. Of course, the engineer did have good technical capabilities; but the real reason Gerwick hired him was that he felt that someone who would make the effort to find a firm specializing in his area of interest would make an enthusiastic and productive member of the organization.

The book *What Color Is Your Parachute, a Practical Manual for Job Hunters and Career Changers* by Bolles offers some excellent ideas on resumés and seeking the proper employment.*

20.6. Office Politics and Diplomacy

There often seem to be six stages in a career, as well as in a project. Recall the witticism often found posted in the project office.

1. Exultation
2. Disenchantment
3. Confusion
4. Search for the guilty
5. Punishment for the innocent
6. Reward for the uninvolved

More seriously, one's ability to work with people is vital to a successful career. Effective construction management demands cooperation and teamwork. It is best not to get too involved with company social activities and politics—stay friendly with all but don't get entwined.

Engineers are judged on letters, charts, and reports. It is imperative to be neat and to write well. Also, large mistakes in reports must be avoided. Be sure to look over a report or letter before it is sent. In making a decision on whom to promote, management will almost always give preference to the person who can write a good letter or report.

Writing papers and giving papers is an effective way to establish one's professional reputation. This will be of help in one's own company.

Also, Parkinson's laws sometimes seem to describe one's career.

1. Work grows to fill the time and staff available.
2. A man is judged by the size of the staff he manages.
3. The accomplishment of a committee is inversely proportional to the number of members.

---

* R. H. Bolles, *What Color Is Your Parachute, a Practical Manual for Job Hunters and Career Changers* (Berkeley, CA: Ten Speed Press, 1972).

4. The effort spent on an item is inversely proportional to its importance. (This law appears particularly relevant to engineering.)

At some time in one's career there usually is a problem with middle people who did not make it. The young engineer comes along and the middle person tries to keep the young engineer from getting above him. It can become almost a goal in his life to try to inhibit and frustrate the young engineer. In this case, the young engineer needs a thick skin and needs to realize that quite likely management is aware of this problem too. It is a difficult situation, but one needs to keep his sense of humor and use some of his own marketing capabilities within his own company. Management is always looking for good people, and they are not going to ignore a good individual who is stuck below someone whom they know to be a problem.

One way to get ahead in a career is to volunteer when an opportunity arises. A brilliant young engineer had just been hired by a West Coast contractor and, as was customary with new hires, had just been assigned to the estimating department to do take-offs, a very routine job.

On Saturday morning, the executives would review the planning of the construction methods and the pricing. The young engineer would voluntarily come in and sit in a corner, listening and learning.

The company was successful bidder on the project. The young engineer went to the vice president of operations and told him he had been working on the estimate for that job and wanted to now go out on it as a field engineer. The vice president knew the young engineer had been in the meetings, so although he theoretically was not ready, he sent him out as "assistant project engineer." A few months later the project engineer became ill, and the young engineer took over. He did a highly creditable job, so well performed that the project manager asked for him on his next job.

Halfway through the next job, the project manager had to return home on an emergency, and the young engineer took over while the company lined up a replacement. A telex came from the client: the young engineer was doing so well it wanted him as permanent project manager. This was a case of graduate student rising to project manager of a multimillion dollar job in three years.

There have been similar cases: a young field engineer who correctly diagnosed the need on a large-bore tunneling job as a materials-handling problem and proposed a solution. It was so well thought out that the company president assigned him the job to implement it, and from that success he rapidly moved to higher responsibilities.

One should not forget that he has to live, as well as work, and that his family exists. Therefore, at some time, one has to request a vacation or days off. In the construction and engineering business, there is a tendency to do a lot of overtime work—one works 14 days straight or one works 6 nights in a row preparing a bid. It is easy to never get to a vacation. Companies tend to be a little shortsighted sometimes. When a person starts running down, looks haggard and tired, and starts making mistakes, they will say, "He used to be a good man; he is overworked; we will give him a month off and see whether he can come back." This is the wrong way to get a vacation. No good engineer or company handles equipment maintenance this way. The best way to handle this matter is to be positive when requesting a vacation or days off (one should be realistic, however; if there is an emergency, one does have to work). Do not just accept the continual pressures of the job, or one will never get any time off. When requesting days off for personal matters, it is always best to discuss the matter openly but briefly and firmly.

One of the most difficult matters is how to get a raise. Usually the company has the same general idea that you do, only it tends to lag behind. Sometimes it becomes necessary for the employee to take the initiative. Do not beat about the bush; do not go on about inflation and costs, and so forth. If you have decided to ask, be polite, direct, specific: "I'd like you and the company to consider my salary; I feel that I am now doing work that justifies a raise." Note that you are emphasizing greater value to the company, not greater need to you. Never threaten to quit. Consider the case of Quinby who had been doing a good job with frequently increasing responsibilities but no increase in salary. One day, in frustration, he came to the president and blurted out, "If I don't get a good raise, I am going to quit!" The president reacted as could be expected: "You can't quit; you're fired!"

Peter's principle often applies to growth in one's career. "A man is advanced to the level of his incompetence" when he is considered for promotion. Hopefully, the readers of this book will not end up at too low a level of incompetence! Finally, never underestimate another person's ability to rise to an emergency or a major problem, and never underestimate your own ability.

20.7. Handling of an Emergency

In one's career, Murphy's Laws seem to apply.

"What can go wrong will go wrong." "They will go wrong at the worst possible time." Note O'Brien's corollary: "Murphy is an optimist."

Murphy's Laws raise the question of how to handle an emergency. Emergencies do happen; and if handled badly, they can hurt or destroy one's career as well as one's self-confidence. One should plan ahead as to how he will handle potential accidents. When an accident happens, report the accident with no unnecessary alarm. Report what action was taken, and then ask for help if needed. If there were no injuries, report that there were no injuries.

An experienced construction manager was in charge of a large terminal project in the Mideast. Unfortunately, some of the indigenous workers were killed. It was a volatile situation, occurring at a time of political turmoil and might have gotten out of hand. The other Westerners on the job ran off the pier. The local workers were being incited to mob action. But the experienced manager handled it properly. He lit his pipe as he calmly walked out on the pier to survey the damage and situation. He talked calmly and slowly with the Arab workers. He kept the situation in hand. His calming influence cooled down the situation.

Compare two stories. In one, Phil Hart, president of the Pacific Bridge Company, who was erecting the deck on the Golden Gate Bridge, was on the witness stand during an investigation. A safety net had failed to save some workers who had fallen with a girder. The question was, "Who designed the net?" Phil Hart said, "I did." The interrogation was repeated: "I mean what individuals designed it?" Phil Hart replied, "I review all the drawings and I approve them. They are all my responsibility entirely." Incidentally, he was completely cleared of negligence.

On another bridge disaster the managing partner of the design firm was asked who was responsible. He replied, "Well, it is really the contractor's responsibility to check on design and, if he finds something wrong, to notify us. So I guess he is responsible." We do not know if the results were directly attributable to this remark or not, but the design firm was ultimately held responsible.

20.8. Give Your Ideas or Die with Them

We are all marketing ideas: concepts, plans, services. Either give your ideas or die with them. Otherwise, your competition discovers them independently and gets the benefit. Do not worry about "brain picking." Of course, sometimes your idea will be used without direct payment. But usually the benefits come back. Once a subordinate, with his boss, made a call on a client. The client pressed eagerly for ideas about how to solve a problem, and the subordinate suggested many good ideas. The boss subsequently criticized him for doing this when the client put the project out for bids with the good ideas right there in the plans for everyone to bid on: "Do not give ideas until you get a contract and get paid for them." However, when the bids went in, their firm was awarded the job even though it was second bidder. The client remembered where he got the ideas.

20.9. Changing Jobs

At some point in one's career, it is quite likely that changing jobs will be considered. One has the feeling that he is in the wrong branch of work and would like to make a change. One can change jobs within the company (particularly if it is a large company) or one can change companies. Often one's own company has many opportunities. And there are a certain amount of benefits that one builds up within a company—not just a pension fund or vacation time but the benefits one builds with having been with a company for 8 or 10 years and knowing how to make his part of the organization work. Therefore, one of the things one should look at is the possibility of changing jobs within one's company—maybe a lateral move to another job. If one does change companies, it should be with no bitterness. He may some day come back to that company, or the old company may become a joint-venture partner with the new one or even become a client. So do not get tempted into justifying your decision by citing all the bad things you can remember.

20.10. Accepting or Rejecting New Assignments

In a recent study, successful construction executives and managers were asked to analyze the decisions that they had made which most affected their careers. The overall concensus was that they made the right decision 80 percent of the time when they said "yes" to an opportunity and that most of their mistakes were made when they said "no" to an opportunity. One must learn to always say "no" on matters of principle—but on matters of venture, the odds are better than 60 percent that "yes" will be the right answer.

When it comes to that time in one's career that it is time to say "yes" to an offer for the new job assignment that has been long waited for, one good approach in business is to say "yes" and then request right at this point the other resources that may be needed to make the job a success. For example, if an administrative assistant and a good accountant will be needed to make your project successful, this is the time to ask for them. Also, one should settle questions as to what one's actual authority will be. Perhaps the authority to make certain decisions within general rules or guidelines is required by the new job. Now is the time to settle those questions with one's superiors.

One brilliant young engineer was offered and accepted his first big job assignment overseas. But he did not ask for assistance. About two months later, he realized that he needed lots of help. He wrote several low-key plaintive letters back to his company asking for a person or persons to help "if it is at all possible." These letters were misplaced; they did not seem urgent, and no action was taken. Eventually, when the job was nearing an end, the administration was in such a mess that the main office became alarmed. The resultant losses were blamed on the project manager, and he was eventually terminated. This came at the height of his career; he did not make it on his first big job. Fortunately, he eventually did come back and rebuilt a very successful career.

In hindsight, he should have asked for help when he started the job. As soon as he saw he had a problem, he should have demanded help. If he needed to call the company president to get help, he should have called him.

20.11. Career Growth

Civil Engineering, November 1975, contained an article, "Career Growth for the Practicing Engineer," which stressed several key points which lead to a growing and successful career. First is outstanding performance in the present job. One must perform his present job well and not spend all his time worrying about the job ahead of him. Second is preparation, both technically and psychologically, for a new, more advanced job. What is required for the technical preparation is obvious. The psychological preparation, however, involves the fact that one will be acting in a different way. The professional can be caught in the dichotomy between the demands that he administer and manage the technical work of others and the personal demand that he perform the work himself. That dichotomy or conflict will stay with him all his life as long as he is in the engineering field. It is extremely complex and difficult to manage effectively other people in professional or technical work. It is much easier to manage people when they are doing physical or business work. Third is learning how to continue learning—the "capability of developing your intellect through the years." An engineer who has been out of college 5 to 10 years can become fossilized. The recent graduate in his first job will learn everything he can—both the good and bad of the profession. However, at some point in his career he tends to settle down in his job and become comfortable and complacent. At this point there may be little incentive to continue to learn and to grow—one becomes stagnated. It is at this point that one needs to learn how to continuously learn.

One should use "down cycles" creatively—prepare, sell, develop, think. It is quite common for people to experience discouragement at age 40; if this happens, it is a good time to be creative. At age 40, Stonewall Jackson was out of a job as an army officer, so he went back to teaching. He used this down cycle to plan the tactics which made him one of the most brilliant generals of the Civil War.

Finally, there is Gerwick's law: "Man creates problems for himself just hard enough to stretch himself to the limit."

20.12. In Conclusion

Following are ten precepts for successful marketing in your career:

1. *Service*: This is the heart of marketing. It is responding to another's needs.
2. *Empathy*: To try to sell or communicate, you should put yourself in the other person's position. What can you do for him? In what is he interested?
3. *Imagination*: Imagination, creativity, or innovation can provide the ideas that are essential to marketing.
4. *Conviction*: If you really believe in what you are doing or offering, the client tends to believe you also.
5. *Preparation*: You cannot sell, market, or do anything if you are not prepared and do not know the details—do your homework.
6. *Perseverance*: You have got to keep going—do not give up. Can you lose a big, important job and still want to go on to the next one?

7. *Respect*: Recognize the other person as an individual—show aggressive goodwill. (I'm O.K.—you're O.K.)
8. *Enthusiasm*: This has often been called "the priceless ingredient," both in marketing and in your career.
9. *Sincerity*: Speak and live the truth, as you know and believe it.
10. *Integrity*: This is the wholeness or unity of character, life, job, and goals.

A metaphor can be drawn between pursuing a career and climbing a mountain. As in climbing a mountain, the usual goal is getting to the top. As in climbing a mountain, there are many routes to the top—the professional route or the management route, for example. Some people spend all their lives going around the base of the mountain looking for an easy route to the top—trying different paths. One can make one or two false starts, but to be successful, somewhere along the line, one needs to make a decision and start climbing. This is the only way to reach the top.

References

Bonny, J. B. and Frein, J. P. *Handbook of Construction Management and Organization.* New York: Van Nostrand-Reinhold, 1973.

Booth, W. D. *Selling Commercial and Industrial Construction.* New York: Van Nostrand-Reinhold, 1981.

Coxe, W. *Marketing Architectural and Engineering Services.* Van Nostrand-Reinhold, 1971.

Drucker, P. F. *Managing for Results.* New York: Harper & Row, 1964.

──────. *Age of Discontinuity.* New York: Harper & Row, 1978.

Meyers, V. "Business Acquisition," in O'Brien and Zilly, *Contractor's Management Handbook.* New York: McGraw-Hill, 1971.

Nierenberg, G. I. *The Art of Negotiating.* New York: Cornerstone Library, 1968.

Packard, V. *The Hidden Persuaders.* New York: D. McKay Co., 1957.

Sibson, R. E. *Managing Professional Services Enterprises.* New York: Pitman, 1971.

Wilson, A. *The Marketing of Professional Services.* London: McGraw-Hill, 1972.

Appendix 1

Sample Exercises

Set One

1. Write a two- or three-page letter to the instructor outlining your own personal objectives in your career and university program. Where did you come from, why are you here, and where do you hope to go?
2. Select a corporate business identity that is compatible with your career plans. For your assumed corporate identity, prepare corporate marketing objectives for the next two fiscal years.
3. Write a memorandum to your company's president in which you propose that the company enter a new market. Explain why and how.

Suggested project (4-10). Westinghouse has just started a 10,000-home development project south of Sonora, California, and has requested proposals for the sewage collection, treatment, and discharge system on a design and construct basis. *Note:* Some students may wish to substitute another project.

4. Draw a diagram showing the multiple offices and interrelated clients for a typical industrial project. Show those involved and number (in order) those whom you think are primarily involved in the new project.
5. Prepare a proposal letter to Westinghouse for performance of the above work on an alternative basis (alternative to lump sum) that you believe might be attractive and profitable. *Note:* You may take a specific portion of the work, if you wish. If you are a general contractor, take it all; if a soils engineer, take the soils design portion; if an architect, take the buildings and/or overall management.
6. Submit a prequalification letter as a contractor (or engineer) for a proposed project, listing past history and experience of your company and

personnel, management capabilities, and your special qualifications for this project.
7. Prepare a brochure for your company, designed to assist one or several of your company's marketing objectives. Use cutouts from magazines or photocopies of photos. You will be judged on effectiveness, clarity, and relevance. (You will not be judged on the quality of the pictures.)
8. Write a letter to the owner (for example, Westinghouse) when you find out that his architect (construction manager or staff) has recommended award to a competitor when you know that your proposal is best.
9. Write an initial claim letter. You have been awarded the (Westinghouse) project on a fixed price basis and have just encountered seriously changed conditions. In the case of the sewer project, where you expected firm clay, you have run into fine sand that flows like quicksand under a head of water.
10. Assume Westinghouse has answered your letter in this way:

We have received your letter of _____ , claiming changed conditions and asking for adjustments to the contract price and extension of time for completion. In view of the fact that this is a fixed-price contract, this request is denied. No adjustment in either time or compensation will be made. You are hereby directed to take whatever steps are necessary to complete the project without delay and without extra cost to Westinghouse, Inc."

Prepare a suitable letter for reply.

Set Two

1. Write a two- or three-page letter to your instructor outlining your own personal objectives in your university program and in your career. Where did you come from, why are you here, and where do you hope to go. (This will not be graded.)
2. Select a corporate business identity that is compatible with your career plans. Examples: (a) A large construction-engineering firm; (b) A small, specialized construction firm; (c) A geotechnical engineering firm. For your assumed company identity, prepare company marketing objectives for the next fiscal year and also for the five-year period ahead.
3. Write a memorandum to your company executive in which you propose that the company enter a new market. Explain why and how.

Suggested project (4-10). Pacific Gas and Electric (P.G. & E.) plans to build a large coal-fired power plant at Collinsville, 40 mi northeast of Berkeley on the Sacramento River. It has received a permit for site preparation only but expects to obtain further permits in the coming year.

You have heard through your friends at P.G. & E. that the company has decided to get the site work (fill, drainage, and so forth) started as quickly as possible, under a separate contract. Therefore, it will call for bids to grade the area about February 1, with award to be made March 1. This will involve cutting from weak sandstone hills and filling over low-lying marsh; about 3,000,000 cu yd of ripper, scraper, and compaction work, with short haul inside its property. Also included will be clearing and grubbing to remove the marsh vegetation, drainage ditches and culverts, fencing, and erection of two temporary steel warehouses (furnished by P.G. & E.).

This contract will be let separately.

Meanwhile, P.G. & E. will be selecting a large CM firm (Bechtel, Stone and Webster, or Fluor) to manage the overall facility. It expects to select that firm about April 1, and the site preparation contract will then be placed under the CM for management.

4. Who is the client? Prepare a diagram showing the multiple offices and "clients" within P.G. & E. for this project. Indicate (by priority) who will be the decision makers and who will have the significant advisory influence.

5. Submit a prequalification letter for your company, for the proposed project, listing past history and experience of your company relevant to the project, your present capabilities and resources (financial, equipment, personnel, management), and your special qualification for this project.

6. Prepare a proposal letter to P.G. & E. for performance of the above work on an alternative basis (other than the fixed unit prices which it asked for), which you believe will be more favorable to it and you.

7. Prepare a brochure for your company, designed to assist your company's marketing objectives. Use cutouts from magazines or photocopies of photographs for your illustrations. You will be judged on effectiveness and relevance. (You will not be judged on the quality of the pictures.)

8. Write a letter to the owner when you have just learned that the new construction manager being most seriously considered for selection (you do not know whether it is Bechtel or Fluor as it is still a secret) has recommended that the work, on which you proposed (exercise 6) above, be given to your competitor even though from everything you have been able to learn your proposal was the best.

9. Write an appropriate initial claim letter to P.G. & E. You have been awarded the P.G. & E. job on a contract that has a lump-sum price for the clearing and grubbing; the rest of the contract is in accord with your alternative proposal. You discover seriously changed conditions. There is a deep, ancient slough channel, which was missed by the borings, winding right through the middle of your fill area. It is now filled with peat. The resident engineer from the CM has directed you to remove it (about 80,000 cu yd) and has stated that this will be done as part of the clearing and grubbing item. (No separate payment.) The replacement

fill is to be placed under water, and he directs you to compact it to 90 percent relative density. Write your claim notifications letter. To whom do you write it? There is no specific "changed subsurface conditions" clause in the contract.

10. You have received an answer from the CM, indicating that a copy was also sent to P.G. & E., denying your claim:

We have received your letter of March 11, claiming changed conditions and asking for adjustment to the contract price and an extension of time for completion. In view of the fact that this removal is clearly part of item B-4, clearing and grubbing, which includes 'removal of all unsatisfactory material including decayed vegetation, and so forth,' and that since item B-4 is a fixed lump-sum price, your request for additional compensation and time extension is denied.

You are hereby directed to complete the project without delay, taking whatever steps are necessary to stay on schedule, without extra cost to P.G. & E.

Prepare a suitable letter for reply.

Set Three

1. Your company was awarded the Watergate North Project-foundation contract, to furnish and install 2000 prestressed concrete piles, 40–80 ft in length, in accordance with the soils engineer's (Woodward-Lundgren) predetermined geotechnical profile. The contract is a lump-sum contract, with unit prices for directed (ordered) changes in pile length and additional piles. Piles are to be driven to predetermined tip elevations and as much further as necessary to develop 100 tons bearing by the Engineering News Record Formula:—$P = 2E/(S + 0.1)$ where S is in in. per blow, E in ft. lbs of energy, and P is the safe bearing load in lb. Borings disclose fill, with lumps of small rock fragments, then organic material (garbage), "foreign material, such as wood fragments," then bay mud, and finally sand. During the driving of the first group of piling you get absolute and sudden refusal of the piles at 16 ft of penetration. This happens on the first six piles. Then 10 piles in a row drive to 62 ft, as specified for this location. Then you encounter refusal at 20 ft on the next three piles. All this happens in the first two days on the job. The contract contains no special clauses of any kind, that is, no changed condition clause. Write an initial letter.

2. You receive an answer, four days later, stating that this is your problem and that you have to drill, using spuds or whatever is necessary, to achieve the required penetration. Meanwhile, you have driven 46 more piles, of which 16 scattered piles came to refusal at 18–22 ft, while 30 performed as originally planned. You tried to put down a boring but encountered firm material, like rock. You have dug a hole with a clamshell and brought up one large slab of concrete from −13 ft; however, this is at a location 50 ft away from the piles. Write a second letter.

Appendix 2

Sample Correspondence (Memoranda and Letters)

MEMORANDUM

To: Myron K. Jones, President
From: Anton Ghissard
Date: May 29, 1979
Subject: 1979-1980 Corporate Marketing Objectives

You will find herein a description of our corporation, the service we offer, our potential market, and our marketing objectives for the fiscal year 1979-1980.

The name of our company is NASP, Limited, which stands for "Numerical Analysis of Soil Problems." We are equipped to deal with a large variety of problems including slope stability, horizontally loaded piles and dynamics. Furthermore, the availability of sophisticated finite element programs enables us to analyze dams, foundations, roads, excavations, cutoff walls, and so on for predicting stress, strain, deformation, and pore pressure in a two- or three-dimensional analysis. Consolidation and seepage can also be simulated by this type of program.

Our potential market is constituted of firms and companies that occasionally need simple or sophisticated numerical analysis services. It includes:

(a) Small- and medium-sized engineering firms: to carry out, check, or refine a part of their design.
(b) Contractors: to challenge the engineer's design (to modify it for potential economies).
(c) Owner or project supervisor: to check the engineer's design or the contractor's procedure before approval.
(d) Regulatory bodies: for example, to satisfy the dam safety act.

Our firm gives to those potential clients access to the expertise they need within a reasonable period of time.

Our firm is based in Montreal and our geographical market will be first limited to Eastern Canada. The construction industry is presently in poor condition, but we have few competitors, if any. For the years to come, the James Bay and the NBR hydroelectric projects (16 and 10 billion dollars, respectively) will provide a number of opportunities. Our firm will use those opportunities to establish a good reputation. The emphasis will therefore be on volume rather than on immediate profitability.

Consequently, our first objective for the fiscal year 1979-1980 is to get contracts from the James Bay Energy Society (supervisor of the design and construction of the two projects) which badly needs the kind of expertise we offer. In this regard, we will meet with the authorities to convince them that specialists are more qualified to perform those types of analyses than their own staff and that they will be done faster than in any university.

Our second objective is to contact all engineering firms and contractors on the James Bay Project in order to eventually diversify our clientele. This could be achieved by personal meetings, letters, and brochures emphasizing our past experience on this project.

Finally, a certain effort should be made to market our services to the engineering firms in Montreal, not so much to obtain contracts this year, but rather to be ready to enter the new market that will open when the economy gets better and the construction starts again. This could also be done by means of letters, brochures, and lectures in engineering clubs.

After this fiscal year, the results of our marketing strategy shall be submitted to a critical review. If the goals are not reached, our course of action shall be modified or the objective oriented in another direction.

Anton Ghissard

AG:pd

Corporate marketing objectives.

MEMORANDUM

To: M. B. Astute, President
From: Janis Q. Roberts
Date: May 29, 1979
Subject: Proposed Sacramento Market Expansion

The results of my recent analysis of the current market outlook for the industrial and commercial building services provided by our company strongly indicate that the greatest opportunity now exists in the Sacramento area.

Although we have always considered Sacramento to be a part of our service area, serious expansion into this market at this time is seen to be desirable for the following reasons:

1. Sacramento represents one of the fastest growing areas in the state. This rapid growth in population fosters a demand for service industries, that is, stores, restaurants, banks, and so on.

2. Many of the businesses that will be locating in this area will be of the "chain" type, with headquarters and outlets in other cities. They will expect buildings of comparable quality to others in their chains. However, current construction in Sacramento is, on the average, lower in quality than that found in the Bay Area. We can provide the required standard of quality.

3. The demand for our particular services can be expected to be high. Firms will find it desirable to obtain a guaranteed maximum cost, or guaranteed lease cost, on which to base their final decision to locate in a new area. Also, a contractor who will accept the full responsibility for design and construction of a facility will be providing an owner, with headquarters elsewhere and no particular expertise in construction, a valuable service.
4. This market represents a logical extension to our existing market in the Bay Area. In addition to providing a growing source of new opportunities with new clients, better service to our current clients will be available through a wider geographical coverage.

The emphasis in our approach to this market should be placed on providing a superior service to the owner, in terms of completeness and guaranteed results, rather than the lowest initial cost.

Inasmuch as we are a union contractor entering a predominantly open-shop area, we cannot be expected to be in a highly competitive position on jobs that are put out to bid. Thus we must offer owners comprehensive proposals to design and construct facilities to their specifications, while emphasizing the advantages of this approach over competitive bidding.

Initial market penetration may best be achieved by contacting our established clients in the Bay Area to determine if any expansion into the Sacramento area is planned on their part in the near future. The eventual market for our services is seen to be the larger chain, or branch, businesses contemplating location in Sacramento.

Please let me know if you should require any additional information or would like to discuss this matter in further detail.

Janis Q. Roberts

JR:dn

How and why to enter a new market.

MEMORANDUM

To: B. C. Gerwick, Jr., Vice President, Operations
From: Saad Abkher
Date: May 29, 1979

Reference is made to your memorandum dated January 30, 1979, concerning the diversification of our firm to work in Saudi Arabia. A project has been selected to be studied and considered.

Project

Bell has signed an S. R. 11 billion contract with the Saudi Arabian government to develop and improve the telecommunication system in the country. Recently, Bell asked contractors to bid on a project for building service stations in thirty-five different locations. The stations are composed of five main structures each:

1. Receiving and transmitting station
2. Computer center

3. Training school
4. Accommodation quarters
5. Recreational facilities (outdoor and indoor)

Bell headquarters are in Toronto; their field project office is in Riyadh, Saudi Arabia.

Our firm is considering the possibility of providing the construction management services for this project. In principle, we are planning to start an advertising campaign to serve the concerned parties. This will include a brief presentation about our past and present work in this field, a description of some successfully completed jobs, and a brief description of our future goals.

Since Bell is a multiheaded organization, we will approach those who are considered the key individuals for this project. The key individuals include but are not limited to:

1. *Project Manager—Riyadh*: He is in charge of evaluating and analyzing all bids received, and he will write recommendations for the committee.
2. *Chief Engineer—Toronto*: He is in charge of the actual design of the project and can help in clearing up some points as well as pointing out different important design elements which are considered sensitive.
3. *Procurement Managers—Toronto and Riyadh*: They are the main people to start with in our advertising campaign, since they are in charge of prequalifications of all contractors and they write recommendations in terms of technical and financial capabilities of bidders.
4. *Vice President, Projects—Toronto*: He is the head of the selection committee. Bids will be evaluated in Riyadh and sent to Toronto together with recommendations for the selection. A brief presentation about our firm will help to introduce us to him.
5. *Resident Engineer—Riyadh*: The resident engineer is the representative of the Saudi Arabian Ministry of Telegraph, Post, and Telephone. He recommends the bidders from the owner's point of view.

Moreover, the engineering department in the Ministry of Telegraph, Post, and Telephone will be contacted through our Saudi sponsor to present our firm as being qualified and legally registered in the country.

Finally, we hope that our approach for contacting the concerned people for the above mentioned project meets your approval, and we will be very glad to discuss any point with you at your convenience.

Saad M. Abkher

Diagram of multiple offices and interrelated clients.

MBA CONSTRUCTION COMPANY

777 Carpenter Avenue
Lathe, California 98765
Telephone: (200) 333-6666
February 13, 1979

Mr. Jerry Mathers
Vice President, Property Management
A to Z Stores
1427 Beaver Boulevard
Dublin, California 94318

Dear Mr. Mathers:

We were very pleased to receive your request for a proposal to design and construct a new freight consolidation center in the Sacramento area for A to Z Stores.

As you may know, MBA Construction Company has successfully completed several similar projects in the recent past. During the course of these projects, we have had the opportunity to gain an insight into some of the special requirements of the retail store business and how we may best serve it. As a result, we are prepared to submit, at this time, a comprehensive proposal to design and construct, including financing, the freight terminal facility with guaranteed results on quality and cost. The advantages of this proposal to A to Z Stores, above and beyond the typical open-ended, cost plus arrangement will be self-apparent.

The key points of our proposal are as follows:

1. We will provide a management team consisting of our trained personnel and real estate broker to assist your people in selecting a final site by providing relative developmental costs of possible sites.
2. Through our contacts with investors, pension funds, and so on we will arrange complete financing for the proposed project, including acquisition of the land.
3. We will provide you, prior to incurring design or construction costs, a guaranteed maximum net lease cost, based upon a 20-year lease. The lease agreement shall also give you the option to purchase the facility after a specific number of years.
4. The facility itself shall be designed and constructed to meet your specific needs.

It has always been our philosophy to offer clients the best possible service that we can provide to meet their special requirements, and to guarantee the results. In this case, our particular expertise on this type of project has enabled us to make a proposal to provide a single source of total responsibility, including financing, for completion of the project, which we believe may be of interest to your company. Should you care to discuss any of the points of our proposal in greater detail, we remain at your service.

Sincerely yours,

Edward Haskell, President

Proposal letter.

WILLIAM PARKER AND ASSOCIATES

60 Long Street
Here, California 99999
Telephone: (444) 555-6666
May 31, 1979

Mr. John Q. Smith, Vice President
Chicago Title Company
348 Allan Boulevard
Chicago, Illinois 12345

Dear Mr. Smith:

We are pleased that we are being considered for the dynamic analysis of your proposed 50-story office building in Chicago. William Parker and Associates have been involved in the dynamic analyses of high-rise buildings since our inception in 1965. Some of the high-rise structures we have been involved in include One Shell Place in New Orleans, the Hilton Hotel in Miami, and the Pacific Bank Building in Seattle. Currently, we are performing a parametric study of a 47-story building subjected to gusty wind for the New York City Bank.

William Parker and Associates, founded in 1965, specializes in dynamic analyses of large structures in which we have been involved include One Shell Place in New Orleans, the Hilton Hotel in Miami, and the Pacific Bank Building in Seattle. Currently, we are performing a parametric study of a 47-story building subjected to gusty wind for the New York City Bank.

The professional staff of William Parker and Associates consists of 43 engineers and 8 draftsmen. Of the 43 engineers, 35 are civil and structural, 5 are mechanical, 2 are naval architects, and 1 is electrical. Biographies of our senior engineers are attached.

The computer progam library at William Parker and Associates consists of programs capable of both linear and nonlinear analyses, finite element analyses, and dynamic analyses of three-dimensional structural systems.

Model testing is subcontracted to one of the testing laboratories listed in an attached sheet.

All of the financing for William Parker and Associates is handled through Crocker National Bank. Attached is a sheet summarizing our financial position. Further questions should be directed to Mr. Frank Johnson at Crocker National Bank, 111 California Street, San Francisco, California 94111.

We look forward to the opportunity of working with your firm on this challenging project.

Very truly yours,

WILLIAM PARKER AND ASSOCIATES

by _____
Mark Adamson, Senior Engineer

Prequalification letter.

MBA CONSTRUCTION COMPANY

777 Carpenter Avenue
Lathe, California 98765
Telephone: (200) 333-6666
March 6, 1979

Mr. Jerry Mathers
Vice President, Property Management
A to Z Stores
1427 Beaver Boulevard
Dublin, California 94318

Dear Mr. Mathers:

   Since our meeting of February 27, 1979, regarding our proposal to design and construct your company's new freight consolidation center on a guaranteed maximum net lease cost basis, we have investigated several potential sources of financing for the project with rather encouraging results.
   In particular, we have located a New York–based life insurance trust fund and a Washington, D.C.–based pension fund, both of whom would be interested in providing construction and permanent financing for your project. They have indicated the availability of funds at rather attractive interest rates in comparison with the high level of current bank short- and long-term interest rates.
   The additional benefits of obtaining this single source of interim and permanent external financing are twofold:

1. Your company's capital will become available to pursue more profitable ventures.
2. You will receive tax advantages generally associated with leasing and still obtain a facility designed and constructed to meet your specific requirements.

   We believe that our comprehensive proposal provides a single source of responsibility, including financing, for the successful completion of your project with guaranteed results on cost, schedule, and quality.
   We look forward to the opportunity to work with your company on this major project and will contact you within the next few days to answer any remaining questions that you may have concerning our proposal.

Very truly yours,

Mark Adamson, President

Response to recommended award to a competitor.

MARINE RESOURCES INCORPORATED

600 Dock Street
Seawall, California 99999
Telephone: (999) 000-5555
March 13, 1979

Mr. John W. Dean, III, Chief Engineer
Bethlehem Steel Corporation
Shipbuilding Division
Pittsburgh, Pennsylvania 55529

Dear Mr. Dean:

Following our conference with you on March 1, it was my understanding that we were to proceed with all due speed toward completion of the design of membrane storage tanks for Exxon's proposed LNG ships. At that time, it was decided to drop the parallel development of a design for spherical tanks so that we could concentrate our resources in an effort to attain a completed design by the deadline of April 15, 1979.

Yesterday, however, I received a call from your project engineer, Mr. Smith, seeking to confirm that we are on schedule for both the membrane tank design and the spherical tank design. At the time I recounted my impression of the decision made at the March 1 meeting, but Mr. Smith reiterated his request for the two alternative designs.

I hope in this instance that you will refer to our exchange of conference follow-up letters (on March 3 and March 6), in which we both state that our company will proceed solely with the design of the membrane tanks. I do not believe that this is an insurmountable problem; in fact, I think we can complete both designs. However, because we stopped development of the spherical tanks after March 1, we will require an extension of time to complete this design, and an increase in fees to compensate for the many extra man-hours required to finish a satisfactory design by April 15.

At this point, our company is prepared to immediately resurrect the spherical tank design so as to minimize time delays, but I think we should have a meeting as soon as possible in order to come to an agreement on how best to handle this change. Please let me know how you wish to proceed in this situation.

Sincerely yours,

Gregg Visineau

GV:pd

Initial claim letter.

HJB GENERAL CONTRACTORS, INC.

7 Alphabet Street
Anytown, California 56789
Telephone: (111) 222-3333
March 30, 1979

Mr. Ben Robertson
Vice President, Real Estate Development
Sonora Development Company, Incorporated
123 B Street
Sonora, California 44444

Dear Mr. Robertson:

Your letter, dated March 14, 1979, is acknowledged. Your points, denying our requests made in our letter, dated March 12, 1979, are well taken when latent subsurface conditions are considered. We wish to point out, however, that your borings, No. 33.10 and No. 34.60, are shown as being located directly on the centerline of the main sewer line right-of-way. Those two borings show stiff, silty clay (unified classification CL) to exist at their locations (stations 33+10 and 34+60), the same as 45 other borings drilled for this project. Actual conditions encountered by us show loose, silty sand (unified classification SM). Please refer to the photographs we sent you with our March 12 letter. They were taken of the excavation at stations 33+10 and 34+60.

We feel that our requests made in our letter, dated March 12, 1979, were not based on latent subsurface conditions, but on changed conditions. The two borings in question, being incorrect, were obviously misleading. Our lump-sum bid was predicated upon the fact that minimum trench shoring would be required and that a daily progress of 300 feet could be made during one eight-hour shift. We had actually exceeded that rate by averaging 400 feet per day before we encountered the unexpected water and silty sand.

We feel that our claim for the required time extension and for reimbursement of the extra expenses is reasonable and certainly justified, in view of the incorrect information which was furnished with the plans and specifications.

We would like to meet with you to settle this issue as soon as possible. We believe that this would be in the best interest of both parties, and facilitate the administration of this important project.

Very truly yours,

Bob Johnson
President

BJ:pd

Second claim letter.

Appendix 3

Conference Problems

The conference problems which follow may be used to practice the presentation of construction and engineering ideas in a conference. During the marketing course that is taught at Berkeley, each student participates in a conference (in groups of six to eight) representing a situation in which we (the students plus the instructor) are all members of a large firm meeting to plan a marketing approach to an important project. This conference is one to two hours in length.

The essentials of the situation are distributed in advance so that each student may prepare to play his role. The conference problems, which follow, are examples of different situations that may be used. It is recommended that a student assume a role along the lines of his specialized interests or experience or planned career. The instructor assumes the role of chairperson.

At Berkeley 20 percent of the student's grade is based on his conference performance. Therefore, there is considerable pressure on the student to do well. Each student is competing with his co-students to get his ideas across during the conference along with the important objective of working in and with a group. A student could just as likely be downgraded for not working in the best interest of the group as for failing to communicate and sell his good ideas to the group. After the conference, each student and the instructor meet to evaluate the student's performance privately.

Problem 1

A. Situation. Your firm is a moderate-sized, local engineer-construction firm doing industrial facilities in the San Francisco Bay Area. You have good local knowledge, contacts, an excellent reputation, and so forth. You have good structural and civil engineers as well as good construction ability for industrial and harbor facilities.

B. Intelligence (heard at a luncheon). Kaiser Fabricators (of Napa, California), a wholly owned subsidiary of Kaiser Steel Corporation, has just received a $100 million contract from Standard Oil of Ohio, a partially owned subsidiary of British Petroleum (United States), to fabricate and assemble complex modules of processing equipment for shipment to the North Slope. It is a crash program.

It has just obtained a lease on 80 acres adjoining Richmond Inner Harbor from Santa Fe Railroad and proposes to set up an assembly facility. All fabricated steel will come from Napa. All equipment will come from various suppliers. Final shipment will be by 6000-ton seagoing barge.

Also overheard: Kaiser Fabricators is thinking about contacting Bechtel to do the engineering and construction management associated with getting the project underway. However, Kaiser is swamped with the problem of organizing itself for this huge crash project. That is why it needs help.

C. Preparation. Your president has called a meeting of the top people in the company to see whether your company can come up with some way of getting a job.

The project is really right in your company's scope.

Demolition of old facilities
Grading and fill
Slope protection along the harbor side
Railroad spur, roads, paving
Utilities, power, lighting
Warehouses
Fencing
Office building
Shops for mechanical and electrical work
Crane tracks
Load-out wharf

Total estimated cost is $20 million.

Your president has already called his friend in the city of Richmond, California, who told him that the facilities are under lease by the city with the condition that Kaiser employ 50 percent minority workers. Kaiser's present plant in Napa is 100 percent union.

You are invited to the meeting one week from now. What can you do to get ready? What ideas do you think can be useful? How will you present them?

Problem 2

We are a general contractor, engaged in heavy-concrete construction, earthwork, and so forth, with a background of sewage treatment plants, pumping stations, and industrial facilities. While last year was a good year, we need business for next year. I (the vice president) have just sent you a note as follows:

I have just learned that a group of investors is backing a salmon ranch facility on the Oregon coast and that they have raised $15 million for construction. Apparently one of the investors is an engineering firm that is also designing the facility.

I made a call to the engineering firm, AMU Engineers, that told me the facility includes rearing ponds and smolt ponds (both of reinforced concrete); an intake structure for saltwater (right out in the surf zone); a flume; roads and grading; freshwater wells; and considerable mechanical piping, pumps, and so forth.

The engineers plan to call for bids from six contractors, when the plans are finished (in about three months), and will have to have completion in one year, as that is when the first fish will be brought to the smolt ponds (about 5 million smolt). Otherwise, they could lose an entire year.

Please come to a meeting at 5 P.M. tomorrow so we can discuss ways to approach this project and hopefully negotiate a contract.

Problem 3

A developer from Texas calls you (a large building contractor) to tell you he has an exciting project for a large restaurant and import shop to be built at Piers 1-5 in front of the Ferry Building in Seattle, Washington.

In subsequent conversations, you and he agree on a general negotiated contract, cost plus fixed fee, for the largest job you have ever had (about $37 million).

Then he tells you he has run into a snag. He has been approached by individuals on the Port Commission, the Planning Commission, and the Building Department (one each). They want "compensation" for the extra work they are doing in helping to process his project. They will accept a special lease on a restaurant site. One would like a new home at Lake Tahoe. He has the lot, and "you are a building contractor with spare materials and labor that will not really cost you anything." To top it all, even the banker (who is arranging the interim loan) says he needs some lumber and timbers for his ranch. He would like a few of the selected timbers from the demolition—perhaps 200 thousand board feet—delivered to his ranch.

What to do? That is the subject of the conference. The cost of any of these matters is not critical at all. You could easily do it from a financial point of view. The gain to you from the project would be just what you need to put your firm on a sound basis.

Problem 4

We are a utility contractor, with extensive background in both mechanical and electrical services in urban areas, including work in city streets, underground, and so forth.

The City of San Francisco Waste Water Treatment plan has been altered, so as to require a considerable section of open cut in place of the original cross-town tunnel. Since the new cut and cover sections run along city streets in developed sections of the city (although generally low-income housing), it is obvious that there will be a lot of utility work.

We know that there will be lots of competition among general contractors but very little among subs. Our "competition" is that the general contractor will do the utility work himself, rather than sub it out. How do we bid as a subcontractor; how do we set it up so that we can get a job?

Problem 5

WXYZ Engineers has a contract to design an extension of BARTD to the San Francisco Airport. It is to be entirely overhead viaduct, a very large job.

You are a prestressed concrete manufacturer. You have come up with a different design for girders than was used on the first BARTD work; yours eliminates the interior diaphragms and uses pretensioning instead of post-tensioning.

If you take your idea to WXYZ, it may accept and incorporate it into the plans, which means that all your competitors will have an equal opportunity to bid on your idea.

If you don't take it to WXYZ, then you have to bid on the old design and, if successful, have to build it that way. Even if there is a value-engineering clause, experience with WXYZ in the past shows that it demands essentially all the savings. (Theoretically, only half but, in practice, it has proved to be an unreasonably hard negotiator). Besides, there may not be time to run all the design checks for a value-engineering change. What to do?

Problem 6

A large shopping center is proposed at Fifth and Market Streets in San Francisco.

Sheldon Gordon, a Los Angeles shopping center entrepreneur, is arranging a deal with Bullock's to take over J. C. Penney's empty building on the southwest corner of Fifth and Market. (Bullock's is owned by Cincinnati-based Federated Department Stores, Inc.) It will be demolished and a new building will be erected. Then he will build a shopping mall on the southeast corner of Fifth and Market to connect with the Emporium. (Carter-Hawley-Hales owns the Emporium and will help finance the project.) There will be a pedestrian bridge over Fifth Street. The San Francisco School District owns the building on the southeast corner and is anxious to increase its revenue because of Proposition 13.

Our firm is a construction management firm, with in-house design capabilities. We do not have large experience in department stores; our previous work has concentrated on integrated hospitals with complex mechanical-electrical systems. We originated the idea of designing hospitals so as to minimize the costs of modification as new systems were developed. We also specialized in minimizing the use of nurses and orderlies (that is, maximizing their ability to control large areas through TV and so forth). Should we consider bidding on this work?

Problem 7

Precast concrete architectural panels are being specified for a number of the new buildings being constructed in San Francisco. Up to now, all large buildings have been done by Basalt Rock Company which has done a good job. Some of the smaller jobs have been carried out by two or three small firms.

You have a large precast concrete plant at Pleasanton, California, and need work. You believe, with good reason, that Basalt's price is high and will be even higher, due to the surge in demand.

So here is an opportunity. However, you have never made architectural panels before. One job is particularly attractive since the owner is a past client on other types of work. How do we go after this job?

Problem 8

Pacific Coast Engineering Company (Paceco) is a 50-year-old firm in Alameda, California specializing in heavy-machinery fabrication with a substantial mechanical engineering input. It was a world leader in specialized dredges, then in container cranes. Its plant is efficient but small, constricted, and old.

It has just received an order for 60 container cranes for Saudi Arabia to be delivered over the next five years. You have learned from the Port of Oakland that Paceco has taken an option on 20 acres of Port of Oakland land, across the estuary and downstream 1 mile. Your informant tells you that Paceco has applied for a U.S. loan based on setting up a plant in an area of high minority unemployment and will build a new plant on a crash basis.

It will start engineering and part fabrication in its old plant, move to the new plant in about 12 months, and complete the move in 24. Also, it is discussing within the company how to go about this—its own people are over their heads with current work plus the new project. One obligation of the government loan is that 50 percent of the employees at the new plant must be minority workers. The present plant is unionized.

The new plant project includes demolition, grading, paving, utilities, fabrication buildings, machinery procurement and installation, railroad and crane trackage, a wharf for barge shipments, mechanical and electrical, and an office building. Total estimated cost is $20 million.

Our company has engineering and construction capabilities and construction management. We hear Paceco is considering Kaiser and Atkinson as potential turnkey contractors. We're only a fraction of their size, with no particular political clout, but we need the work badly. How can we get this one?

Problem 9

"Marine World"—the recreation park in Belmont, California, originally built in the mud flats—has become a very successful amusement facility. It proposes to build a monorail of prestressed concrete, covering a total length of 3 mi, all elevated to 30–50 ft above ground, and almost all on curves. They want spans of about 80 ft minimum, 100 ft preferred.

Alweg, which will provide the cars and operate the system, demands extremely close tolerances ($\pm 1/8$ in. on surfaces) and minimum camber/deflection, that is, control creep to an absolute minimum. It will pay for quality if they are assured of it. They want a full package: design and construct.

I, as vice president of the XYZ Prestressed Concrete Manufacturing Company, have written you a note, telling you of this new project and asking you to meet with me at 5 P.M. We need this job, but it is something we have never tackled before, either as to quality or scope.

Problem 10

The Dow Chemical Company tried several years ago to build a petrochemical plant near Rio Vista, California, on the Sacramento River. It ran into so many obstructions from environmentalists that it gave up on the project. Now the climate has changed. The state has apparently indicated to Dow that perhaps things could be worked out. Dow has announced that it will revive the project.

Your firm is a large grading contractor located in Antioch. You have done all the earthwork for the Humble Refinery in Benicia.

Your first contact with Dow has indicated that if the project is approved, it will be a rush job. Dow plans to invite proposals from C. F. Braun, Bechtel, and Parsons.

What can you do to get the $3 million grading job?

Problem 11

Your firm has entered into the cooling tower market, furnishing and erecting a completely prefabricated, preengineered cooling tower for power plants, both fossil-fueled and nuclear. You have the license for a German system and have successfully built one such cooling tower in eastern Washington. Now you have found a wide-open market ahead. Every plant around the country needs to install a cooling tower. The only problem is that the conditions differ!

One company has offered you two contracts at your asking price, and you signed quickly. Now you have discovered that one of the plants will use a saltwater spray, instead of freshwater, and your sales engineer verbally agreed to accept this, saying the concrete would be durable for 30 years. As you get into the contract, you discover that the concrete aggregates, being supplied by the owner, have questionable durability. Further, your design has never been applied to saltwater.

You discuss this with the owner, who takes the hard-nosed attitude that you have promised a fully durable design for his conditions. He is unwilling to pay any more for imported aggregates or special treatments.

To do what we think is proper (conservative) would cost us our full estimated profit on this one job. What do we do?

Appendix 4

Presentation Exercise for Sales Calls

During the quarter, each student is required to make two 15-minute sales calls on the instructor. It is normally best that the student assume the role of representative of a firm or company along the lines of his specialized interests or experience or his planned career.

The day before the exercise, the student sets the stage by providing the instructor with a one-page explanation of what assumed firm or company he represents and what role the instructor shoud enact. For example, the student might represent a structural engineering firm (if he is a structural engineer), and the instructor might be asked to assume the role of a partner in an architectural firm. Also, the stage could be set where the representative of the structural engineering firm is trying to sell his firm's services for a specific project that the architectural firm has just been awarded.

The next day, the student comes to the instructor's office to make the call. The instructor portrays the role set the day before and assumes they have never met. In the sales call presentation we try to create the real-life pressures or stress an individual must face in various situations. For instance, a representative must walk into the office of an owner he has never met, there is a lot of competition to get the work, and the company he represents badly needs work.

At Berkeley, 20 percent of the student's grade is based on his sales call presentations. As in the conference problem, there is considerable pressure on the student to do well. Each student is competing with his fellow students. If the student is nervous, young, or maybe a foreign student, he might find the instructor very cooperative in his role and helpful in getting the student through the situation. However, if the student is mature or has some work experience, the instructor is quite likely to pull some of the "dirty tricks" that are used in too many real-life situations. For example, the telephone might ring, and the instructor will answer the telephone just as the student is making a key point, talk for five minutes, and proceed to forget what was said before. Or the instructor might announce he has another meeting in five minutes and does not have much time to spend with the student and to hurry it up. Perhaps the instructor will simply tell the student he has already chosen another consultant for the services to see how the student reacts. Possibly to keep the student off balance and nervous, the instructor will not be agreeable but will be like an owner in real life who has 10 other people interested in providing this service. He might very well ask the question, "Well, why should we give the job to your company as opposed to the other 10 competitors?" The student should count on some thing(s) happening that he is not expecting. Also, he should expect to be asked at least one hard question about his services which he is unlikely to be able to answer. It is often said that young engineers are very poor in sales because they do not listen to the client's questions and are not willing to give the client what he is asking for. They will simply go blindly or straight ahead with the ideas they want to present and ignore the questions from the client. One of the purposes of the sales call presentation is to train students to overcome this weakness. Very often, the student will find the instructor not interested at all in his services and will need to find some way to leave on a friendly basis and keep the door open to make the call at another time. At the end of the call the student shakes hands, says goodbye, leaves the door open to make a second call, and leaves the room.

A few minutes later the student is called back into the room and his presentation is evaluated. If he was squirming around in the chair, appearing tense, he will be told so. If he was too relaxed and played with the papers on the client's desk, he will be told so. The good, along with the weak, points of his presentation will be discussed. He is also judged on the technical aspects of his presentation. If, for example, he was asked what he thought of a certain competitor he will be evaluated on the way in which he responded to that question.

"A CONSTRUCTION PLAY"
by an unknown playwright

The actors:

Mr. Dyno—president of the Dynopile Company.
Mr. Crandall—manager of South Georgia Edison Power.

The play:

Mr. Crandall is holding a board meeting in his office; the door bursts open and in strides Mr. Dyno.

Dyno—Excuse me for interrupting you, Crandall, but I have to talk to you before you proceed further with this meeting.

Crandall—What! What are you doing here? This is a closed board meeting.

Dyno—This is important! I got up at five o'clock to catch a flight down here. I know you are having a board meeting but this is urgent. I can only stay a few minutes: I have to fly to New York today.

CRANDALL—Good (*muttered*).

DYNO—Now Ed called me last night and told me about the terrible mistake you are about to make. Here you are, about to award the piling contract to that little contractor in San Francisco—what's his name? Ace Pile Company. Ed called me last night . . ."

CRANDALL—Ed? Our Ed Looselips?

DYNO (*taking out some pictures*)—Look at these pictures of broken piles I took on Ace Pile's last job. They are a terrible contractor. Look, Crandall, we have done all your work for 20 years. We have been helping . . .

CRANDALL (*breaking in*)—Oh, you must be Mr. Dyno.

DYNO—Yes, I'm Dyno, president of Dynopile, largest pile-driving contractor in the West. Your awarding this job to Ace Pile is just not fair; it's not playing square with us, after all the work we've done for you in the past years. Why, you didn't even tell us there was going to be competition this time.

CRANDALL—Now I remember this case. You were second bidder, and as I recall, your bid was grossly unbalanced. Why was your bid so unbalanced? You bid one cent per foot to furnish the piles and $1000 each to drive them.

DYNO—Oh, that, that's technical; it has to do with soil mechanics. You wouldn't understand. Anyway, this is just a little nuisance job.

Look, I am absolutely sincere. If we don't get this job, we won't do business with you again. Ace Pile is too small to help you; they'll probably go bankrupt or move away. On your next job we are going to be too busy. We have other clients, clients who work with us and honor their long-term relationships.

CRANDALL—Would you please leave your name and number with my secretary on your way out? The door is over there.

The play review: How many errors in salesmanship did Mr. Dyno make? Mr. Dyno is truly remarkable. He not only interrupted the client but also was arrogant and talked down to the client. He ran down the competition—this is unprofessional and bad salesmanship. He used name dropping and, in the process, revealed a confidential source of information (Ed Looselips). Mr. Dyno also stated he was not interested in this "little nuisance job." Finally, Mr. Dyno threatened Mr. Crandall. Mr. Dyno is truly remarkable.

In a subsequent development, one year later, Dynopile and Ace Pile form a joint venture, and Mr. Dyno, undaunted, appears before the same Mr. Crandall to argue that "we are the two best piling companies in the West."

Appendix 5

Sample Contracts

THE AMERICAN INSTITUTE OF ARCHITECTS

AIA Document A101

Standard Form of Agreement Between Owner and Contractor

where the basis of payment is a
STIPULATED SUM

1977 EDITION

THIS DOCUMENT HAS IMPORTANT LEGAL CONSEQUENCES; CONSULTATION WITH
AN ATTORNEY IS ENCOURAGED WITH RESPECT TO ITS COMPLETION OR MODIFICATION

Use only with the 1976 Edition of AIA Document A201, General Conditions of the Contract for Construction.

This document has been approved and endorsed by The Associated General Contractors of America.

AGREEMENT

made as of the   day of   in the year of Nineteen Hundred and

BETWEEN the Owner:

and the Contractor:

The Project:

FIGURE A5.1. American Institute of Architects (AIA) Document A101

The Architect:

> AIA copyrighted material has been reproduced with permission of the American Institute of Architects under permission number 81067. Further reproduction is prohibited.
>
> Because AIA Documents are revised from time to time, users should ascertain from the AIA the current edition(s) of the Document(s) reproduced herein.

The Owner and the Contractor agree as set forth below.

Copyright 1915, 1918, 1925, 1937, 1951, 1958, 1961, 1963, 1967, 1974, © 1977 by the American Institute of Architects, 1735 New York Avenue, N.W., Washington, D. C. 20006. Reproduction of the material herein or substantial quotation of its provisions without permission of the AIA violates the copyright laws of the United States and will be subject to legal prosecution.

AIA DOCUMENT A101 • OWNER-CONTRACTOR AGREEMENT • ELEVENTH EDITION • JUNE 1977 • AIA®
©1977 • THE AMERICAN INSTITUTE OF ARCHITECTS, 1735 NEW YORK AVE., N.W., WASHINGTON, D. C. 20006   A101-1977

ARTICLE 1

THE CONTRACT DOCUMENTS

The Contract Documents consist of this Agreement, the Conditions of the Contract (General, Supplementary and other Conditions), the Drawings, the Specifications, all Addenda issued prior to and all Modifications issued after execution of this Agreement. These form the Contract, and all are as fully a part of the Contract as if attached to this Agreement or repeated herein. An enumeration of the Contract Documents appears in Article 7.

ARTICLE 2

THE WORK

The Contractor shall perform all the Work required by the Contract Documents for
(Here insert the caption descriptive of the Work as used on other Contract Documents.)

ARTICLE 3

TIME OF COMMENCEMENT AND SUBSTANTIAL COMPLETION

The Work to be performed under this Contract shall be commenced

and, subject to authorized adjustments, Substantial Completion shall be achieved not later than

(Here insert any special provisions for liquidated damages relating to failure to complete on time.)

AIA DOCUMENT A101 • OWNER-CONTRACTOR AGREEMENT • ELEVENTH EDITION • JUNE 1977 • AIA®
©1977 • THE AMERICAN INSTITUTE OF ARCHITECTS, 1735 NEW YORK AVE., N.W., WASHINGTON, D. C. 20006   A101-1977

FIGURE A5.1. *(continued)*

ARTICLE 4

CONTRACT SUM

The Owner shall pay the Contractor in current funds for the performance of the Work, subject to additions and deductions by Change Order as provided in the Contract Documents, the Contract Sum of

The Contract Sum is determined as follows:
(State here the base bid or other lump sum amount, accepted alternates, and unit prices, as applicable.)

ARTICLE 5

PROGRESS PAYMENTS

Based upon Applications for Payment submitted to the Architect by the Contractor and Certificates for Payment issued by the Architect, the Owner shall make progress payments on account of the Contract Sum to the Contractor as provided in the Contract Documents for the period ending the   day of the month as follows:

Not later than   days following the end of the period covered by the Application for Payment   percent (%) of the portion of the Contract Sum properly allocable to labor, materials and equipment incorporated in the Work and   percent (%) of the portion of the Contract Sum properly allocable to materials and equipment suitably stored at the site or at some other location agreed upon in writing, for the period covered by the Application for Payment, less the aggregate of previous payments made by the Owner; and upon Substantial Completion of the entire Work, a sum sufficient to increase the total payments to   percent (%) of the Contract Sum, less such amounts as the Architect shall determine for all incomplete Work and unsettled claims as provided in the Contract Documents.

(If not covered elsewhere in the Contract Documents, here insert any provision for limiting or reducing the amount retained after the Work reaches a certain stage of completion.)

Payments due and unpaid under the Contract Documents shall bear interest from the date payment is due at the rate entered below, or in the absence thereof, at the legal rate prevailing at the place of the Project.
(Here insert any rate of interest agreed upon.)

(Usury laws and requirements under the Federal Truth in Lending Act, similar state and local consumer credit laws and other regulations at the Owner's and Contractor's principal places of business, the location of the Project and elsewhere may affect the validity of this provision. Specific legal advice should be obtained with respect to deletion, modification, or other requirements such as written disclosures or waivers.)

AIA DOCUMENT A101 • OWNER-CONTRACTOR AGREEMENT • ELEVENTH EDITION • JUNE 1977 • AIA®
©1977 • THE AMERICAN INSTITUTE OF ARCHITECTS, 1735 NEW YORK AVE., N.W., WASHINGTON, D. C. 20006   A101-1977

FIGURE A5.1. *(continued)*

ARTICLE 6
FINAL PAYMENT

Final payment, constituting the entire unpaid balance of the Contract Sum, shall be paid by the Owner to the Contractor when the Work has been completed, the Contract fully performed, and a final Certificate for Payment has been issued by the Architect.

ARTICLE 7
MISCELLANEOUS PROVISIONS

7.1 Terms used in this Agreement which are defined in the Conditions of the Contract shall have the meanings designated in those Conditions.

7.2 The Contract Documents, which constitute the entire agreement between the Owner and the Contractor, are listed in Article 1 and, except for Modifications issued after execution of this Agreement, are enumerated as follows:
(List below the Agreement, the Conditions of the Contract (General, Supplementary, and other Conditions), the Drawings, the Specifications, and any Addenda and accepted alternates, showing page or sheet numbers in all cases and dates where applicable.)

This Agreement entered into as of the day and year first written above.

OWNER   CONTRACTOR

_____   _____

_____   _____

_____   _____

AIA DOCUMENT A101 • OWNER-CONTRACTOR AGREEMENT • ELEVENTH EDITION • JUNE 1977 • AIA®   A101-1977
©1977 • THE AMERICAN INSTITUTE OF ARCHITECTS, 1735 NEW YORK AVE., N.W., WASHINGTON, D. C. 20006

FIGURE A5.1. *(continued)*

THE AMERICAN INSTITUTE OF ARCHITECTS

AIA Document A201

General Conditions of the Contract for Construction

THIS DOCUMENT HAS IMPORTANT LEGAL CONSEQUENCES; CONSULTATION WITH AN ATTORNEY IS ENCOURAGED WITH RESPECT TO ITS MODIFICATION

1976 EDITION
TABLE OF ARTICLES

1. CONTRACT DOCUMENTS
2. ARCHITECT
3. OWNER
4. CONTRACTOR
5. SUBCONTRACTORS
6. WORK BY OWNER OR BY SEPARATE CONTRACTORS
7. MISCELLANEOUS PROVISIONS
8. TIME
9. PAYMENTS AND COMPLETION
10. PROTECTION OF PERSONS AND PROPERTY
11. INSURANCE
12. CHANGES IN THE WORK
13. UNCOVERING AND CORRECTION OF WORK
14. TERMINATION OF THE CONTRACT

This document has been approved and endorsed by The Associated General Contractors of America.

Copyright 1911, 1915, 1918, 1925, 1937, 1951, 1958, 1961, 1963, 1966, 1967, 1970, © 1976 by The American Institute of Architects, 1735 New York Avenue, N.W., Washington, D. C. 20006. Reproduction of the material herein or substantial quotation of its provisions without permission of the AIA violates the copyright laws of the United States and will be subject to legal prosecution.

> AIA copyrighted material has been reproduced with permission of the American Institute of Architects under permission number 81067. Further reproduction is prohibited.
>
> Because AIA Documents are revised from time to time, users should ascertain from the AIA the current edition(s) of the Document(s) reproduced herein.

AIA DOCUMENT A201 • GENERAL CONDITIONS OF THE CONTRACT FOR CONSTRUCTION • THIRTEENTH EDITION • AUGUST 1976
AIA® • © 1976 • THE AMERICAN INSTITUTE OF ARCHITECTS, 1735 NEW YORK AVENUE, N.W., WASHINGTON, D.C. 20006   **A201-1976**

FIGURE A5.2. American Institute of Architects (AIA) Document A201

INDEX

Acceptance of Defective or Non-Conforming Work . . 6.2.2, **13.3**
Acceptance of Work 5.4.2, 9.5.5, 9.8.1, 9.9.1, 9.9.3
Access to Work 2.2.5, 6.2.1
Accident Prevention 2.2.4, 10
Acts and Omissions 2.2.4, 4.18.3, 7.4, 7.6.2, 8.3.1, 10.2.5
Additional Costs, Claims for 12.3
Administration of the Contract **2.2**, 4.3.3
All Risk Insurance 11.3.1
Allowances **4.8**
Applications for Payment 2.2.6, 9.2, **9.3**, 9.4,
9.5.3, 9.6.1, 9.8.2, 9.9.1, 9.9.3, 14.2.2
Approvals . . . . 2.2.14, 3.4, 4.3.3, 4.5, 4.12.4 through
4.12.6, 4.12.8, 4.18.3, 7.7, 9.3.2
Arbitration . . . . 2.2.7 through 2.2.13, 2.2.19, 6.2.5,
7.9, 8.3.1, 11.3.7, 11.3.8
ARCHITECT **2**
Architect, **Definition** of **2.1**
Architect, Extent of Authority 2.2, 3.4, 4.12.8, 5.2, 6.3, 7.7.2,
8.1.3, 8.3.1, 9.2, 9.3.1, 9.4, 9.5.3, 9.6, 9.8, 9.9.1, 9.9.3, 12.1.1,
12.1.4, 12.3.1, 12.4.1, 13.1, 13.2.1, 13.2.5, 14.2
Architect, Limitations of Authority and Responsibility 2.2.2
through 2.2.4, 2.2.10 through 2.2.14, 2.2.17, 2.2.18,
4.3.3, 4.12.6, 5.2.1, 9.4.2, 9.5.4, 9.5.5, 12.4
Architect's Additional Services . . 3.4, 7.7.2, 13.2.1, 13.2.5, 14.2.2
Architect's Approvals 2.2.14, 3.4, 4.5, 4.12.6, 4.12.8, 4.18.3
Architect's Authority to Reject Work 2.2.13, 4.5, 13.1.2, 13.2
Architect's Copyright 1.3
Architect's Decisions . . . . 2.2.7 through 2.2.13, 6.3, 7.7.2,
7.9.1, 8.3.1, 9.2, 9.4, 9.6.1, 9.8.1, 12.1.4, 12.3.1
Architect's Inspections 2.2.13, 2.2.16, 9.8.1, 9.9.1
Architect's Instructions 2.2.13, 2.2.15, 7.7.2, 12.4, 13.1
Architect's Interpretations . . . . 2.2.7 through 2.2.10, 12.3.2
Architect's On-Site Observations 2.2.3, 2.2.5, 2.2.6, 2.2.17,
7.7.1, 7.7.4, 9.4.2, 9.6.1, 9.9.1
Architect's Project Representative 2.2.17, 2.2.18
Architect's Relationship with Contractor 1.1.2, 2.2.4, 2.2.5,
2.2.10, 2.2.13, 4.3.3, 4.5, 4.7.3, 4.12.6, 4.18, 11.3.6
Architect's Relationship with
Subcontractors 1.1.2, 2.2.13, 9.5.3, 9.5.4
Architect's Representations 9.4.2, 9.6.1, 9.9.1
Artistic Effect 1.2.3, 2.2.11, 2.2.12, 7.9.1
Attorneys' Fees 4.18.1, 6.2.5, 9.9.2
Award of Separate Contracts 6.1.1
Award of Subcontracts and Other Contracts for
Portions of the Work **5.2**
Bonds, Lien 9.9.2
Bonds, Performance, Labor and Material Payment 7.5, 9.9.3
Building Permit 4.7
Certificate of Substantial Completion 9.8.1
Certificates of Inspection, Testing or Approval 7.7.3
Certificates of Insurance 9.3.2, 11.1.4
Certificates for Payment 2.2.6, 2.2.16, **9.4**, 9.5.1, 9.5.5, 9.6.1,
9.7.1, 9.8.2, 9.9.1, 9.9.3, 12.1.4, 14.2.2
Change Orders 1.1.1, 2.2.15, 3.4, 4.8.2.3, 5.2.3, 7.7.2,
8.3.1, 9.7, 9.9.3, 11.3.1, 11.3.5, 11.3.7,
12.1, 13.1.2, 13.2.5, 13.3.1
Change Orders, Definition of 12.1.1
CHANGES IN THE WORK 2.2.15, 4.1.1, **12**
Claims for Additional Cost or Time 8.3.2, 8.3.3, 12.2.1, **12.3**
Claims for Damages 6.1.1, 6.2.5, **7.4**, 8.3, 9.6.1.1

Cleaning Up **4.15**, 6.3
Commencement of the Work, Conditions Relating to . . 3.2.1, 4.2,
4.7.1, 4.10, 5.2.1, 6.2.2, 7.5, 9.2, 11.1.4, 11.3.4
Commencement of the Work, Definition of 8.1.2
Communications 2.2.2, 3.2.6, 4.9.1, **4.16**
Completion,
Conditions Relating to . . . 2.2.16, 4.11, 4.15, 9.4.2, 9.9, 13.2.2
COMPLETION, PAYMENTS AND **9**
Completion, Substantial . . . 2.2.16, 8.1.1, 8.1.3, 8.2.2, 9.8, 13.2.2
Compliance with Laws 1.3, 2.1.1, 4.6, 4.7, 4.13,
7.1, 7.7, 10.2.2, 14
Concealed Conditions **12.2**
Consent,
Written . . . 2.2.18, 4.14.2, 7.2, 7.6.2, 9.8.1, 9.9.2, 9.9.3, 11.3.9
Contract, Definition of **1.1.2**
Contract Administration 2.2, 4.3.3
Contract Award and Execution, Conditions
Relating to 4.7.1, 4.10, 5.2, 7.5, 9.2, 11.1.4, 11.3.4
CONTRACT DOCUMENTS **1**
Contract Documents,
Copies Furnished and Use of 1.3, 3.2.5, 5.3
Contract Documents, Definition of **1.1.1**
Contract Sum, Definition of **9.1.1**
Contract Termination 14
Contract Time, Definition of 8.1.1
CONTRACTOR **4**
Contractor, **Definition** of **4.1**, 6.1.2
Contractor's Employees 4.3.2, 4.4.2, 4.8.1, 4.9, 4.18, 10.2.1
through 10.2.4, 10.2.6, 10.3, 11.1.1
Contractor's Liability Insurance **11.1**
Contractor's Relationship with
Separate Contractors and Owner's Forces 3.2.7, 6
Contractor's Relationship with
Subcontractors 1.2.4, 5.2, 5.3, 9.5.2, 11.3.3, 11.3.6
Contractor's Relationship with the Architect 1.1.2, 2.2.4,
2.2.5, 2.2.10, 2.2.13, 4.3.3, 4.5, 4.7.3, 4.12.6, 4.18, 11.3.6
Contractor's Representations 1.2.2, 4.5, 4.12.5, 9.3.3
Contractor's Responsibility for
Those Performing the Work 4.3.2, 4.18, 10
Contractor's Review of Contract Documents 1.2.2, 4.2, 4.7.3
Contractor's Right to Stop the Work 9.7
Contractor's Right to Terminate the Contract 14.1
Contractor's Submittals 2.2.14, 4.10, 4.12.5, 5.2.1,
5.2.3, 9.2, 9.3.1, 9.8.1, 9.9.2, 9.9.3
Contractor's Superintendent 4.9, 10.2.6
Contractor's Supervision and
Construction Procedures 1.2.4, 2.2.4, 4.3, 4.4, 10
Contractual Liability Insurance 11.1.3
Coordination and
Correlation 1.2.2, 1.2.4, 4.3.1, 4.10.1, 4.12.5, 6.1.3, 6.2.1
Copies Furnished of Drawings and Specifications . . 1.3, 3.2.5, 5.3
Correction of Work 3.3, 3.4, 10.2.5, **13.2**
Cost, Definition of 12.1.4
Costs 3.4, 4.8.2, 4.15.2, 5.2.3, 6.1.1, 6.2.3, 6.2.5, 6.3, 7.7.1,
7.7.2, 9.7, 11.3.1, 11.3.5, 12.1.3, 12.1.4, 12.3, 13.1.2, 13.2, 14
Cutting and Patching of Work **4.14**, 6.2
Damage to the Work 6.2.4, 6.2.5, 9.6.1.5, 9.8.1,
10.2.1.2, 10.3, 11.3, 13.2.6
Damages, Claims for 6.1.1, 6.2.5, 7.4, 8.3.4, 9.6.1.2
Damages for Delay 6.1.1, 8.3.4, 9.7
Day, Definition of 8.1.4

FIGURE A5.2. *(continued)*

Decisions of the Architect ......2.2.9 through 2.2.12, 6.3, 7.7.2,
            7.9.1, 8.3.1, 9.2, 9.4, 9.6.1, 9.8.1, 12.1.4, 12.3.1, 14.2.1
Defective or Non-Conforming Work, Acceptance, Rejection
      and Correction of 2.2.3, 2.2.13, 3.3, 3.4, 4.5, 6.2.2, 6.2.3,
                                                 9.6.1.1, 9.9.4.2, 13
Definitions ....**1.1, 2.1, 3.1, 4.1,** 4.12.1 through 4.12.3, **5.1,**
                                         6.1.2, **8.1,** 9.1.1, 12.1.1, 12.1.4
Delays and Extensions of Time **8.3**
Disputes 2.2.9, 2.2.12, 2.2.19, 6.2.5, 6.3, 7.9.1
Documents and Samples at the Site **4.11**
Drawings and Specifications, Use and
      Ownership of 1.1.1, 1.3, 3.2.5, 5.3
Emergencies **10.3**
Employees, Contractor's 4.3.2, 4.4.2, 4.8.1, 4.9, 4.18, 10.2.1
                                           through 10.2.4, 10.2.6, 10.3, 11.1.1
Equipment, Labor, Materials and 1.1.3, 4.4, 4.5, 4.12, 4.13,
                              4.15.1, 6.2.1, 9.3.2, 9.3.3, 11.3, 13.2.2, 13.2.5, 14
Execution and Progress of the Work 1.1.3, 1.2.3, 2.2.3, 2.2.4,
                                             2.2.8, 4.2, 4.4.1, 4.5, 6.2.2, 7.9.3, 8.2,
                                                   8.3, 9.6.1, 10.2.3, 10.2.4, 14.2
Execution, Correlation and Intent of the
      Contract Documents **1.2,** 4.7.1
Extensions of Time 8.3, 12.1
Failure of Payment by Owner **9.7,** 14.1
Failure of Payment of Subcontractors ..9.5.2, 9.6.1.3, 9.9.2, 14.2.1
Final Completion and Final Payment ..2.2.12, 2.2.16, **9.9,** 13.3.1
Financial Arrangements, Owner's 3.2.1
Fire and Extended Coverage Insurance 11.3.1
Governing Law **7.1**
Guarantees (See Warranty
      and Warranties) 2.2.16, 4.5, 9.3.3, 9.8.1, 9.9.4, 13.2.2
Indemnification 4.17, **4.18,** 6.2.5, 9.9.2
Identification of Contract Documents 1.2.1
Identification of Subcontractors and Suppliers 5.2.1
**Information and
      Services Required of the Owner** **3.2,** 6, 9, 11.2, 11.3
Inspections 2.2.13, 2.2.16, 4.3.3, 7.7, 9.8.1, 9.9.1
Instructions to Bidders 1.1.1, 7.5
Instructions to the
      Contractor 2.2.2, 3.2.6, 4.8.1, 7.7.2, 12.1.2, 12.1.4
INSURANCE **9.8.1,** 11
Insurance, Contractor's Liability 11.1
Insurance, Loss of Use 11.4
Insurance, Owner's Liability 11.2
Insurance, Property 11.3
Insurance, Stored Materials 9.3.2, 11.3.1
Insurance Companies, Consent to Partial Occupancy 11.3.9
Insurance Companies, Settlement With 11.3.8
Intent of
      the Contract Documents ...1.2.3, 2.2.10, 2.2.13, 2.2.14, 12.4
Interest **7.8**
Interpretations, Written 1.1.1, 2.2.7, 2.2.8, 2.2.10, 12.4
Labor and Materials, Equipment 1.1.3, **4.4,** 4.5, 4.12, 4.13,
                              4.15.1, 6.2.1, 9.3.2, 9.3.3, 11.3, 13.2.2, 13.2.5, 14
Labor and Material Payment Bond 7.5
Labor Disputes 8.3.1
Laws and Regulations 1.3, 2.1., 4.6, 4.7, 4.13, 7.1,
                                                            7.7, 10.2.2, 14
Liens 9.3.3, 9.9.2, 9.9.4.1
Limitations of Authority 2.2.2, 2.2.17, 2.2.18, 11.3.8, 12.4.1
Limitations of Liability 2.2.10, 2.2.13, 2.2.14, 3.3, 4.2, 4.7.3,

                                 4.12.6, 4.17, 4.18.3, 6.2.2, 7.6.2, 9.4.2,
                                    9.9.4, 9.9.5, 10.2.5, 11.1.2, 11.3.6
Limitations of Time, General 2.2.8, 2.2.14, 3.2.4, 4.2, 4.7.3,
                              4.12.4, 4.15, 5.2.1, 5.2.3, 7.4, 7.7, 8.2, 9.5.2, 9.6,
                              9.8, 9.9, 11.3.4, 12.1.4, 12.4, 13.2.1, 13.2.2, 13.2.5
Limitations of Time, Specific 2.2.8, 2.2.12, 3.2.1, 3.4,
                              4.10, 5.3, 6.2.2, 7.9.2, 8.2, 8.3.2, 8.3.3, 9.2, 9.3.1, 9.4.1, 9.5.1,
                                9.7, 11.1.4, 11.3.1, 11.3.8, 11.3.9, 12.2, 12.3, 13.2.2,
                                                13.2.5, 13.2.7, 14.1, 14.2.1
Limitations, Statutes of 7.9.2, 13.2.2, 13.2.7
Loss of Use Insurance **11.4**
Materials, Labor, Equipment and 1.1.3, 4.4, 4.5, 4.12, 4.13,
                              4.15.1, 6.2.1, 9.3.2, 9.3.3, 11.3.1, 13.2.2, 13.2.5, 14
Materials Suppliers 4.12.1, 5.2.1, 9.3.3
Means, Methods, Techniques, Sequences and
      Procedures of Construction 2.2.4, 4.3.1, 9.4.2
Minor Changes in the Work 1.1.1, 2.2.15, **12.4**
MISCELLANEOUS PROVISIONS **7**
Modifications, Definition of 1.1.1
Modifications to the Contract 1.1.1, 1.1.2, 2.2.2, 2.2.18,
                                                               4.7.3, 7.9.3, 12
Mutual Responsibility **6.2**
Non-Conforming Work, Acceptance of Defective or 13.3.1
Notice, Written 2.2.8, 2.2.12, 3.4, 4.2, 4.7.3, 4.7.4, 4.9,
                              4.12.6, 4.12.7, 4.17, 5.2.1, 7.3, 7.4, 7.7, 7.9.2, 8.1.2, 8.3.2,
                                8.3.3, 9.4.1, 9.6.1, 9.7, 9.9.1, 9.9.5, 10.2.6, 11.1.4, 11.3.1,
                                   11.3.4, 11.3.5, 11.3.7, 11.3.8, 12.2, 12.3, 13.2.2, 13.2.5, 14
Notices, Permits, Fees and 4.7, 10.2.2
Notice of Testing and Inspections 7.7
Notice to Proceed 8.1.2
Observations, Architect's On-Site 2.2.3, 7.7.1, 7.7.4, 9.4.2
Observations, Contractor's 1.2.2, 4.2.1, 4.7.3
Occupancy 8.1.3, 9.5.5, 11.3.9
On-Site Inspections by the Architect 2.2.3, 2.2.16, 9.4.2,
                                                                9.8.1, 9.9.1
On-Site Observations by the Architect 2.2.3, 2.2.6, 2.2.17,
                                                  7.7.1, 7.7.4, 9.4.2, 9.6.1, 9.9.1
Orders, Written 3.3, 4.9, 12.1.4, 12.4.1, 13.1
OWNER **3**
Owner, **Definition** of **3.1**
Owner, Information and Services Required of the 3.2, 6.1.3,
                                                               6.2, 9, 11.2, 11.3
Owner's Authority 2.2.16, 4.8.1, 7.7.2, 9.3.1, 9.3.2,
                                                       9.8.1, 11.3.8, 12.1.2, 12.1.4
Owner's Financial Capability 3.2.1
Owner's Liability Insurance **11.2**
Owner's Relationship with Subcontractors 1.1.2, 9.5.4
Owner's Right to Carry Out the Work **3.4,** 13.2.4
Owner's Right to Clean Up 4.15.2, **6.3**
**Owner's Right to Perform Work and to Award
      Separate Contracts** **6.1**
Owner's Right to Terminate the Contract 14.2
Owner's Right to Stop the Work **3.3**
Ownership and Use of Documents 1.1.1, **1.3,** 3.2.5, 5.2.3
Patching of Work, Cutting and 4.14, 6.2.2
Patents, Royalties and 4.17.1
Payment Bond, Labor and Material 7.5
Payment, Applications for 2.2.6, 9.2, 9.3, 9.4, 9.5.3,
                                                   9.6.1, 9.8.2, 9.9.1, 9.9.3, 14.2.2
Payment, Certificates for 2.2.6, 2.2.16, 9.4, 9.5.1,
                                           9.5.5, 9.6.1, 9.7.1, 9.8.2, 9.9.1, 9.9.3, 12.1.4, 14.2.2

FIGURE A5.2. (continued)

Payment, Failure of 9.5.2, 9.6.1.3, 9.7, 9.9.2, 14	SUBCONTRACTORS 5
Payment, Final 2.2.12, 2.2.16, 9.9, 13.3.1	Subcontractors, **Definition** of **5.1**
Payments, Progress 7.8, 7.9.3, 9.5.5, 9.8.2, 9.9.3, 12.1.4	Subcontractors, Work by 1.2.4, 2.2.4, 4.3.1, 4.3.2
PAYMENTS AND COMPLETION **9**	**Subcontractual Relations** **5.3**
Payments to Subcontractors 9.5.2, 9.5.3, 9.5.4, 9.6.1.3, 11.3.3, 14.2.1	Submittals .1.3, 4.10, 4.12, 5.2.1, 5.2.3, 9.2, 9.3.1, 9.8.1, 9.9.2, 9.9.3
Payments Withheld **9.6**	Subrogation, Waiver of 11.3.6
Performance Bond and Labor and Material Payment Bond . .**7.5**	Substantial Completion 2.2.16, 8.1.1, 8.1.3, 8.2.2, **9.8**, 13.2.2
Permits, Fees and Notices 3.2.3, **4.7**, 4.13	Substantial Completion, Definition of 8.1.3
PERSONS AND PROPERTY, PROTECTION OF **10**	Substitution of Subcontractors 5.2.3, 5.2.4
Product Data, Definition of 4.12.2	Substitution of the Architect 2.2.19
Product Data, Shop Drawings, Samples and . . .2.2.14, 4.2.1, 4.12	Substitutions of Materials 4.5, 12.1.4
Progress and Completion .2.2.3, 7.9.3, **8.2**	Sub-subcontractors, Definition of 5.1.2
Progress Payments 7.8, 7.9.3, 9.5.5, 9.8.2, 9.9.3, 12.1.4	Subsurface Conditions 12.2.1
Progress Schedule . . . .-. **4.10**	Successors and Assigns 7.2
Project, Definition of **1.1.4**	**Supervision and Construction Procedures** .1.2.4, 2.2.4, **4.3**, 4.4, 10
Project Representative 2.2.17	**Superintendent**, Contractor's **4.9**, 10.2.6
Property Insurance **11.3**	Surety, Consent of 9.9.2, 9.9.3
PROTECTION OF PERSONS AND PROPERTY **10**	Surveys 3.2.2, 4.18.3
Regulations and Laws 1.3, 2.1.1, 4.6, 4.7, 4.13, 7.1, 10.2.2, 14	**Taxes** **4.6**
Rejection of Work 2.2.13, 4.5.1, 13.2	**Termination by the Contractor** **14.1**
Releases of Waivers and Liens 9.9.2, 9.9.4	**Termination by the Owner** **14.2**
Representations 1.2.2, 4.5, 4.12.5, 9.4.2, 9.6.1, 9.9.1	Termination of the Architect 2.2.19
Representatives 2.1, 2.2.2, 2.2.17, 2.2.18, 3.1, 4.1, 4.9, 5.1, 9.3.3	**TERMINATION OF THE CONTRACT** **14**
	Tests 2.2.13, 4.3.3, **7.7**, 9.4.2
Responsibility for Those Performing the Work 2.2.4, 4.3.2, 6.1.3, 6.2, 9.8.1	**Time** **8**
Retainage 9.3.1, 9.5.2, 9.8.2, 9.9.2, 9.9.3	Time, **Definition** of **8.1**
Review of Contract Documents	Time, Delays and Extensions of 8.3, 12.1, 12.3, 13.2.7
by the Contractor 1.2.2, **4.2**, 4.7.3	Time Limits, Specific .2.2.8, 2.2.12, 3.2.1, 3.4, 4.10, 5.3, 6.2.2, 7.9.2, 8.2, 8.3.2, 8.3.3, 9.2, 9.3.1, 9.4.1, 9.5.1, 9.7, 11.1.4, 11.3.1, 11.3.8, 11.3.9, 12.2, 12.3, 13.2.2, 13.2.5, 13.2.7, 14.1, 14.2.1
Reviews of Contractor's Submittals by Owner and Architect 2.2.14, 4.10, 4.12, 5.2.1, 5.2.3, 9.2	
Rights and Remedies 1.1.2, 2.2.12, 2.2.13, 3.3, 3.4, 5.3, 6.1, 6.3, **7.6**, 7.9, 8.3.1, 9.6.1, 9.7, 10.3, 12.1.2, 12.2, 13.2.2, 14	Title to Work 9.3.2, 9.3.3
	UNCOVERING AND CORRECTION OF WORK **13**
Royalties and Patents **4.17**	Uncovering of Work **13.1**
Safety of Persons and Property **10.2**	Unforseen Conditions 8.3, 12.2
Safety Precautions and Programs 2.2.4, **10.1**	Unit Prices 12.1.3, 12.1.5
Samples, Definition of 4.12.3	Use of Documents 1.1.1, 1.3, 3.2.5, 5.3
Samples, Shop Drawings, Product Data and 2.2.14, 4.2, 4.12	**Use of Site** **4.13**, 6.2.1
Samples at the Site, Documents and .4.11	Values, Schedule of 9.2
Schedule of Values **9.2**	Waiver of Claims by the Contractor 7.6.2, 8.3.2, 9.9.5, 11.3.6
Schedule, Progress 4.10	Waiver of Claims by the Owner 7.6.2, 9.9.4, 11.3.6, 11.4.1
Separate Contracts and Contractors 4.14.2, 6, 11.3.6, 13.1.2	Waiver of Liens 9.9.2
Shop Drawings, Definition of 4.12.1	**Warranty** and Warranties 2.2.16, **4.5**, 9.3.3, 9.8.1, 9.9.4, 13.2.2
Shop Drawings, Product Data and Samples . . . 2.2.14, 4.2, **4.12**	Weather Delays 8.3.1
Site, Use of 4.13, 6.2.1	**Work**, Definition of **1.1.3**
Site Visits, Architect's 2.2.3, 2.2.5, 2.2.6, 2.2.17, 7.7.1, 7.7.4, 9.4.2, 9.6.1, 9.9.1	**Work by Owner or by Separate Contractors** **6**
	Written Consent 2.2.18, 4.14.2, 7.2, 7.6.2, 9.8.1, 9.9.3, 9.9.4
Site Inspections 1.2.2, 2.2.3, 2.2.16, 7.7, 9.8.1, 9.9.1	Written Interpretations 1.1.1, 1.2.4, 2.2.8, 12.3.2
Special Inspection and Testing 2.2.13, 7.7	**Written Notice** 2.2.8, 2.2.12, 3.4, 4.2, 4.7.3, 4.7.4, 4.9, 4.12.6, 4.12.7, 4.17, 5.2.1, **7.3**, 7.4, 7.7, 7.9.2, 8.1.2, 8.3.2, 8.3.3, 9.4.1, 9.6.1, 9.7, 9.9.1, 9.9.5, 10.2.6, 11.1.4, 11.3.1, 11.3.4, 11.3.5, 11.3.7, 11.3.8, 12.2, 12.3, 13.2.2, 13.2.5, 14
Specifications 1.1.1, 1.2.4, 1.3	
Statutes of Limitations 7.9.2, 13.2.2, 13.2.7	
Stopping the Work 3.3, 9.7.1, 10.3, 14.1	
Stored Materials 6.2.1, 9.3.2, 10.2.1.2, 11.3.1, 13.2.5	Written Orders 3.3, 4.9, 12.1.4, 12.4.1, 13.1

A201-1976   AIA DOCUMENT A201 • GENERAL CONDITIONS OF THE CONTRACT FOR CONSTRUCTION • THIRTEENTH EDITION • AUGUST 1976
AIA® • © 1976 • THE AMERICAN INSTITUTE OF ARCHITECTS, 1735 NEW YORK AVENUE, N.W., WASHINGTON, D.C. 20006

FIGURE A5.2. *(continued)*

GENERAL CONDITIONS OF THE CONTRACT FOR CONSTRUCTION

ARTICLE 1

CONTRACT DOCUMENTS

1.1 DEFINITIONS

1.1.1 THE CONTRACT DOCUMENTS

The Contract Documents consist of the Owner-Contractor Agreement, the Conditions of the Contract (General, Supplementary and other Conditions), the Drawings, the Specifications, and all Addenda issued prior to and all Modifications issued after execution of the Contract. A Modification is (1) a written amendment to the Contract signed by both parties, (2) a Change Order, (3) a written interpretation issued by the Architect pursuant to Subparagraph 2.2.8, or (4) a written order for a minor change in the Work issued by the Architect pursuant to Paragraph 12.4. The Contract Documents do not include Bidding Documents such as the Advertisement or Invitation to Bid, the Instructions to Bidders, sample forms, the Contractor's Bid or portions of Addenda relating to any of these, or any other documents, unless specifically enumerated in the Owner-Contractor Agreement.

1.1.2 THE CONTRACT

The Contract Documents form the Contract for Construction. This Contract represents the entire and integrated agreement between the parties hereto and supersedes all prior negotiations, representations, or agreements, either written or oral. The Contract may be amended or modified only by a Modification as defined in Subparagraph 1.1.1. The Contract Documents shall not be construed to create any contractual relationship of any kind between the Architect and the Contractor, but the Architect shall be entitled to performance of obligations intended for his benefit, and to enforcement thereof. Nothing contained in the Contract Documents shall create any contractual relationship between the Owner or the Architect and any Subcontractor or Sub-subcontractor.

1.1.3 THE WORK

The Work comprises the completed construction required by the Contract Documents and includes all labor necessary to produce such construction, and all materials and equipment incorporated or to be incorporated in such construction.

1.1.4 THE PROJECT

The Project is the total construction of which the Work performed under the Contract Documents may be the whole or a part.

1.2 EXECUTION, CORRELATION AND INTENT

1.2.1 The Contract Documents shall be signed in not less than triplicate by the Owner and Contractor. If either the Owner or the Contractor or both do not sign the Conditions of the Contract, Drawings, Specifications, or any of the other Contract Documents, the Architect shall identify such Documents.

1.2.2 By executing the Contract, the Contractor represents that he has visited the site, familiarized himself with the local conditions under which the Work is to be performed, and correlated his observations with the requirements of the Contract Documents.

1.2.3 The intent of the Contract Documents is to include all items necessary for the proper execution and completion of the Work. The Contract Documents are complementary, and what is required by any one shall be as binding as if required by all. Work not covered in the Contract Documents will not be required unless it is consistent therewith and is reasonably inferable therefrom as being necessary to produce the intended results. Words and abbreviations which have well-known technical or trade meanings are used in the Contract Documents in accordance with such recognized meanings.

1.2.4 The organization of the Specifications into divisions, sections and articles, and the arrangement of Drawings shall not control the Contractor in dividing the Work among Subcontractors or in establishing the extent of Work to be performed by any trade.

1.3 OWNERSHIP AND USE OF DOCUMENTS

1.3.1 All Drawings, Specifications and copies thereof furnished by the Architect are and shall remain his property. They are to be used only with respect to this Project and are not to be used on any other project. With the exception of one contract set for each party to the Contract, such documents are to be returned or suitably accounted for to the Architect on request at the completion of the Work. Submission or distribution to meet official regulatory requirements or for other purposes in connection with the Project is not to be construed as publication in derogation of the Architect's common law copyright or other reserved rights.

ARTICLE 2

ARCHITECT

2.1 DEFINITION

2.1.1 The Architect is the person lawfully licensed to practice architecture, or an entity lawfully practicing architecture identified as such in the Owner-Contractor Agreement, and is referred to throughout the Contract Documents as if singular in number and masculine in gender. The term Architect means the Architect or his authorized representative.

2.2 ADMINISTRATION OF THE CONTRACT

2.2.1 The Architect will provide administration of the Contract as hereinafter described.

2.2.2 The Architect will be the Owner's representative during construction and until final payment is due. The Architect will advise and consult with the Owner. The Owner's instructions to the Contractor shall be forwarded

FIGURE A5.2. (continued)

through the Architect. The Architect will have authority to act on behalf of the Owner only to the extent provided in the Contract Documents, unless otherwise modified by written instrument in accordance with Subparagraph 2.2.18.

2.2.3 The Architect will visit the site at intervals appropriate to the stage of construction to familiarize himself generally with the progress and quality of the Work and to determine in general if the Work is proceeding in accordance with the Contract Documents. However, the Architect will not be required to make exhaustive or continuous on-site inspections to check the quality or quantity of the Work. On the basis of his on-site observations as an architect, he will keep the Owner informed of the progress of the Work, and will endeavor to guard the Owner against defects and deficiencies in the Work of the Contractor.

2.2.4 The Architect will not be responsible for and will not have control or charge of construction means, methods, techniques, sequences or procedures, or for safety precautions and programs in connection with the Work, and he will not be responsible for the Contractor's failure to carry out the Work in accordance with the Contract Documents. The Architect will not be responsible for or have control or charge over the acts or omissions of the Contractor, Subcontractors, or any of their agents or employees, or any other persons performing any of the Work.

2.2.5 The Architect shall at all times have access to the Work wherever it is in preparation and progress. The Contractor shall provide facilities for such access so the Architect may perform his functions under the Contract Documents.

2.2.6 Based on the Architect's observations and an evaluation of the Contractor's Applications for Payment, the Architect will determine the amounts owing to the Contractor and will issue Certificates for Payment in such amounts, as provided in Paragraph 9.4.

2.2.7 The Architect will be the interpreter of the requirements of the Contract Documents and the judge of the performance thereunder by both the Owner and Contractor.

2.2.8 The Architect will render interpretations necessary for the proper execution or progress of the Work, with reasonable promptness and in accordance with any time limit agreed upon. Either party to the Contract may make written request to the Architect for such interpretations.

2.2.9 Claims, disputes and other matters in question between the Contractor and the Owner relating to the execution or progress of the Work or the interpretation of the Contract Documents shall be referred initially to the Architect for decision which he will render in writing within a reasonable time.

2.2.10 All interpretations and decisions of the Architect shall be consistent with the intent of and reasonably inferable from the Contract Documents and will be in writing or in the form of drawings. In his capacity as interpreter and judge, he will endeavor to secure faithful performance by both the Owner and the Contractor, will not show partiality to either, and will not be liable for the result of any interpretation or decision rendered in good faith in such capacity.

2.2.11 The Architect's decisions in matters relating to artistic effect will be final if consistent with the intent of the Contract Documents.

2.2.12 Any claim, dispute or other matter in question between the Contractor and the Owner referred to the Architect, except those relating to artistic effect as provided in Subparagraph 2.2.11 and except those which have been waived by the making or acceptance of final payment as provided in Subparagraphs 9.9.4 and 9.9.5, shall be subject to arbitration upon the written demand of either party. However, no demand for arbitration of any such claim, dispute or other matter may be made until the earlier of (1) the date on which the Architect has rendered a written decision, or (2) the tenth day after the parties have presented their evidence to the Architect or have been given a reasonable opportunity to do so, if the Architect has not rendered his written decision by that date. When such a written decision of the Architect states (1) that the decision is final but subject to appeal, and (2) that any demand for arbitration of a claim, dispute or other matter covered by such decision must be made within thirty days after the date on which the party making the demand receives the written decision, failure to demand arbitration within said thirty days' period will result in the Architect's decision becoming final and binding upon the Owner and the Contractor.. If the Architect renders a decision after arbitration proceedings have been initiated, such decision may be entered as evidence but will not supersede any arbitration proceedings unless the decision is acceptable to all parties concerned.

2.2.13 The Architect will have authority to reject Work which does not conform to the Contract Documents. Whenever, in his opinion, he considers it necessary or advisable for the implementation of the intent of the Contract Documents, he will have authority to require special inspection or testing of the Work in accordance with Subparagraph 7.7.2 whether or not such Work be then fabricated, installed or completed. However, neither the Architect's authority to act under this Subparagraph 2.2.13, nor any decision made by him in good faith either to exercise or not to exercise such authority, shall give rise to any duty or responsibility of the Architect to the Contractor, any Subcontractor, any of their agents or employees, or any other person performing any of the Work.

2.2.14 The Architect will review and approve or take other appropriate action upon Contractor's submittals such as Shop Drawings, Product Data and Samples, but only for conformance with the design concept of the Work and with the information given in the Contract Documents. Such action shall be taken with reasonable promptness so as to cause no delay. The Architect's approval of a specific item shall not indicate approval of an assembly of which the item is a component.

2.2.15 The Architect will prepare Change Orders in accordance with Article 12, and will have authority to order minor changes in the Work as provided in Subparagraph 12.4.1.

A201-1976   AIA DOCUMENT A201 • GENERAL CONDITIONS OF THE CONTRACT FOR CONSTRUCTION • THIRTEENTH EDITION • AUGUST 1976
AIA® • © 1976 • THE AMERICAN INSTITUTE OF ARCHITECTS, 1735 NEW YORK AVENUE, N.W., WASHINGTON, D.C. 20006

FIGURE A5.2. *(continued)*

2.2.16 The Architect will conduct inspections to determine the dates of Substantial Completion and final completion, will receive and forward to the Owner for the Owner's review written warranties and related documents required by the Contract and assembled by the Contractor, and will issue a final Certificate for Payment upon compliance with the requirements of Paragraph 9.9.

2.2.17 If the Owner and Architect agree, the Architect will provide one or more Project Representatives to assist the Architect in carrying out his responsibilities at the site. The duties, responsibilities and limitations of authority of any such Project Representative shall be as set forth in an exhibit to be incorporated in the Contract Documents.

2.2.18 The duties, responsibilities and limitations of authority of the Architect as the Owner's representative during construction as set forth in the Contract Documents will not be modified or extended without written consent of the Owner, the Contractor and the Architect.

2.2.19 In case of the termination of the employment of the Architect, the Owner shall appoint an architect against whom the Contractor makes no reasonable objection whose status under the Contract Documents shall be that of the former architect. Any dispute in connection with such appointment shall be subject to arbitration.

ARTICLE 3

OWNER

3.1 DEFINITION

3.1.1 The Owner is the person or entity identified as such in the Owner-Contractor Agreement and is referred to throughout the Contract Documents as if singular in number and masculine in gender. The term Owner means the Owner or his authorized representative.

3.2 INFORMATION AND SERVICES REQUIRED OF THE OWNER

3.2.1 The Owner shall, at the request of the Contractor, at the time of execution of the Owner-Contractor Agreement, furnish to the Contractor reasonable evidence that he has made financial arrangements to fulfill his obligations under the Contract. Unless such reasonable evidence is furnished, the Contractor is not required to execute the Owner-Contractor Agreement or to commence the Work.

3.2.2 The Owner shall furnish all surveys describing the physical characteristics, legal limitations and utility locations for the site of the Project, and a legal description of the site.

3.2.3 Except as provided in Subparagraph 4.7.1, the Owner shall secure and pay for necessary approvals, easements, assessments and charges required for the construction, use or occupancy of permanent structures or for permanent changes in existing facilities.

3.2.4 Information or services under the Owner's control shall be furnished by the Owner with reasonable promptness to avoid delay in the orderly progress of the Work.

3.2.5 Unless otherwise provided in the Contract Documents, the Contractor will be furnished, free of charge, all copies of Drawings and Specifications reasonably necessary for the execution of the Work.

3.2.6 The Owner shall forward all instructions to the Contractor through the Architect.

3.2.7 The foregoing are in addition to other duties and responsibilities of the Owner enumerated herein and especially those in respect to Work by Owner or by Separate Contractors, Payments and Completion, and Insurance in Articles 6, 9 and 11 respectively.

3.3 OWNER'S RIGHT TO STOP THE WORK

3.3.1 If the Contractor fails to correct defective Work as required by Paragraph 13.2 or persistently fails to carry out the Work in accordance with the Contract Documents, the Owner, by a written order signed personally or by an agent specifically so empowered by the Owner in writing, may order the Contractor to stop the Work, or any portion thereof, until the cause for such order has been eliminated; however, this right of the Owner to stop the Work shall not give rise to any duty on the part of the Owner to exercise this right for the benefit of the Contractor or any other person or entity, except to the extent required by Subparagraph 6.1.3.

3.4 OWNER'S RIGHT TO CARRY OUT THE WORK

3.4.1 If the Contractor defaults or neglects to carry out the Work in accordance with the Contract Documents and fails within seven days after receipt of written notice from the Owner to commence and continue correction of such default or neglect with diligence and promptness, the Owner may, after seven days following receipt by the Contractor of an additional written notice and without prejudice to any other remedy he may have, make good such deficiencies. In such case an appropriate Change Order shall be issued deducting from the payments then or thereafter due the Contractor the cost of correcting such deficiencies, including compensation for the Architect's additional services made necessary by such default, neglect or failure. Such action by the Owner and the amount charged to the Contractor are both subject to the prior approval of the Architect. If the payments then or thereafter due the Contractor are not sufficient to cover such amount, the Contractor shall pay the difference to the Owner.

ARTICLE 4

CONTRACTOR

4.1 DEFINITION

4.1.1 The Contractor is the person or entity identified as such in the Owner-Contractor Agreement and is referred to throughout the Contract Documents as if singular in number and masculine in gender. The term Contracto means the Contractor or his authorized representative.

4.2 REVIEW OF CONTRACT DOCUMENTS

4.2.1 The Contractor shall carefully study and compare the Contract Documents and shall at once report to the Architect any error, inconsistency or omission he may discover. The Contractor shall not be liable to the Owner or

FIGURE A5.2. *(continued)*

the Architect for any damage resulting from any such errors, inconsistencies or omissions in the Contract Documents. The Contractor shall perform no portion of the Work at any time without Contract Documents or, where required, approved Shop Drawings, Product Data or Samples for such portion of the Work.

4.3 SUPERVISION AND CONSTRUCTION PROCEDURES

4.3.1 The Contractor shall supervise and direct the Work, using his best skill and attention. He shall be solely responsible for all construction means, methods, techniques, sequences and procedures and for coordinating all portions of the Work under the Contract.

4.3.2 The Contractor shall be responsible to the Owner for the acts and omissions of his employees, Subcontractors and their agents and employees, and other persons performing any of the Work under a contract with the Contractor.

4.3.3 The Contractor shall not be relieved from his obligations to perform the Work in accordance with the Contract Documents either by the activities or duties of the Architect in his administration of the Contract, or by inspections, tests or approvals required or performed under Paragraph 7.7 by persons other than the Contractor.

4.4 LABOR AND MATERIALS

4.4.1 Unless otherwise provided in the Contract Documents, the Contractor shall provide and pay for all labor, materials, equipment, tools, construction equipment and machinery, water, heat, utilities, transportation, and other facilities and services necessary for the proper execution and completion of the Work, whether temporary or permanent and whether or not incorporated or to be incorporated in the Work.

4.4.2 The Contractor shall at all times enforce strict discipline and good order among his employees and shall not employ on the Work any unfit person or anyone not skilled in the task assigned to him.

4.5 WARRANTY

4.5.1 The Contractor warrants to the Owner and the Architect that all materials and equipment furnished under this Contract will be new unless otherwise specified, and that all Work will be of good quality, free from faults and defects and in conformance with the Contract Documents. All Work not conforming to these requirements, including substitutions not properly approved and authorized, may be considered defective. If required by the Architect, the Contractor shall furnish satisfactory evidence as to the kind and quality of materials and equipment. This warranty is not limited by the provisions of Paragraph 13.2.

4.6 TAXES

4.6.1 The Contractor shall pay all sales, consumer, use and other similar taxes for the Work or portions thereof provided by the Contractor which are legally enacted at the time bids are received, whether or not yet effective.

4.7 PERMITS, FEES AND NOTICES

4.7.1 Unless otherwise provided in the Contract Documents, the Contractor shall secure and pay for the building permit and for all other permits and governmental fees, licenses and inspections necessary for the proper execution and completion of the Work which are customarily secured after execution of the Contract and which are legally required at the time the bids are received.

4.7.2 The Contractor shall give all notices and comply with all laws, ordinances, rules, regulations and lawful orders of any public authority bearing on the performance of the Work.

4.7.3 It is not the responsibility of the Contractor to make certain that the Contract Documents are in accordance with applicable laws, statutes, building codes and regulations. If the Contractor observes that any of the Contract Documents are at variance therewith in any respect, he shall promptly notify the Architect in writing, and any necessary changes shall be accomplished by appropriate Modification.

4.7.4 If the Contractor performs any Work knowing it to be contrary to such laws, ordinances, rules and regulations, and without such notice to the Architect, he shall assume full responsibility therefor and shall bear all costs attributable thereto.

4.8 ALLOWANCES

4.8.1 The Contractor shall include in the Contract Sum all allowances stated in the Contract Documents. Items covered by these allowances shall be supplied for such amounts and by such persons as the Owner may direct, but the Contractor will not be required to employ persons against whom he makes a reasonable objection.

4.8.2 Unless otherwise provided in the Contract Documents:
- .1 these allowances shall cover the cost to the Contractor, less any applicable trade discount, of the materials and equipment required by the allowance delivered at the site, and all applicable taxes;
- .2 the Contractor's costs for unloading and handling on the site, labor, installation costs, overhead, profit and other expenses contemplated for the original allowance shall be included in the Contract Sum and not in the allowance;
- .3 whenever the cost is more than or less than the allowance, the Contract Sum shall be adjusted accordingly by Change Order, the amount of which will recognize changes, if any, in handling costs on the site, labor, installation costs, overhead, profit and other expenses.

4.9 SUPERINTENDENT

4.9.1 The Contractor shall employ a competent superintendent and necessary assistants who shall be in attendance at the Project site during the progress of the Work. The superintendent shall represent the Contractor and all communications given to the superintendent shall be as binding as if given to the Contractor. Important communications shall be confirmed in writing. Other communications shall be so confirmed on written request in each case.

4.10 PROGRESS SCHEDULE

4.10.1 The Contractor, immediately after being awarded the Contract, shall prepare and submit for the Owner's and Architect's information an estimated progress sched-

---

A201-1976 · AIA DOCUMENT A201 · GENERAL CONDITIONS OF THE CONTRACT FOR CONSTRUCTION · THIRTEENTH EDITION · AUGUST 1976
AIA® · © 1976 · THE AMERICAN INSTITUTE OF ARCHITECTS, 1735 NEW YORK AVENUE, N.W., WASHINGTON, D.C. 20006

FIGURE A5.2. *(continued)*

ule for the Work. The progress schedule shall be related to the entire Project to the extent required by the Contract Documents, and shall provide for expeditious and practicable execution of the Work.

4.11 DOCUMENTS AND SAMPLES AT THE SITE

4.11.1 The Contractor shall maintain at the site for the Owner one record copy of all Drawings, Specifications, Addenda, Change Orders and other Modifications, in good order and marked currently to record all changes made during construction, and approved Shop Drawings, Product Data and Samples. These shall be available to the Architect and shall be delivered to him for the Owner upon completion of the Work.

4.12 SHOP DRAWINGS, PRODUCT DATA AND SAMPLES

4.12.1 Shop Drawings are drawings, diagrams, schedules and other data specially prepared for the Work by the Contractor or any Subcontractor, manufacturer, supplier or distributor to illustrate some portion of the Work.

4.12.2 Product Data are illustrations, standard schedules, performance charts, instructions, brochures, diagrams and other information furnished by the Contractor to illustrate a material, product or system for some portion of the Work.

4.12.3 Samples are physical examples which illustrate materials, equipment or workmanship and establish standards by which the Work will be judged.

4.12.4 The Contractor shall review, approve and submit, with reasonable promptness and in such sequence as to cause no delay in the Work or in the work of the Owner or any separate contractor, all Shop Drawings, Product Data and Samples required by the Contract Documents.

4.12.5 By approving and submitting Shop Drawings, Product Data and Samples, the Contractor represents that he has determined and verified all materials, field measurements, and field construction criteria related thereto, or will do so, and that he has checked and coordinated the information contained within such submittals with the requirements of the Work and of the Contract Documents.

4.12.6 The Contractor shall not be relieved of responsibility for any deviation from the requirements of the Contract Documents by the Architect's approval of Shop Drawings, Product Data or Samples under Subparagraph 2.2.14 unless the Contractor has specifically informed the Architect in writing of such deviation at the time of submission and the Architect has given written approval to the specific deviation. The Contractor shall not be relieved from responsibility for errors or omissions in the Shop Drawings, Product Data or Samples by the Architect's approval thereof.

4.12.7 The Contractor shall direct specific attention, in writing or on resubmitted Shop Drawings, Product Data or Samples, to revisions other than those requested by the Architect on previous submittals.

4.12.8 No portion of the Work requiring submission of a Shop Drawing, Product Data or Sample shall be commenced until the submittal has been approved by the Architect as provided in Subparagraph 2.2.14. All such portions of the Work shall be in accordance with approved submittals.

4.13 USE OF SITE

4.13.1 The Contractor shall confine operations at the site to areas permitted by law, ordinances, permits and the Contract Documents and shall not unreasonably encumber the site with any materials or equipment.

4.14 CUTTING AND PATCHING OF WORK

4.14.1 The Contractor shall be responsible for all cutting, fitting or patching that may be required to complete the Work or to make its several parts fit together properly.

4.14.2 The Contractor shall not damage or endanger any portion of the Work or the work of the Owner or any separate contractors by cutting, patching or otherwise altering any work, or by excavation. The Contractor shall not cut or otherwise alter the work of the Owner or any separate contractor except with the written consent of the Owner and of such separate contractor. The Contractor shall not unreasonably withhold from the Owner or any separate contractor his consent to cutting or otherwise altering the Work.

4.15 CLEANING UP

4.15.1 The Contractor at all times shall keep the premises free from accumulation of waste materials or rubbish caused by his operations. At the completion of the Work he shall remove all his waste materials and rubbish from and about the Project as well as all his tools, construction equipment, machinery and surplus materials.

4.15.2 If the Contractor fails to clean up at the completion of the Work, the Owner may do so as provided in Paragraph 3.4 and the cost thereof shall be charged to the Contractor.

4.16 COMMUNICATIONS

4.16.1 The Contractor shall forward all communications to the Owner through the Architect.

4.17 ROYALTIES AND PATENTS

4.17.1 The Contractor shall pay all royalties and license fees. He shall defend all suits or claims for infringement of any patent rights and shall save the Owner harmless from loss on account thereof, except that the Owner shall be responsible for all such loss when a particular design, process or the product of a particular manufacturer or manufacturers is specified, but if the Contractor has reason to believe that the design, process or product specified is an infringement of a patent, he shall be responsible for such loss unless he promptly gives such information to the Architect.

4.18 INDEMNIFICATION

4.18.1 To the fullest extent permitted by law, the Contractor shall indemnify and hold harmless the Owner and the Architect and their agents and employees from and against all claims, damages, losses and expenses, including but not limited to attorneys' fees, arising out of or resulting from the performance of the Work, provided that any such claim, damage, loss or expense (1) is attributable to bodily injury, sickness, disease or death, or to injury to or destruction of tangible property (other than the Work itself) including the loss of use resulting therefrom,

FIGURE A5.2. (continued)

and (2) is caused in whole or in part by any negligent act or omission of the Contractor, any Subcontractor, anyone directly or indirectly employed by any of them or anyone for whose acts any of them may be liable, regardless of whether or not it is caused in part by a party indemnified hereunder. Such obligation shall not be construed to negate, abridge, or otherwise reduce any other right or obligation of indemnity which would otherwise exist as to any party or person described in this Paragraph 4.18.

4.18.2 In any and all claims against the Owner or the Architect or any of their agents or employees by any employee of the Contractor, any Subcontractor, anyone directly or indirectly employed by any of them or anyone for whose acts any of them may be liable, the indemnification obligation under this Paragraph 4.18 shall not be limited in any way by any limitation on the amount or type of damages, compensation or benefits payable by or for the Contractor or any Subcontractor under workers' or workmen's compensation acts, disability benefit acts or other employee benefit acts.

4.18.3 The obligations of the Contractor under this Paragraph 4.18 shall not extend to the liability of the Architect, his agents or employees, arising out of (1) the preparation or approval of maps, drawings, opinions, reports, surveys, change orders, designs or specifications, or (2) the giving of or the failure to give directions or instructions by the Architect, his agents or employees provided such giving or failure to give is the primary cause of the injury or damage.

ARTICLE 5
SUBCONTRACTORS

5.1 DEFINITION

5.1.1 A Subcontractor is a person or entity who has a direct contract with the Contractor to perform any of the Work at the site. The term Subcontractor is referred to throughout the Contract Documents as if singular in number and masculine in gender and means a Subcontractor or his authorized representative. The term Subcontractor does not include any separate contractor or his subcontractors.

5.1.2 A Sub-subcontractor is a person or entity who has a direct or indirect contract with a Subcontractor to perform any of the Work at the site. The term Sub-subcontractor is referred to throughout the Contract Documents as if singular in number and masculine in gender and means a Sub-subcontractor or an authorized representative thereof.

5.2 AWARD OF SUBCONTRACTS AND OTHER CONTRACTS FOR PORTIONS OF THE WORK

5.2.1 Unless otherwise required by the Contract Documents or the Bidding Documents, the Contractor, as soon as practicable after the award of the Contract, shall furnish to the Owner and the Architect in writing the names of the persons or entities (including those who are to furnish materials or equipment fabricated to a special design) proposed for each of the principal portions of the Work. The Architect will promptly reply to the Contractor in writing stating whether or not the Owner or the Architect, after due investigation, has reasonable objection to any such proposed person or entity. Failure of the Owner or Architect to reply promptly shall constitute notice of no reasonable objection.

5.2.2 The Contractor shall not contract with any such proposed person or entity to whom the Owner or the Architect has made reasonable objection under the provisions of Subparagraph 5.2.1. The Contractor shall not be required to contract with anyone to whom he has a reasonable objection.

5.2.3 If the Owner or the Architect has reasonable objection to any such proposed person or entity, the Contractor shall submit a substitute to whom the Owner or the Architect has no reasonable objection, and the Contract Sum shall be increased or decreased by the difference in cost occasioned by such substitution and an appropriate Change Order shall be issued; however, no increase in the Contract Sum shall be allowed for any such substitution unless the Contractor has acted promptly and responsively in submitting names as required by Subparagraph 5.2.1.

5.2.4 The Contractor shall make no substitution for any Subcontractor, person or entity previously selected if the Owner or Architect makes reasonable objection to such substitution.

5.3 SUBCONTRACTUAL RELATIONS

5.3.1 By an appropriate agreement, written where legally required for validity, the Contractor shall require each Subcontractor, to the extent of the Work to be performed by the Subcontractor, to be bound to the Contractor by the terms of the Contract Documents, and to assume toward the Contractor all the obligations and responsibilities which the Contractor, by these Documents, assumes toward the Owner and the Architect. Said agreement shall preserve and protect the rights of the Owner and the Architect under the Contract Documents with respect to the Work to be performed by the Subcontractor so that the subcontracting thereof will not prejudice such rights, and shall allow to the Subcontractor, unless specifically provided otherwise in the Contractor-Subcontractor agreement, the benefit of all rights, remedies and redress against the Contractor that the Contractor, by these Documents, has against the Owner. Where appropriate, the Contractor shall require each Subcontractor to enter into similar agreements with his Sub-subcontractors. The Contractor shall make available to each proposed Subcontractor, prior to the execution of the Subcontract, copies of the Contract Documents to which the Subcontractor will be bound by this Paragraph 5.3, and identify to the Subcontractor any terms and conditions of the proposed Subcontract which may be at variance with the Contract Documents. Each Subcontractor shall similarly make copies of such Documents available to his Sub-subcontractors.

ARTICLE 6
WORK BY OWNER OR BY SEPARATE CONTRACTORS

6.1 OWNER'S RIGHT TO PERFORM WORK AND TO AWARD SEPARATE CONTRACTS

6.1.1 The Owner reserves the right to perform work related to the Project with his own forces, and to award

FIGURE A5.2. (continued)

separate contracts in connection with other portions of the Project or other work on the site under these or similar Conditions of the Contract. If the Contractor claims that delay or additional cost is involved because of such action by the Owner, he shall make such claim as provided elsewhere in the Contract Documents.

6.1.2 When separate contracts are awarded for different portions of the Project or other work on the site, the term Contractor in the Contract Documents in each case shall mean the Contractor who executes each separate Owner-Contractor Agreement.

6.1.3 The Owner will provide for the coordination of the work of his own forces and of each separate contractor with the Work of the Contractor, who shall cooperate therewith as provided in Paragraph 6.2.

6.2 MUTUAL RESPONSIBILITY

6.2.1 The Contractor shall afford the Owner and separate contractors reasonable opportunity for the introduction and storage of their materials and equipment and the execution of their work, and shall connect and coordinate his Work with theirs as required by the Contract Documents.

6.2.2 If any part of the Contractor's Work depends for proper execution or results upon the work of the Owner or any separate contractor, the Contractor shall, prior to proceeding with the Work, promptly report to the Architect any apparent discrepancies or defects in such other work that render it unsuitable for such proper execution and results. Failure of the Contractor so to report shall constitute an acceptance of the Owner's or separate contractors' work as fit and proper to receive his Work, except as to defects which may subsequently become apparent in such work by others.

6.2.3 Any costs caused by defective or ill-timed work shall be borne by the party responsible therefor.

6.2.4 Should the Contractor wrongfully cause damage to the work or property of the Owner, or to other work on the site, the Contractor shall promptly remedy such damage as provided in Subparagraph 10.2.5.

6.2.5 Should the Contractor wrongfully cause damage to the work or property of any separate contractor, the Contractor shall upon due notice promptly attempt to settle with such other contractor by agreement, or otherwise to resolve the dispute. If such separate contractor sues or initiates an arbitration proceeding against the Owner on account of any damage alleged to have been caused by the Contractor, the Owner shall notify the Contractor who shall defend such proceedings at the Owner's expense, and if any judgment or award against the Owner arises therefrom the Contractor shall pay or satisfy it and shall reimburse the Owner for all attorneys' fees and court or arbitration costs which the Owner has incurred.

6.3 OWNER'S RIGHT TO CLEAN UP

6.3.1 If a dispute arises between the Contractor and separate contractors as to their responsibility for cleaning up as required by Paragraph 4.15, the Owner may clean up and charge the cost thereof to the contractors responsible therefor as the Architect shall determine to be just.

ARTICLE 7

MISCELLANEOUS PROVISIONS

7.1 GOVERNING LAW

7.1.1 The Contract shall be governed by the law of the place where the Project is located.

7.2 SUCCESSORS AND ASSIGNS

7.2.1 The Owner and the Contractor each binds himself, his partners, successors, assigns and legal representatives to the other party hereto and to the partners, successors, assigns and legal representatives of such other party in respect to all covenants, agreements and obligations contained in the Contract Documents. Neither party to the Contract shall assign the Contract or sublet it as a whole without the written consent of the other, nor shall the Contractor assign any moneys due or to become due to him hereunder, without the previous written consent of the Owner.

7.3 WRITTEN NOTICE

7.3.1 Written notice shall be deemed to have been duly served if delivered in person to the individual or member of the firm or entity or to an officer of the corporation for whom it was intended, or if delivered at or sent by registered or certified mail to the last business address known to him who gives the notice.

7.4 CLAIMS FOR DAMAGES

7.4.1 Should either party to the Contract suffer injury or damage to person or property because of any act or omission of the other party or of any of his employees, agents or others for whose acts he is legally liable, claim shall be made in writing to such other party within a reasonable time after the first observance of such injury or damage.

7.5 PERFORMANCE BOND AND LABOR AND MATERIAL PAYMENT BOND

7.5.1 The Owner shall have the right to require the Contractor to furnish bonds covering the faithful performance of the Contract and the payment of all obligations arising thereunder if and as required in the Bidding Documents or in the Contract Documents.

7.6 RIGHTS AND REMEDIES

7.6.1 The duties and obligations imposed by the Contract Documents and the rights and remedies available thereunder shall be in addition to and not a limitation of any duties, obligations, rights and remedies otherwise imposed or available by law.

7.6.2 No action or failure to act by the Owner, Architect or Contractor shall constitute a waiver of any right or duty afforded any of them under the Contract, nor shall any such action or failure to act constitute an approval of or acquiescence in any breach thereunder, except as may be specifically agreed in writing.

FIGURE A5.2. (*continued*)

7.7 TESTS

7.7.1 If the Contract Documents, laws, ordinances, rules, regulations or orders of any public authority having jurisdiction require any portion of the Work to be inspected, tested or approved, the Contractor shall give the Architect timely notice of its readiness so the Architect may observe such inspection, testing or approval. The Contractor shall bear all costs of such inspections, tests or approvals conducted by public authorities. Unless otherwise provided, the Owner shall bear all costs of other inspections, tests or approvals.

7.7.2 If the Architect determines that any Work requires special inspection, testing, or approval which Subparagraph 7.7.1 does not include, he will, upon written authorization from the Owner, instruct the Contractor to order such special inspection, testing or approval, and the Contractor shall give notice as provided in Subparagraph 7.7.1. If such special inspection or testing reveals a failure of the Work to comply with the requirements of the Contract Documents, the Contractor shall bear all costs thereof, including compensation for the Architect's additional services made necessary by such failure; otherwise the Owner shall bear such costs, and an appropriate Change Order shall be issued.

7.7.3 Required certificates of inspection, testing or approval shall be secured by the Contractor and promptly delivered by him to the Architect.

7.7.4 If the Architect is to observe the inspections, tests or approvals required by the Contract Documents, he will do so promptly and, where practicable, at the source of supply.

7.8 INTEREST

7.8.1 Payments due and unpaid under the Contract Documents shall bear interest from the date payment is due at such rate as the parties may agree upon in writing or, in the absence thereof, at the legal rate prevailing at the place of the Project.

7.9 ARBITRATION

7.9.1 All claims, disputes and other matters in question between the Contractor and the Owner arising out of, or relating to, the Contract Documents or the breach thereof, except as provided in Subparagraph 2.2.11 with respect to the Architect's decisions on matters relating to artistic effect, and except for claims which have been waived by the making or acceptance of final payment as provided by Subparagraphs 9.9.4 and 9.9.5, shall be decided by arbitration in accordance with the Construction Industry Arbitration Rules of the American Arbitration Association then obtaining unless the parties mutually agree otherwise. No arbitration arising out of or relating to the Contract Documents shall include, by consolidation, joinder or in any other manner, the Architect, his employees or consultants except by written consent containing a specific reference to the Owner-Contractor Agreement and signed by the Architect, the Owner, the Contractor and any other person sought to be joined. No arbitration shall include by consolidation, joinder or in any other manner, parties other than the Owner, the Contractor and any other persons substantially involved in a common question of fact or law, whose presence is required if complete relief is to be accorded in the arbitration. No person other than the Owner or Contractor shall be included as an original third party or additional third party to an arbitration whose interest or responsibility is insubstantial. Any consent to arbitration involving an additional person or persons shall not constitute consent to arbitration of any dispute not described therein or with any person not named or described therein. The foregoing agreement to arbitrate and any other agreement to arbitrate with an additional person or persons duly consented to by the parties to the Owner-Contractor Agreement shall be specifically enforceable under the prevailing arbitration law. The award rendered by the arbitrators shall be final, and judgment may be entered upon it in accordance with applicable law in any court having jurisdiction thereof.

7.9.2 Notice of the demand for arbitration shall be filed in writing with the other party to the Owner-Contractor Agreement and with the American Arbitration Association, and a copy shall be filed with the Architect. The demand for arbitration shall be made within the time limits specified in Subparagraph 2.2.12 where applicable, and in all other cases within a reasonable time after the claim, dispute or other matter in question has arisen, and in no event shall it be made after the date when institution of legal or equitable proceedings based on such claim, dispute or other matter in question would be barred by the applicable statute of limitations.

7.9.3 Unless otherwise agreed in writing, the Contractor shall carry on the Work and maintain its progress during any arbitration proceedings, and the Owner shall continue to make payments to the Contractor in accordance with the Contract Documents.

ARTICLE 8

TIME

8.1 DEFINITIONS

8.1.1 Unless otherwise provided, the Contract Time is the period of time allotted in the Contract Documents for Substantial Completion of the Work as defined in Subparagraph 8.1.3, including authorized adjustments thereto.

8.1.2 The date of commencement of the Work is the date established in a notice to proceed. If there is no notice to proceed, it shall be the date of the Owner-Contractor Agreement or such other date as may be established therein.

8.1.3 The Date of Substantial Completion of the Work or designated portion thereof is the Date certified by the Architect when construction is sufficiently complete, in accordance with the Contract Documents, so the Owner can occupy or utilize the Work or designated portion thereof for the use for which it is intended.

8.1.4 The term day as used in the Contract Documents shall mean calendar day unless otherwise specifically designated.

8.2 PROGRESS AND COMPLETION

8.2.1 All time limits stated in the Contract Documents are of the essence of the Contract.

A201-1976 · AIA DOCUMENT A201 · GENERAL CONDITIONS OF THE CONTRACT FOR CONSTRUCTION · THIRTEENTH EDITION · AUGUST 1976
AIA® · © 1976 · THE AMERICAN INSTITUTE OF ARCHITECTS, 1735 NEW YORK AVENUE, N.W., WASHINGTON, D.C. 20006

FIGURE A5.2. *(continued)*

8.2.2 The Contractor shall begin the Work on the date of commencement as defined in Subparagraph 8.1.2. He shall carry the Work forward expeditiously with adequate forces and shall achieve Substantial Completion within the Contract Time.

8.3 DELAYS AND EXTENSIONS OF TIME

8.3.1 If the Contractor is delayed at any time in the progress of the Work by any act or neglect of the Owner or the Architect, or by any employee of either, or by any separate contractor employed by the Owner, or by changes ordered in the Work, or by labor disputes, fire, unusual delay in transportation, adverse weather conditions not reasonably anticipatable, unavoidable casualties, or any causes beyond the Contractor's control, or by delay authorized by the Owner pending arbitration, or by any other cause which the Architect determines may justify the delay, then the Contract Time shall be extended by Change Order for such reasonable time as the Architect may determine.

8.3.2 Any claim for extension of time shall be made in writing to the Architect not more than twenty days after the commencement of the delay; otherwise it shall be waived. In the case of a continuing delay only one claim is necessary. The Contractor shall provide an estimate of the probable effect of such delay on the progress of the Work.

8.3.3 If no agreement is made stating the dates upon which interpretations as provided in Subparagraph 2.2.8 shall be furnished, then no claim for delay shall be allowed on account of failure to furnish such interpretations until fifteen days after written request is made for them, and not then unless such claim is reasonable.

8.3.4 This Paragraph 8.3 does not exclude the recovery of damages for delay by either party under other provisions of the Contract Documents.

ARTICLE 9

PAYMENTS AND COMPLETION

9.1 CONTRACT SUM

9.1.1 The Contract Sum is stated in the Owner-Contractor Agreement and, including authorized adjustments thereto, is the total amount payable by the Owner to the Contractor for the performance of the Work under the Contract Documents.

9.2 SCHEDULE OF VALUES

9.2.1 Before the first Application for Payment, the Contractor shall submit to the Architect a schedule of values allocated to the various portions of the Work, prepared in such form and supported by such data to substantiate its accuracy as the Architect may require. This schedule, unless objected to by the Architect, shall be used only as a basis for the Contractor's Applications for Payment.

9.3 APPLICATIONS FOR PAYMENT

9.3.1 At least ten days before the date for each progress payment established in the Owner-Contractor Agreement, the Contractor shall submit to the Architect an itemized Application for Payment, notarized if required, supported by such data substantiating the Contractor's right to payment as the Owner or the Architect may require, and reflecting retainage, if any, as provided elsewhere in the Contract Documents.

9.3.2 Unless otherwise provided in the Contract Documents, payments will be made on account of materials or equipment not incorporated in the Work but delivered and suitably stored at the site and, if approved in advance by the Owner, payments may similarly be made for materials or equipment suitably stored at some other location agreed upon in writing. Payments for materials or equipment stored on or off the site shall be conditioned upon submission by the Contractor of bills of sale or such other procedures satisfactory to the Owner to establish the Owner's title to such materials or equipment or otherwise protect the Owner's interest, including applicable insurance and transportation to the site for those materials and equipment stored off the site.

9.3.3 The Contractor warrants that title to all Work, materials and equipment covered by an Application for Payment will pass to the Owner either by incorporation in the construction or upon the receipt of payment by the Contractor, whichever occurs first, free and clear of all liens, claims, security interests or encumbrances, hereinafter referred to in this Article 9 as "liens"; and that no Work, materials or equipment covered by an Application for Payment will have been acquired by the Contractor, or by any other person performing Work at the site or furnishing materials and equipment for the Project, subject to an agreement under which an interest therein or an encumbrance thereon is retained by the seller or otherwise imposed by the Contractor or such other person.

9.4 CERTIFICATES FOR PAYMENT

9.4.1 The Architect will, within seven days after the receipt of the Contractor's Application for Payment, either issue a Certificate for Payment to the Owner, with a copy to the Contractor, for such amount as the Architect determines is properly due, or notify the Contractor in writing his reasons for withholding a Certificate as provided in Subparagraph 9.6.1.

9.4.2 The issuance of a Certificate for Payment will constitute a representation by the Architect to the Owner, based on his observations at the site as provided in Subparagraph 2.2.3 and the data comprising the Application for Payment, that the Work has progressed to the point indicated; that, to the best of his knowledge, information and belief, the quality of the Work is in accordance with the Contract Documents (subject to an evaluation of the Work for conformance with the Contract Documents upon Substantial Completion, to the results of any subsequent tests required by or performed under the Contract Documents, to minor deviations from the Contract Documents correctable prior to completion, and to any specific qualifications stated in his Certificate); and that the Contractor is entitled to payment in the amount certified. However, by issuing a Certificate for Payment, the Architect shall not thereby be deemed to represent that he has made exhaustive or continuous on-site inspections to check the quality or quantity of the Work or that he has reviewed the construction means, methods, techniques,

FIGURE A5.2. (*continued*)

sequences or procedures, or that he has made any examination to ascertain how or for what purpose the Contractor has used the moneys previously paid on account of the Contract Sum.

9.5 PROGRESS PAYMENTS

9.5.1 After the Architect has issued a Certificate for Payment, the Owner shall make payment in the manner and within the time provided in the Contract Documents.

9.5.2 The Contractor shall promptly pay each Subcontractor, upon receipt of payment from the Owner, out of the amount paid to the Contractor on account of such Subcontractor's Work, the amount to which said Subcontractor is entitled, reflecting the percentage actually retained, if any, from payments to the Contractor on account of such Subcontractor's Work. The Contractor shall, by an appropriate agreement with each Subcontractor, require each Subcontractor to make payments to his Subsubcontractors in similar manner.

9.5.3 The Architect may, on request and at his discretion, furnish to any Subcontractor, if practicable, information regarding the percentages of completion or the amounts applied for by the Contractor and the action taken thereon by the Architect on account of Work done by such Subcontractor.

9.5.4 Neither the Owner nor the Architect shall have any obligation to pay or to see to the payment of any moneys to any Subcontractor except as may otherwise be required by law.

9.5.5 No Certificate for a progress payment, nor any progress payment, nor any partial or entire use or occupancy of the Project by the Owner, shall constitute an acceptance of any Work not in accordance with the Contract Documents.

9.6 PAYMENTS WITHHELD

9.6.1 The Architect may decline to certify payment and may withhold his Certificate in whole or in part, to the extent necessary reasonably to protect the Owner, if in his opinion he is unable to make representations to the Owner as provided in Subparagraph 9.4.2. If the Architect is unable to make representations to the Owner as provided in Subparagraph 9.4.2 and to certify payment in the amount of the Application, he will notify the Contractor as provided in Subparagraph 9.4.1. If the Contractor and the Architect cannot agree on a revised amount, the Architect will promptly issue a Certificate for Payment for the amount for which he is able to make such representations to the Owner. The Architect may also decline to certify payment or, because of subsequently discovered evidence or subsequent observations, he may nullify the whole or any part of any Certificate for Payment previously issued, to such extent as may be necessary in his opinion to protect the Owner from loss because of:

.1 defective Work not remedied,

.2 third party claims filed or reasonable evidence indicating probable filing of such claims,

.3 failure of the Contractor to make payments properly to Subcontractors or for labor, materials or equipment,

.4 reasonable evidence that the Work cannot be completed for the unpaid balance of the Contract Sum,

.5 damage to the Owner or another contractor,

.6 reasonable evidence that the Work will not be completed within the Contract Time, or

.7 persistent failure to carry out the Work in accordance with the Contract Documents.

9.6.2 When the above grounds in Subparagraph 9.6.1 are removed, payment shall be made for amounts withheld because of them.

9.7 FAILURE OF PAYMENT

9.7.1 If the Architect does not issue a Certificate for Payment, through no fault of the Contractor, within seven days after receipt of the Contractor's Application for Payment, or if the Owner does not pay the Contractor within seven days after the date established in the Contract Documents any amount certified by the Architect or awarded by arbitration, then the Contractor may, upon seven additional days' written notice to the Owner and the Architect, stop the Work until payment of the amount owing has been received. The Contract Sum shall be increased by the amount of the Contractor's reasonable costs of shut-down, delay and start-up, which shall be effected by appropriate Change Order in accordance with Paragraph 12.3.

9.8 SUBSTANTIAL COMPLETION

9.8.1 When the Contractor considers that the Work, or a designated portion thereof which is acceptable to the Owner, is substantially complete as defined in Subparagraph 8.1.3, the Contractor shall prepare for submission to the Architect a list of items to be completed or corrected. The failure to include any items on such list does not alter the responsibility of the Contractor to complete all Work in accordance with the Contract Documents. When the Architect on the basis of an inspection determines that the Work or designated portion thereof is substantially complete, he will then prepare a Certificate of Substantial Completion which shall establish the Date of Substantial Completion, shall state the responsibilities of the Owner and the Contractor for security, maintenance, heat, utilities, damage to the Work, and insurance, and shall fix the time within which the Contractor shall complete the items listed therein. Warranties required by the Contract Documents shall commence on the Date of Substantial Completion of the Work or designated portion thereof unless otherwise provided in the Certificate of Substantial Completion. The Certificate of Substantial Completion shall be submitted to the Owner and the Contractor for their written acceptance of the responsibilities assigned to them in such Certificate.

9.8.2 Upon Substantial Completion of the Work or designated portion thereof and upon application by the Contractor and certification by the Architect, the Owner shall make payment, reflecting adjustment in retainage, if any, for such Work or portion thereof, as provided in the Contract Documents.

9.9 FINAL COMPLETION AND FINAL PAYMENT

9.9.1 Upon receipt of written notice that the Work is ready for final inspection and acceptance and upon receipt of a final Application for Payment, the Architect will

FIGURE A5.2. (*continued*)

promptly make such inspection and, when he finds the Work acceptable under the Contract Documents and the Contract fully performed, he will promptly issue a final Certificate for Payment stating that to the best of his knowledge, information and belief, and on the basis of his observations and inspections, the Work has been completed in accordance with the terms and conditions of the Contract Documents and that the entire balance found to be due the Contractor, and noted in said final Certificate, is due and payable. The Architect's final Certificate for Payment will constitute a further representation that the conditions precedent to the Contractor's being entitled to final payment as set forth in Subparagraph 9.9.2 have been fulfilled.

9.9.2 Neither the final payment nor the remaining retained percentage shall become due until the Contractor submits to the Architect (1) an affidavit that all payrolls, bills for materials and equipment, and other indebtedness connected with the Work for which the Owner or his property might in any way be responsible, have been paid or otherwise satisfied, (2) consent of surety, if any, to final payment and (3), if required by the Owner, other data establishing payment or satisfaction of all such obligations, such as receipts, releases and waivers of liens arising out of the Contract, to the extent and in such form as may be designated by the Owner. If any Subcontractor refuses to furnish a release or waiver required by the Owner, the Contractor may furnish a bond satisfactory to the Owner to indemnify him against any such lien. If any such lien remains unsatisfied after all payments are made, the Contractor shall refund to the Owner all moneys that the latter may be compelled to pay in discharging such lien, including all costs and reasonable attorneys' fees.

9.9.3 If, after Substantial Completion of the Work, final completion thereof is materially delayed through no fault of the Contractor or by the issuance of Change Orders affecting final completion, and the Architect so confirms, the Owner shall, upon application by the Contractor and certification by the Architect, and without terminating the Contract, make payment of the balance due for that portion of the Work fully completed and accepted. If the remaining balance for Work not fully completed or corrected is less than the retainage stipulated in the Contract Documents, and if bonds have been furnished as provided in Paragraph 7.5, the written consent of the surety to the payment of the balance due for that portion of the Work fully completed and accepted shall be submitted by the Contractor to the Architect prior to certification of such payment. Such payment shall be made under the terms and conditions governing final payment, except that it shall not constitute a waiver of claims.

9.9.4 The making of final payment shall constitute a waiver of all claims by the Owner except those arising from:
  .1 unsettled liens,
  .2 faulty or defective Work appearing after Substantial Completion,
  .3 failure of the Work to comply with the requirements of the Contract Documents, or
  .4 terms of any special warranties required by the Contract Documents.

9.9.5 The acceptance of final payment shall constitute a waiver of all claims by the Contractor except those previously made in writing and identified by the Contractor as unsettled at the time of the final Application for Payment.

ARTICLE 10
PROTECTION OF PERSONS AND PROPERTY

10.1 SAFETY PRECAUTIONS AND PROGRAMS

10.1.1 The Contractor shall be responsible for initiating, maintaining and supervising all safety precautions and programs in connection with the Work.

10.2 SAFETY OF PERSONS AND PROPERTY

10.2.1 The Contractor shall take all reasonable precautions for the safety of, and shall provide all reasonable protection to prevent damage, injury or loss to:
  .1 all employees on the Work and all other persons who may be affected thereby;
  .2 all the Work and all materials and equipment to be incorporated therein, whether in storage on or off the site, under the care, custody or control of the Contractor or any of his Subcontractors or Sub-subcontractors; and
  .3 other property at the site or adjacent thereto, including trees, shrubs, lawns, walks, pavements, roadways, structures and utilities not designated for removal, relocation or replacement in the course of construction.

10.2.2 The Contractor shall give all notices and comply with all applicable laws, ordinances, rules, regulations and lawful orders of any public authority bearing on the safety of persons or property or their protection from damage, injury or loss.

10.2.3 The Contractor shall erect and maintain, as required by existing conditions and progress of the Work, all reasonable safeguards for safety and protection, including posting danger signs and other warnings against hazards, promulgating safety regulations and notifying owners and users of adjacent utilities.

10.2.4 When the use or storage of explosives or other hazardous materials or equipment is necessary for the execution of the Work, the Contractor shall exercise the utmost care and shall carry on such activities under the supervision of properly qualified personnel.

10.2.5 The Contractor shall promptly remedy all damage or loss (other than damage or loss insured under Paragraph 11.3) to any property referred to in Clauses 10.2.1.2 and 10.2.1.3 caused in whole or in part by the Contractor, any Subcontractor, any Sub-subcontractor, or anyone directly or indirectly employed by any of them, or by anyone for whose acts any of them may be liable and for which the Contractor is responsible under Clauses 10.2.1.2 and 10.2.1.3, except damage or loss attributable to the acts or omissions of the Owner or Architect or anyone directly or indirectly employed by either of them, or by anyone for whose acts either of them may be liable, and not attributable to the fault or negligence of the Contractor. The foregoing obligations of the Contractor are in addition to his obligations under Paragraph 4.18.

FIGURE A5.2. (*continued*)

10.2.6 The Contractor shall designate a responsible member of his organization at the site whose duty shall be the prevention of accidents. This person shall be the Contractor's superintendent unless otherwise designated by the Contractor in writing to the Owner and the Architect.

10.2.7 The Contractor shall not load or permit any part of the Work to be loaded so as to endanger its safety.

10.3 EMERGENCIES

10.3.1 In any emergency affecting the safety of persons or property, the Contractor shall act, at his discretion, to prevent threatened damage, injury or loss. Any additional compensation or extension of time claimed by the Contractor on account of emergency work shall be determined as provided in Article 12 for Changes in the Work.

ARTICLE 11

INSURANCE

11.1 CONTRACTOR'S LIABILITY INSURANCE

11.1.1 The Contractor shall purchase and maintain such insurance as will protect him from claims set forth below which may arise out of or result from the Contractor's operations under the Contract, whether such operations be by himself or by any Subcontractor or by anyone directly or indirectly employed by any of them, or by anyone for whose acts any of them may be liable:

- .1 claims under workers' or workmen's compensation, disability benefit and other similar employee benefit acts;
- .2 claims for damages because of bodily injury, occupational sickness or disease, or death of his employees;
- .3 claims for damages because of bodily injury, sickness or disease, or death of any person other than his employees;
- .4 claims for damages insured by usual personal injury liability coverage which are sustained (1) by any person as a result of an offense directly or indirectly related to the employment of such person by the Contractor, or (2) by any other person;
- .5 claims for damages, other than to the Work itself, because of injury to or destruction of tangible property, including loss of use resulting therefrom; and
- .6 claims for damages because of bodily injury or death of any person or property damage arising out of the ownership, maintenance or use of any motor vehicle.

11.1.2 The insurance required by Subparagraph 11.1.1 shall be written for not less than any limits of liability specified in the Contract Documents, or required by law, whichever is greater.

11.1.3 The insurance required by Subparagraph 11.1.1 shall include contractual liability insurance applicable to the Contractor's obligations under Paragraph 4.18.

11.1.4 Certificates of Insurance acceptable to the Owner shall be filed with the Owner prior to commencement of the Work. These Certificates shall contain a provision that coverage afforded under the policies will not be cancelled until at least thirty days' prior written notice has been given to the Owner.

11.2 OWNER'S LIABILITY INSURANCE

11.2.1 The Owner shall be responsible for purchasing and maintaining his own liability insurance and, at his option, may purchase and maintain such insurance as will protect him against claims which may arise from operations under the Contract.

11.3 PROPERTY INSURANCE

11.3.1 Unless otherwise provided, the Owner shall purchase and maintain property insurance upon the entire Work at the site to the full insurable value thereof. This insurance shall include the interests of the Owner, the Contractor, Subcontractors and Sub-subcontractors in the Work and shall insure against the perils of fire and extended coverage and shall include "all risk" insurance for physical loss or damage including, without duplication of coverage, theft, vandalism and malicious mischief. If the Owner does not intend to purchase such insurance for the full insurable value of the entire Work, he shall inform the Contractor in writing prior to commencement of the Work. The Contractor may then effect insurance which will protect the interests of himself, his Subcontractors and the Sub-subcontractors in the Work, and by appropriate Change Order the cost thereof shall be charged to the Owner. If the Contractor is damaged by failure of the Owner to purchase or maintain such insurance and to so notify the Contractor, then the Owner shall bear all reasonable costs properly attributable thereto. If not covered under the all risk insurance or otherwise provided in the Contract Documents, the Contractor shall effect and maintain similar property insurance on portions of the Work stored off the site or in transit when such portions of the Work are to be included in an Application for Payment under Subparagraph 9.3.2.

11.3.2 The Owner shall purchase and maintain such boiler and machinery insurance as may be required by the Contract Documents or by law. This insurance shall include the interests of the Owner, the Contractor, Subcontractors and Sub-subcontractors in the Work.

11.3.3 Any loss insured under Subparagraph 11.3.1 is to be adjusted with the Owner and made payable to the Owner as trustee for the insureds, as their interests may appear, subject to the requirements of any applicable mortgagee clause and of Subparagraph 11.3.8. The Contractor shall pay each Subcontractor a just share of any insurance moneys received by the Contractor, and by appropriate agreement, written where legally required for validity, shall require each Subcontractor to make payments to his Sub-subcontractors in similar manner.

11.3.4 The Owner shall file a copy of all policies with the Contractor before an exposure to loss may occur.

11.3.5 If the Contractor requests in writing that insurance for risks other than those described in Subparagraphs 11.3.1 and 11.3.2 or other special hazards be included in the property insurance policy, the Owner shall, if possible, include such insurance, and the cost thereof shall be charged to the Contractor by appropriate Change Order.

FIGURE A5.2. (*continued*)

11.3.6 The Owner and Contractor waive all rights against (1) each other and the Subcontractors, Sub-subcontractors, agents and employees each of the other, and (2) the Architect and separate contractors, if any, and their subcontractors, sub-subcontractors, agents and employees, for damages caused by fire or other perils to the extent covered by insurance obtained pursuant to this Paragraph 11.3 or any other property insurance applicable to the Work, except such rights as they may have to the proceeds of such insurance held by the Owner as trustee. The foregoing waiver afforded the Architect, his agents and employees shall not extend to the liability imposed by Subparagraph 4.18.3. The Owner or the Contractor, as appropriate, shall require of the Architect, separate contractors, Subcontractors and Sub-subcontractors by appropriate agreements, written where legally required for validity, similar waivers each in favor of all other parties enumerated in this Subparagraph 11.3.6.

11.3.7 If required in writing by any party in interest, the Owner as trustee shall, upon the occurrence of an insured loss, give bond for the proper performance of his duties. He shall deposit in a separate account any money so received, and he shall distribute it in accordance with such agreement as the parties in interest may reach, or in accordance with an award by arbitration in which case the procedure shall be as provided in Paragraph 7.9. If after such loss no other special agreement is made, replacement of damaged work shall be covered by an appropriate Change Order.

11.3.8 The Owner as trustee shall have power to adjust and settle any loss with the insurers unless one of the parties in interest shall object in writing within five days after the occurrence of loss to the Owner's exercise of this power, and if such objection be made, arbitrators shall be chosen as provided in Paragraph 7.9. The Owner as trustee shall, in that case, make settlement with the insurers in accordance with the directions of such arbitrators. If distribution of the insurance proceeds by arbitration is required, the arbitrators will direct such distribution.

11.3.9 If the Owner finds it necessary to occupy or use a portion or portions of the Work prior to Substantial Completion thereof, such occupancy or use shall not commence prior to a time mutually agreed to by the Owner and Contractor and to which the insurance company or companies providing the property insurance have consented by endorsement to the policy or policies. This insurance shall not be cancelled or lapsed on account of such partial occupancy or use. Consent of the Contractor and of the insurance company or companies to such occupancy or use shall not be unreasonably withheld.

11.4 LOSS OF USE INSURANCE

11.4.1 The Owner, at his option, may purchase and maintain such insurance as will insure him against loss of use of his property due to fire or other hazards, however caused. The Owner waives all rights of action against the Contractor for loss of use of his property, including consequential losses due to fire or other hazards however caused, to the extent covered by insurance under this Paragraph 11.4.

ARTICLE 12

CHANGES IN THE WORK

12.1 CHANGE ORDERS

12.1.1 A Change Order is a written order to the Contractor signed by the Owner and the Architect, issued after execution of the Contract, authorizing a change in the Work or an adjustment in the Contract Sum or the Contract Time. The Contract Sum and the Contract Time may be changed only by Change Order. A Change Order signed by the Contractor indicates his agreement therewith, including the adjustment in the Contract Sum or the Contract Time.

12.1.2 The Owner, without invalidating the Contract, may order changes in the Work within the general scope of the Contract consisting of additions, deletions or other revisions, the Contract Sum and the Contract Time being adjusted accordingly. All such changes in the Work shall be authorized by Change Order, and shall be performed under the applicable conditions of the Contract Documents.

12.1.3 The cost or credit to the Owner resulting from a change in the Work shall be determined in one or more of the following ways:
 .1 by mutual acceptance of a lump sum properly itemized and supported by sufficient substantiating data to permit evaluation;
 .2 by unit prices stated in the Contract Documents or subsequently agreed upon;
 .3 by cost to be determined in a manner agreed upon by the parties and a mutually acceptable fixed or percentage fee; or
 .4 by the method provided in Subparagraph 12.1.4.

12.1.4 If none of the methods set forth in Clauses 12.1.3.1, 12.1.3.2 or 12.1.3.3 is agreed upon, the Contractor, provided he receives a written order signed by the Owner, shall promptly proceed with the Work involved. The cost of such Work shall then be determined by the Architect on the basis of the reasonable expenditures and savings of those performing the Work attributable to the change, including, in the case of an increase in the Contract Sum, a reasonable allowance for overhead and profit. In such case, and also under Clauses 12.1.3.3 and 12.1.3.4 above, the Contractor shall keep and present, in such form as the Architect may prescribe, an itemized accounting together with appropriate supporting data for inclusion in a Change Order. Unless otherwise provided in the Contract Documents, cost shall be limited to the following: cost of materials, including sales tax and cost of delivery; cost of labor, including social security, old age and unemployment insurance, and fringe benefits required by agreement or custom; workers' or workmen's compensation insurance; bond premiums; rental value of equipment and machinery; and the additional costs of supervision and field office personnel directly attributable to the change. Pending final determination of cost to the Owner, payments on account shall be made on the Architect's Certificate for Payment. The amount of credit to be allowed by the Contractor to the Owner for any deletion

AIA DOCUMENT A201 • GENERAL CONDITIONS OF THE CONTRACT FOR CONSTRUCTION • THIRTEENTH EDITION • AUGUST 1976
AIA® • © 1976 • THE AMERICAN INSTITUTE OF ARCHITECTS, 1735 NEW YORK AVENUE, N.W., WASHINGTON, D.C. 20006   **A201-1976**

FIGURE A5.2. (continued)

or change which results in a net decrease in the Contract Sum will be the amount of the actual net cost as confirmed by the Architect. When both additions and credits covering related Work or substitutions are involved in any one change, the allowance for overhead and profit shall be figured on the basis of the net increase, if any, with respect to that change.

12.1.5 If unit prices are stated in the Contract Documents or subsequently agreed upon, and if the quantities originally contemplated are so changed in a proposed Change Order that application of the agreed unit prices to the quantities of Work proposed will cause substantial inequity to the Owner or the Contractor, the applicable unit prices shall be equitably adjusted.

12.2 CONCEALED CONDITIONS

12.2.1 Should concealed conditions encountered in the performance of the Work below the surface of the ground or should concealed or unknown conditions in an existing structure be at variance with the conditions indicated by the Contract Documents, or should unknown physical conditions below the surface of the ground or should concealed or unknown conditions in an existing structure of an unusual nature, differing materially from those ordinarily encountered and generally recognized as inherent in work of the character provided for in this Contract, be encountered, the Contract Sum shall be equitably adjusted by Change Order upon claim by either party made within twenty days after the first observance of the conditions.

12.3 CLAIMS FOR ADDITIONAL COST

12.3.1 If the Contractor wishes to make a claim for an increase in the Contract Sum, he shall give the Architect written notice thereof within twenty days after the occurrence of the event giving rise to such claim. This notice shall be given by the Contractor before proceeding to execute the Work, except in an emergency endangering life or property in which case the Contractor shall proceed in accordance with Paragraph 10.3. No such claim shall be valid unless so made. If the Owner and the Contractor cannot agree on the amount of the adjustment in the Contract Sum, it shall be determined by the Architect. Any change in the Contract Sum resulting from such claim shall be authorized by Change Order.

12.3.2 If the Contractor claims that additional cost is involved because of, but not limited to, (1) any written interpretation pursuant to Subparagraph 2.2.8, (2) any order by the Owner to stop the Work pursuant to Paragraph 3.3 where the Contractor was not at fault, (3) any written order for a minor change in the Work issued pursuant to Paragraph 12.4, or (4) failure of payment by the Owner pursuant to Paragraph 9.7, the Contractor shall make such claim as provided in Subparagraph 12.3.1.

12.4 MINOR CHANGES IN THE WORK

12.4.1 The Architect will have authority to order minor changes in the Work not involving an adjustment in the Contract Sum or an extension of the Contract Time and not inconsistent with the intent of the Contract Documents. Such changes shall be effected by written order, and shall be binding on the Owner and the Contractor.

The Contractor shall carry out such written orders promptly.

ARTICLE 13

UNCOVERING AND CORRECTION OF WORK

13.1 UNCOVERING OF WORK

13.1.1 If any portion of the Work should be covered contrary to the request of the Architect or to requirements specifically expressed in the Contract Documents, it must, if required in writing by the Architect, be uncovered for his observation and shall be replaced at the Contractor's expense.

13.1.2 If any other portion of the Work has been covered which the Architect has not specifically requested to observe prior to being covered, the Architect may request to see such Work and it shall be uncovered by the Contractor. If such Work be found in accordance with the Contract Documents, the cost of uncovering and replacement shall, by appropriate Change Order, be charged to the Owner. If such Work be found not in accordance with the Contract Documents, the Contractor shall pay such costs unless it be found that this condition was caused by the Owner or a separate contractor as provided in Article 6, in which event the Owner shall be responsible for the payment of such costs.

13.2 CORRECTION OF WORK

13.2.1 The Contractor shall promptly correct all Work rejected by the Architect as defective or as failing to conform to the Contract Documents whether observed before or after Substantial Completion and whether or not fabricated, installed or completed. The Contractor shall bear all costs of correcting such rejected Work, including compensation for the Architect's additional services made necessary thereby.

13.2.2 If, within one year after the Date of Substantial Completion of the Work or designated portion thereof or within one year after acceptance by the Owner of designated equipment or within such longer period of time as may be prescribed by law or by the terms of any applicable special warranty required by the Contract Documents, any of the Work is found to be defective or not in accordance with the Contract Documents, the Contractor shall correct it promptly after receipt of a written notice from the Owner to do so unless the Owner has previously given the Contractor a written acceptance of such condition. This obligation shall survive termination of the Contract. The Owner shall give such notice promptly after discovery of the condition.

13.2.3 The Contractor shall remove from the site all portions of the Work which are defective or non-conforming and which have not been corrected under Subparagraphs 4.5.1, 13.2.1 and 13.2.2, unless removal is waived by the Owner.

13.2.4 If the Contractor fails to correct defective or non-conforming Work as provided in Subparagraphs 4.5.1, 13.2.1 and 13.2.2, the Owner may correct it in accordance with Paragraph 3.4.

FIGURE A5.2.   (*continued*)

13.2.5 If the Contractor does not proceed with the correction of such defective or non-conforming Work within a reasonable time fixed by written notice from the Architect, the Owner may remove it and may store the materials or equipment at the expense of the Contractor. If the Contractor does not pay the cost of such removal and storage within ten days thereafter, the Owner may upon ten additional days' written notice sell such Work at auction or at private sale and shall account for the net proceeds thereof, after deducting all the costs that should have been borne by the Contractor, including compensation for the Architect's additional services made necessary thereby. If such proceeds of sale do not cover all costs which the Contractor should have borne, the difference shall be charged to the Contractor and an appropriate Change Order shall be issued. If the payments then or thereafter due the Contractor are not sufficient to cover such amount, the Contractor shall pay the difference to the Owner.

13.2.6 The Contractor shall bear the cost of making good all work of the Owner or separate contractors destroyed or damaged by such correction or removal.

13.2.7 Nothing contained in this Paragraph 13.2 shall be construed to establish a period of limitation with respect to any other obligation which the Contractor might have under the Contract Documents, including Paragraph 4.5 hereof. The establishment of the time period of one year after the Date of Substantial Completion or such longer period of time as may be prescribed by law or by the terms of any warranty required by the Contract Documents relates only to the specific obligation of the Contractor to correct the Work, and has no relationship to the time within which his obligation to comply with the Contract Documents may be sought to be enforced, nor to the time within which proceedings may be commenced to establish the Contractor's liability with respect to his obligations other than specifically to correct the Work.

13.3 ACCEPTANCE OF DEFECTIVE OR NON-CONFORMING WORK

13.3.1 If the Owner prefers to accept defective or non-conforming Work, he may do so instead of requiring its removal and correction, in which case a Change Order will be issued to reflect a reduction in the Contract Sum where appropriate and equitable. Such adjustment shall be effected whether or not final payment has been made.

ARTICLE 14

TERMINATION OF THE CONTRACT

14.1 TERMINATION BY THE CONTRACTOR

14.1.1 If the Work is stopped for a period of thirty days under an order of any court or other public authority having jurisdiction, or as a result of an act of government, such as a declaration of a national emergency making materials unavailable, through no act or fault of the Contractor or a Subcontractor or their agents or employees or any other persons performing any of the Work under a contract with the Contractor, or if the Work should be stopped for a period of thirty days by the Contractor because the Architect has not issued a Certificate for Payment as provided in Paragraph 9.7 or because the Owner has not made payment thereon as provided in Paragraph 9.7, then the Contractor may, upon seven additional days' written notice to the Owner and the Architect, terminate the Contract and recover from the Owner payment for all Work executed and for any proven loss sustained upon any materials, equipment, tools, construction equipment and machinery, including reasonable profit and damages.

14.2 TERMINATION BY THE OWNER

14.2.1 If the Contractor is adjudged a bankrupt, or if he makes a general assignment for the benefit of his creditors, or if a receiver is appointed on account of his insolvency, or if he persistently or repeatedly refuses or fails, except in cases for which extension of time is provided, to supply enough properly skilled workmen or proper materials, or if he fails to make prompt payment to Subcontractors or for materials or labor, or persistently disregards laws, ordinances, rules, regulations or orders of any public authority having jurisdiction, or otherwise is guilty of a substantial violation of a provision of the Contract Documents, then the Owner, upon certification by the Architect that sufficient cause exists to justify such action, may, without prejudice to any right or remedy and after giving the Contractor and his surety, if any, seven days' written notice, terminate the employment of the Contractor and take possession of the site and of all materials, equipment, tools, construction equipment and machinery thereon owned by the Contractor and may finish the Work by whatever method he may deem expedient. In such case the Contractor shall not be entitled to receive any further payment until the Work is finished.

14.2.2 If the unpaid balance of the Contract Sum exceeds the costs of finishing the Work, including compensation for the Architect's additional services made necessary thereby, such excess shall be paid to the Contractor. If such costs exceed the unpaid balance, the Contractor shall pay the difference to the Owner. The amount to be paid to the Contractor or to the Owner, as the case may be, shall be certified by the Architect, upon application, in the manner provided in Paragraph 9.4, and this obligation for payment shall survive the termination of the Contract.

FIGURE A5.2. (*continued*)

INSTRUCTION SHEET *AIA DOCUMENT A101a*

FOR AIA DOCUMENT A101, STANDARD FORM OF AGREEMENT BETWEEN OWNER AND CONTRACTOR — JUNE 1977 EDITION

AIA Document A101, Standard Form of Agreement Between Owner and Contractor, is for use where the basis of payment is a stipulated sum (fixed price). The 1977 Edition has been prepared for use with the 1976 Edition of AIA Document A201, General Conditions of the Contract for Construction. It is suitable for any arrangement between the Owner and the Contractor where the cost has been set in advance either by bidding or by negotiation. Although the Owner has the advantage of advance knowledge of the cost of the Work, increased efforts to assure Contract compliance may be required, in view of the fact that the price is fixed and the Contractor has a financial interest in minimizing the cost of carrying out the Work. A more complete explanation of A101 is provided in Architect's Handbook of Professional Practice, Chapter 17: Owner-Contractor and Contractor-Subcontractor Agreements.

Below is a listing of pertinent provisions revised or added to the 1977 Edition of the Stipulated Sum Owner-Contractor Agreement Form:

Article 3 — Modified to read, "Time of Commencement and *Substantial* Completion." The General Conditions, AIA Document A201, 1976 Edition, make it clear that the Contract Time runs until the Date of Substantial Completion; the Owner should be aware that an additional period of time will be required to reach final completion.

Article 4 — Revised to include reference to the Contract Documents for determination of amounts of Change Orders. Parenthetical instruction describing basis of payment now includes *base bid* and *accepted alternates*.

Article 5 — A sentence has been added at the end of the first paragraph to stipulate a specific day of the month as the end of the period for which progress payments will be made. The Agreement requires that the Owner make progress payments not later than *an agreed-upon number of days* following the end of that period covered by the Application for Payment. (Note that the General Conditions, AIA Document A201, 1976 Edition, require in Subparagraph 9.3.1 that the Contractor apply for payment at least 10 days in advance of the date payment is due.)

The provision for interest on payments due and unpaid has been revised to provide for the entry of a specific rate of interest in accordance with the changes in the interest provision of A201, Paragraph 7.8. A parenthetical statement has been added drawing attention to Truth-in-Lending and other laws which may govern the use and form of an interest provision under certain circumstances.

Article 6 — Modified to provide that final payment is due when the Work has been completed (the reference to an agreed-upon number of days after Substantial Completion of the Work has been deleted). The Certificate of Substantial Completion will provide the time period within which the Contractor will bring the Work to final completion.

Completing the form:
(NOTE: Prospective bidders should be aware of any additional provisions which may be included in A101, such as liquidated damages, retainage, or payment for stored materials, by an appropriate notice in the Bidding Documents.)

Cover Page — The names of the Owner and the Architect should be shown in the same form as in the other Project documents; include the full legal or corporate names under which the Owner and Contractor are entering the Agreement.

Article 1 — The Contract Documents
The Contract Documents must be enumerated in detail under Article 7. If unit prices are incorporated in the Contractor's bid, the bid itself may be incorporated into the Contract; similarly, other bidding documents, bonds, etc. may be incorporated, particularly in public work.

Article 2 — The Work
The general scope of the Work should be carefully defined here, since changes by Change Order, under Paragraph 12.1 of A201, must be within the general scope of the Work contemplated by the Contract. This Article should be used to describe the portions of the Project for which the Contractor is responsible, if separate contracts are used.

Article 3 — Time of Commencement and Substantial Completion
The following items should be included as appropriate:
• Date of commencement of the Work
• Provision for notice to proceed, if any
• Date of Substantial Completion of the Work
• Provision, if any, for liquidated damages if not included in the Supplementary Conditions (see AIA Document A511)

Date of commencement of the Work should not be earlier than the date of execution of the Contract. When time of performance is to be strictly enforced, the statement of starting time should be carefully considered.
A sample provision where a notice to proceed will be used is as follows:
The Work shall commence on the date stipulated in the notice to proceed and shall be substantially completed on ―――――――――――――――――

The Date of Substantial Completion of the Work may be expressed as a number of days (preferably calendar days) or as a specific date. The time requirements will ordinarily have been fulfilled when the Work is Substantially Complete, as defined in A201, Subparagraph 8.1.3, even if a few minor items may remain to be completed or corrected.

AIA DOCUMENT A101a • INSTRUCTION SHEET FOR OWNER-CONTRACTOR AGREEMENT • 1977 EDITION • AIA®
THE AMERICAN INSTITUTE OF ARCHITECTS, 1735 NEW YORK AVENUE, N.W., WASHINGTON, D. C. 20006   A101a-1977

AIA copyrighted material has been reproduced with permission of the American Institute of Architects under permission number 81067. Further reproduction is prohibited.

Because AIA Documents are revised from time to time, users should ascertain from the AIA the current edition(s) of the Document(s) reproduced herein.

FIGURE A5.3. American Institute of Architects (AIA) Document A101a

If liquidated damages are to be assessed because delayed construction will result in the Owner actually suffering loss, the amount per day should be entered in the Supplementary Conditions or the Agreement. Factors such as confidentiality will help determine the choice of location. Liquidated Damages are not a penalty to be inflicted on the Contractor, but must bear an actual and reasonably estimated relationship to the loss to the Owner if the building is not completed on time; for example, the cost per day of renting space to house students if a dormitory cannot be occupied when needed, additional financing costs, loss of profits, etc. This provision, which should be carefully reviewed, if not drafted, by the Owner's attorney, may be as follows:

> The Owner will suffer financial damage if the Project is not Substantially Completed on the date set forth in the Contract Documents. The Contractor (and his Surety) shall pay to the Owner the sums hereinafter stipulated as fixed, agreed and liquidated damages for each calendar day of delay until the Work is Substantially Completed: _____ dollars ($).

A provision for penalty and *bonus*, where such is appropriate, is suggested as follows:

> The Contractor agrees to pay to the Owner a sum of _____ dollars ($) for each calendar day beyond the established completion date that the Work remains uncompleted, in consideration of which the Owner agrees to pay the Contractor a sum of _____ dollars ($) for each calendar day ahead of the established completion date that the Work is determined to be Substantially Completed.

Note that a liquidated damages provision may be placed in the Supplementary Conditions in order to put Subcontractors on notice of this condition.

Article 4 — Contract Sum

The following items should be included as appropriate:
* The Contract Sum
* Unit prices, cash allowances, or cash contingency allowances, if any

If not covered elsewhere in the Contract Documents in more detail, the following provision for unit prices is suggested:

> The unit prices listed below shall determine the value of extra Work or changes, as applicable. They shall be considered complete including all material and equipment, labor, installation costs, overhead and profit, and shall be used uniformly for either additions or deductions.

Specific allowances for overhead and for profit on Change Orders may also be included here.

Article 5 — Progress Payments

The following items should be included as appropriate:
* Due dates for payments
* Retained percentage
* Payment for materials stored off the site

The due date for payment is often arbitrarily set. It should be a date mutually acceptable to both the Owner and the Contractor in consideration of the time required for the Contractor to prepare an Application for Payment, for the Architect to check and certify payment, and for the Owner to make payment, within the time limits set in Subparagraph 9.4.1, of A201, and in this Article of A101.

The last date upon which Work may be included in an Application should be normally not less than fourteen days prior to the payment date to allow seven days for the Architect to evaluate the Application and issue a Certificate for Payment and seven days for the Owner to make payment as provided in Article 9 of AIA Document A201. The Contractor may prefer an additional few days to allow time for preparation of his Application.

Retained percentage: It is a frequent practice to pay the Contractor 90 percent of the earned sum when payments fall due, retaining 10 percent to assure faithful performance of the Contract. These percentages may vary with circumstances and localities. AIA endorses the concept of reducing retainage as rapidly as possible consistent with the continued protection of all affected interests. See AIA Document A511, Guide for Supplementary Conditions, for a complete discussion.

A provision for reducing retainage should provide that the reduction will be made only if, in the judgment of the Architect, satisfactory progress is being made and maintained in the Work. If the Contractor has furnished a bond, he should be required to provide a Consent of Surety to Reduction In or Partial Release of Retainage (AIA Document G707A), before the retainage is reduced.

Payment for materials stored *off* the site should be provided for in a specific agreement and included in Article 7. Provisions regarding transportation to the site and insurance to protect the Owner's interests should be included.

Article 6 — Final Payment

At the time final payment is requested, the Architect should be particularly meticulous in ascertaining that all claims have been settled, in defining any claims that may still be unsettled, in obtaining from the Contractor the certification required in Article 9 of AIA Document A201 that no indebtedness against the Project remains, and in being assured that to the best of his knowledge and belief, based on the final inspection, the Contract requirements have been fulfilled.

Article 7 — Miscellaneous Provisions

An accurate, detailed enumeration of all Documents included in the Contract must be made in this Article.

Signatures — Subparagraph 1.2.1 of AIA Document A201, states that the Contract Documents shall be executed in not less than triplicate by the Owner and the Contractor. The Agreement should be executed by the parties in their capacities as individuals, partners, officers, etc., as appropriate.

A101a-1977   AIA DOCUMENT A101a • INSTRUCTION SHEET FOR OWNER-CONTRACTOR AGREEMENT • 1977 EDITION • AIA®
THE AMERICAN INSTITUTE OF ARCHITECTS, 1735 NEW YORK AVENUE, N.W., WASHINGTON, D.C. 20006

FIGURE A5.3. *(continued)*

THE ASSOCIATED GENERAL CONTRACTORS

STANDARD FORM OF AGREEMENT BETWEEN OWNER AND CONSTRUCTION MANAGER

(GUARANTEED MAXIMUM PRICE OPTION)

(See AGC Document No. 8a for Establishing the Guaranteed Maximum Price)

This Document has important legal and insurance consequences; consultation with an attorney is encouraged with respect to its completion or modification.

AGREEMENT

Made this   day of   in the year of Nineteen Hundred and

BETWEEN
                                                                        the Owner, and
                                                        the Construction Manager

For services in connection with the following described Project: (Include complete Project location and scope)

The Architect/Engineer for the Project is

The Owner and the Construction Manager agree as set forth below:

AIA copyrighted material has been reproduced with permission of the American Institute of Architects under permission number 81067. Further reproduction is prohibited.

Because AIA Documents are revised from time to time, users should ascertain from the AIA the current edition(s) of the Document(s) reproduced herein.

Certain provisions of this document have been derived, with modifications, from the following documents published by The American Institute of Architects: AIA Document A111, Owner Contractor Agreement, © 1974; AIA Document A201, General Conditions, © 1976; AIA Document B801, Owner Construction Manager Agreement, © 1973, by The American Institute of Architects. Usage made of AIA language, with the permission of AIA, does not apply AIA endorsement or approval of this document. Further reproduction of copyrighted AIA materials without separate written permission from AIA is prohibited.

AGC DOCUMENT NO. 8   OWNER CONSTRUCTION MANAGER AGREEMENT   JULY 1980
© 1980 Associated General Contractors of America

FIGURE A5.4. Associated General Contractors (AGC) Document No. 8

TABLE OF CONTENTS

ARTICLES   **PAGE**

1. The Construction Team and Extent of Agreement 1
2. Construction Manager's Services 1
3. The Owner's Responsibilities 4
4. Trade Contracts 5
5. Schedule 5
6. Guaranteed Maximum Price 6
7. Construction Manager's Fee 6
8. Cost of the Project 7
9. Changes in the Project 8
10. Discounts 9
11. Payments to the Construction Manager 10
12. Insurance, Indemnity and Waiver of Subrogation 10
13. Termination of the Agreement and Owner's Right to Perform Construction Manager's Obligations 13
14. Assignment and Governing Law 14
15. Miscellaneous Provisions 14
16. Arbitration 14

FIGURE A5.4. (*continued*)

ARTICLE 1

The Construction Team and Extent of Agreement

The CONSTRUCTION MANAGER accepts the relationship of trust and confidence established between him and the Owner by this Agreement. He covenants with the Owner to furnish his best skill and judgment and to cooperate with the Architect/Engineer in furthering the interests of the Owner. He agrees to furnish efficient business administration and superintendence and to use his best efforts to complete the Project in an expeditious and economical manner consistent with the interest of the Owner.

1.1 *The Construction Team:* The Construction Manager, the Owner, and the Architect/Engineer called the "Construction Team" shall work from the beginning of design through construction completion. The Construction Manager shall provide leadership to the Construction Team on all matters relating to construction.

1.2 *Extent of Agreement:* This Agreement represents the entire agreement between the Owner and the Construction Manager and supersedes all prior negotiations, representations or agreements. When Drawings and Specifications are complete, they shall be identified by amendment to this Agreement. This Agreement shall not be superseded by any provisions of the documents for construction and may be amended only by written instrument signed by both the Owner and the Construction Manager.

1.3 *Definitions:* The Project is the total construction to be performed under this Agreement. The Work is that part of the construction that the Construction Manager is to perform with his own forces or that part of the construction that a particular Trade Contractor is to perform. The term day shall mean calendar day unless otherwise specifically designated.

ARTICLE 2

Construction Manager's Services

The Construction Manager will perform the following services under this Agreement in each of the two phases described below.

2.1 Design Phase

2.1.1 *Consultation During Project Development:* Schedule and attend regular meetings with the Architect/Engineer during the development of conceptual and preliminary design to advise on site use and improvements, selection of materials, building systems and equipment. Provide recommendations on construction feasibility, availability of materials and labor, time requirements for installation and construction, and factors related to cost including costs of alternative designs or materials, preliminary budgets, and possible economies.

2.1.2 *Scheduling:* Develop a Project Time Schedule that coordinates and integrates the Architect/Engineer's design efforts with construction schedules. Update the Project Time Schedule incorporating a detailed schedule for the construction operations of the Project, including realistic activity sequences and durations, allocation of labor and materials, processing of shop drawings and samples, and delivery of products requiring long lead-time procurement. Include the Owner's occupancy requirements showing portions of the Project having occupancy priority.

2.1.3 *Project Construction Budget:* Prepare a Project budget as soon as major Project requirements have been identified, and update periodically for the Owner's approval. Prepare an estimate based on a quantity survey of Drawings and Specifications at the end of the schematic design phase for approval by the Owner as the Project Construction Budget. Update and refine this estimate for the Owner's approval as the development of the Drawings and Specifications proceeds, and advise the Owner and the Architect/Engineer if it appears that the Project Construction Budget will not be met and make recommendations for corrective action.

2.1.4 *Coordination of Contract Documents:* Review the Drawings and Specifications as they are being prepared, recommending alternative solutions whenever design details affect construction feasibility or schedules without, however, assuming any of the Architect/Engineer's responsibilities for design.

FIGURE A5.4. *(continued)*

2.1.5 *Construction Planning:* Recommend for purchase and expedite the procurement of long-lead items to ensure their delivery by the required dates.

2.1.5.1 Make recommendations to the Owner and the Architect/Engineer regarding the division of Work in the Drawings and Specifications to facilitate the bidding and awarding of Trade Contracts, allowing for phased construction taking into consideration such factors as time of performance, availability of labor, overlapping trade jurisdictions, and provisions for temporary facilities.

2.1.5.2 Review the Drawings and Specifications with the Architect/Engineer to eliminate areas of conflict and overlapping in the Work to be performed by the various Trade Contractors and prepare prequalification criteria for bidders.

2.1.5.3 Develop Trade Contractor interest in the Project and as working Drawings and Specifications are completed, take competitive bids on the Work of the various Trade Contractors. After analyzing the bids, either award contracts or recommend to the Owner that such contracts be awarded.

2.1.6 *Equal Employment Opportunity:* Determine applicable requirements for equal emloyment opportunity programs for inclusion in Project bidding documents.

2.2 Construction Phase

2.2.1 *Project Control:* Monitor the Work of the Trade Contractors and coordinate the Work with the activities and responsibilities of the Owner, Architect/Engineer and Construction Manager to complete the Project in accordance with the Owner's objectives of cost, time and quality.

2.2.1.1 Maintain a competent full-time staff at the Project site to coordinate and provide general direction of the Work and progress of the Trade Contractors on the Project.

2.2.1.2 Establish on-site organization and lines of authority in order to carry out the overall plans of the Construction Team.

2.2.1.3 Establish procedures for coordination among the Owner, Architect/Engineer, Trade Contractors and Construction Manager with respect to all aspects of the Project and implement such procedures.

2.2.1.4 Schedule and conduct progress meetings at which Trade Contractors, Owner, Architect/Engineer and Construction Manager can discuss jointly such matters as procedures, progress, problems and scheduling.

2.2.1.5 Provide regular monitoring of the schedule as construction progresses. Identify potential variances between scheduled and probable completion dates. Review schedule for Work not started or incomplete and recommend to the Owner and Trade Contractors adjustments in the schedule to meet the probable completion date. Provide summary reports of each monitoring and document all changes in schedule.

2.2.1.6 Determine the adequacy of the Trade Contractors' personnel and equipment and the availability of materials and supplies to meet the schedule. Recommend courses of action to the Owner when requirements of a Trade Contract are not being met.

2.2.2 *Physical Construction:* Provide all supervision, labor, materials, construction equipment, tools and subcontract items which are necessary for the completion of the Project which are not provided by either the Trade Contractors or the Owner. To the extent that the Construction Manager performs any Work with his own forces, he shall, with respect to such Work, perform in accordance with the Plans and Specifications and in accordance with the procedure applicable to the Project.

2.2.3 *Cost Control:* Develop and monitor an effective system of Project cost control. Revise and refine the initially approved Project Construction Budget, incorporate approved changes as they occur, and develop cash flow reports and forecasts as needed. Identify variances between actual and budgeted or estimated costs and advise Owner and Architect/Engineer whenever projected cost exceeds budgets or estimates.

FIGURE A5.4. (*continued*)

2.2.3.1 Maintain cost accounting records on authorized Work performed under unit costs, actual costs for labor and material, or other bases requiring accounting records. Afford the Owner access to these records and preserve them for a period of three (3) years after final payment.

2.2.4 *Change Orders:* Develop and implement a system for the preparation, review and processing of Change Orders. Recommend necessary or desirable change to the Owner and the Architect/Engineer, review requests for changes, submit recommendations to the Owner and the Architect/Engineer, and assist in negotiating Change Orders.

2.2.5 *Payments to Trade Contractors:* Develop and implement a procedure for the review, processing and payment of applications by Trade Contractors for progress and final payments.

2.2.6 *Permits and Fees:* Assist the Owner and Architect/Engineer in obtaining all building permits and special permits for permanent improvements, excluding permits for inspection or temporary facilities required to be obtained directly by the various Trade Contractors. Assist in obtaining approvals from all the authorities having jurisdiction.

2.2.7 *Owner's Consultants:* If required, assist the Owner in selecting and retaining professional services of a surveyor, testing laboratories and special consultants, and coordinate these services, without assuming any responsibility or liability of or for these consultants.

2.2.8 *Inspection:* Inspect the Work of Trade Contractors for defects and deficiencies in the Work without assuming any of the Architect/Engineer's responsibilities for inspection.

2.2.8.1 Review the safety programs of each of the Trade Contractors and make appropriate recommendations. In making such recommendations and carrying out such reviews, he shall not be required to make exhaustive or continuous inspections to check safety precautions and programs in connection with the Project. The performance of such services by the Construction Manager shall not relieve the Trade Contractors of their responsibilities for the safety of persons and property, and for compliance with all federal, state and local statutes, rules, regulations and orders applicable to the conduct of the Work.

2.2.9 *Document Interpretation:* Refer all questions for interpretation of the documents prepared by the Architect/Engineer to the Architect/Engineer.

2.2.10 *Shop Drawings and Samples:* In collaboration with the Architect/Engineer, establish and implement procedures for expediting the processing and approval of shop drawings and samples.

2.2.11 *Reports and Project Site Documents:* Record the progress of the Project. Submit written progress reports to the Owner and the Architect/Engineer including information on the Trade Contractors' Work, and the percentage of completion. Keep a daily log available to the Owner and the Architect/Engineer.

2.2.11.1 Maintain at the Project site, on a current basis: records of all necessary Contracts, Drawings, samples, purchases, materials, equipment, maintenance and operating manuals and instructions, and other construction related documents, including all revisions. Obtain data from Trade Contractors and maintain a current set of record Drawings, Specifications and operating manuals. At the completion of the Project, deliver all such records to the Owner.

2.2.12 *Substantial Completion:* Determine Substantial Completion of the Work or designated portions thereof and prepare for the Architect/Engineer a list of incomplete or unsatisfactory items and a schedule for their completion.

2.2.13 *Start-Up:* With the Owner's maintenance personnel, direct the checkout of utilities, operations systems and equipment for readiness and assist in their initial start-up and testing by the Trade Contractors.

2.2.14 *Final Completion:* Determine final completion and provide written notice to the Owner and Architect/Engineer that the Work is ready for final inspection. Secure and transmit to the Architect/Engineer required guarantees, affidavits, releases, bonds and waivers. Turn over to the Owner all keys, manuals, record drawings and maintenance stocks.

2.2.15 *Warranty:* Where any Work is performed by the Construction Manager's own forces or by Trade Contractors under contract with the Construction Manager, the Construction Manager shall warrant that all materials and equipment included in such Work will be new, unless otherwise specified, and that such Work will be of good quality, free from improper workmanship and defective materials and in conformance with the Drawings and Specifications. With respect to the same Work, the

FIGURE A5.4. (*continued*)

Construction Manager further agrees to correct all Work defective in material and workmanship for a period of one year from the Date of Substantial Completion or for such longer periods of time as may be set forth with respect to specific warranties contained in the trade sections of the Specifications. The Construction Manager shall collect and deliver to the Owner any specific written warranties given by others.

2.3 Additional Services

2.3.1 At the request of the Owner the Construction Manager will provide the following additional services upon written agreement between the Owner and Construction Manager defining the extent of such additional services and the amount and manner in which the Construction Manager will be compensated for such additional services.

2.3.2 Services related to investigation, appraisals or valuations of existing conditions, facilities or equipment, or verifying the accuracy of existing drawings or other Owner-furnished information.

2.3.3 Services related to Owner-furnished equipment, furniture and furnishings which are not a part of this Agreement.

2.3.4 Services for tenant or rental spaces not a part of this Agreement.

2.3.5 Obtaining or training maintenance personnel or negotiating maintenance service contracts.

ARTICLE 3

Owner's Responsibilities

3.1 The Owner shall provide full information regarding his requirements for the Project.

3.2 The Owner shall designate a representative who shall be fully acquainted with the Project and has authority to issue and approve Project Construction Budgets, issue Change Orders, render decisions promptly and furnish information expeditiously.

3.3 The Owner shall retain an Architect/Engineer for design and to prepare construction documents for the Project. The Architect/Engineer's services, duties and responsibilities are described in the Agreement between the Owner and the Architect/Engineer, a copy of which will be furnished to the Construction Manager. The Agreement between the Owner and the Architect/Engineer shall not be modified without written notification to the Construction Manager.

3.4 The Owner shall furnish for the site of the Project all necessary surveys describing the physical characteristics, soil reports and subsurface investigations, legal limitations, utility locations, and a legal description.

3.5 The Owner shall secure and pay for necessary approvals, easements, assessments and charges required for the construction, use or occupancy of permanent structures or for permanent changes in existing facilities.

3.6 The Owner shall furnish such legal services as may be necessary for providing the items set forth in Paragraph 3.5, and such auditing services as he may require.

3.7 The Construction Manager will be furnished without charge all copies of Drawings and Specifications reasonably necessary for the execution of the Work.

3.8 The Owner shall provide the insurance for the Project as provided in Paragraph 12.4, and shall bear the cost of any bonds required.

3.9 The services, information, surveys and reports required by the above paragraphs or otherwise to be furnished by other consultants employed by the Owner, shall be furnished with reasonable promptness at the Owner's expense and the Construction Manager shall be entitled to rely upon the accuracy and completeness thereof.

3.10 If the Owner becomes aware of any fault or defect in the Project or non-conformance with the Drawings and Specifications, he shall give prompt written notice thereof to the Construction Manager.

FIGURE A5.4. (*continued*)

3.11 The Owner shall furnish, prior to commencing work and at such future times as may be requested, reasonable evidence satisfactory to the Construction Manager that sufficient funds are available and committed for the entire cost of the Project Unless such reasonable evidence is furnished, the Construction Manager is not required to commence or continue any Work, or may, if such evidence is not presented within a reasonable time, stop the Project upon 15 days notice to the Owner The failure of the Construction Manager to insist upon the providing of this evidence at any one time shall not be a waiver of the Owner's obligation to make payments pursuant to this Agreement nor shall it be a waiver of the Construction Manager's right to request or insist that such evidence be provided at a later date.

3.12 The Owner shall communicate with the Trade Contractors only through the Construction Manager.

ARTICLE 4

Trade Contracts

4.1 All portions of the Project that the Construction Manager does not perform with his own forces shall be performed under Trade Contracts. The Construction Manager shall request and receive proposals from Trade Contractors and Trade Contracts will be awarded after the proposals are reviewed by the Architect/Engineer, Construction Manager and Owner.

4.2 If the Owner refuses to accept a Trade Contractor recommended by the Construction Manager, the Construction Manager shall recommend an acceptable substitute and the Guaranteed Maximum Price if applicable shall be increased or decreased by the difference in cost occasioned by such substitution and an appropriate Change Order shall be issued.

4.3 Unless otherwise directed by the Owner, Trade Contracts will be between the Construction Manager and the Trade Contractors. Whether the Trade Contracts are with the Construction Manager or the Owner, the form of the Trade Contracts including the General and Supplementary Conditions shall be satisfactory to the Construction Manager.

4.4 The Construction Manager shall be responsible to the Owner for the acts and omissions of his agents and employees, Trade Contractors performing Work under a contract with the Construction Manager, and such Trade Contractors' agents and employees.

ARTICLE 5

Schedule

5.1 The services to be provided under this Contract shall be in general accordance with the following schedule:

5.2 At the time a Guaranteed Maximum Price is established, as provided for in Article 6, a Date of Substantial Completion of the project shall also be established.

5.3 The Date of Substantial Completion of the Project or a designated portion thereof is the date when construction is sufficiently complete in accordance with the Drawings and Specifications so the Owner can occupy or utilize the Project or designated portion thereof for the use for which it is intended. Warranties called for by this Agreement or by the Drawings and Specifications shall commence on the Date of Substantial Completion of the Project or designated portion thereof.

5.4 If the Construction Manager is delayed at any time in the progress of the Project by any act or neglect of the Owner or the Architect/Engineer or by any employee of either, or by any separate contractor employed by the Owner, or by changes ordered in the Project, or by labor disputes, fire, unusual delay in transportation, adverse weather conditions not reasonably anticipatable, unavoidable casualties or any causes beyond the Construction Manager's control, or by delay authorized by the Owner pending arbitration, the Construction Completion Date shall be extended by Change Order for a reasonable length of time.

AGC DOCUMENT NO. 8 • OWNER-CONSTRUCTION MANAGER AGREEMENT   JULY 1980

FIGURE A5.4. (*continued*)

ARTICLE 6

Guaranteed Maximum Price

6.1 When the design, Drawings and Specifications are sufficiently complete, the Construction Manager will, if desired by the Owner, establish a Guaranteed Maximum Price, guaranteeing the maximum price to the Owner for the Cost of the Project and the Construction Manager's Fee. Such Guaranteed Maximum Price will be subject to modification for Changes in the Project as provided in Article 9, and for additional costs arising from delays caused by the Owner or the Architect/Engineer.

6.2 When the Construction Manager provides a Guaranteed Maximum Price, the Trade Contracts will either be with the Construction Manager or will contain the necessary provisions to allow the Construction Manager to control the performance of the Work. The Owner will also authorize the Construction Manager to take all steps necessary in the name of the Owner, including arbitration or litigation, to assure that the Trade Contractors perform their contracts in accordance with their terms.

6.3 The Guaranteed Maximum Price will only include those taxes in the Cost of the Project which are legally enacted at the time the Guaranteed Maximum Price is established.

ARTICLE 7

Construction Manager's Fee

7.1 In consideration of the performance of the Contract, the Owner agrees to pay the Construction Manager in current funds as compensation for his services a Construction Manager's Fee as set forth in Subparagraphs 7.1.1 and 7.1.2.

7.1.1 For the performance of the Design Phase services, a fee of
which shall be paid monthly, in equal proportions, based on the scheduled Design Phase time.

7.1.2 For work or services performed during the Construction Phase, a fee of
which shall be paid proportionately to the ratio the monthly payment for the Cost of the Project bears to the estimated cost. Any balance of this fee shall be paid at the time of final payment.

7.2 Adjustments in Fee shall be made as follows:

7.2.1 For Changes in the Project as provided in Article 9, the Construction Manager's Fee shall be adjusted as follows:

7.2.2 For delays in the Project not the responsibility of the Construction Manager, there will be an equitable adjustment in the fee to compensate the Constructon Manager for his increased expenses.

7.2.3 The Construction Manager shall be paid an additional fee in the same proportion as set forth in 7.2.1 if the Construction Manager is placed in charge of the reconstruction of any insured or uninsured loss.

7.3 Included in the Construction Manager's Fee are the following:

7.3.1 Salaries or other compensation of the Construction Manager's employees at the principal office and branch offices, except employees listed in Subparagraph 8.2.2.

FIGURE A5.4. (*continued*)

7.3.2 General operating expenses of the Construction Manager's principal and branch offices other than the field office.

7.3.3 Any part of the Construction Manager's capital expenses, including interest on the Construction Manager's capital employed for the project.

7.3.4 Overhead or general expenses of any kind, except as may be expressly included in Article 8.

7.3.5 Costs in excess of the Guaranteed Maximum Price.

ARTICLE 8

Cost of the Project

8.1 The term Cost of the Project shall mean costs necessarily incurred in the Project during either the Design or Construction Phase, and paid by the Construction Manager, or by the Owner if the Owner is directly paying Trade Contractors upon the Construction Manager's approval and direction. Such costs shall include the items set forth below in this Article.

8.1.1 The Owner agrees to pay the Construction Manager for the Cost of the Project as defined in Article 8. Such payment shall be in addition to the Construction Manager's Fee stipulated in Article 7.

8.2 Cost Items

8.2.1 Wages paid for labor in the direct employ of the Construction Manager in the performance of his Work under applicable collective bargaining agreements, or under a salary or wage schedule agreed upon by the Owner and Construction Manager, and including such welfare or other benefits, if any, as may be payable with respect thereto.

8.2.2 Salaries of the Construction Manager's employees when stationed at the field office, in whatever capacity employed, employees engaged on the road in expediting the production or transportation of materials and equipment, and employees in the main or branch office performing the functions listed below:

8.2.3 Cost of all employee benefits and taxes for such items as unemployment compensation and social security, insofar as such cost is based on wages, salaries, or other remuneration paid to employees of the Construction Manager and included in the Cost of the Project under Subparagraphs 8.2.1 and 8.2.2.

8.2.4 Reasonable transportation, traveling, moving, and hotel expenses of the Construction Manager or of his officers or employees incurred in discharge of duties connected with the Project.

8.2.5 Cost of all materials, supplies and equipment incorporated in the Project, including costs of transportation and storage thereof.

8.2.6 Payments made by the Construction Manager or Owner to Trade Contractors for their Work performed pursuant to contract under this Agreement.

8.2.7 Cost, including transportation and maintenance, of all materials, supplies, equipment, temporary facilities and hand tools not owned by the workmen, which are employed or consumed in the performance of the Work, and cost less salvage value on such items used but not consumed which remain the property of the Construciton Manager.

8.2.8 Rental charges of all necessary machinery and equipment, exclusive of hand tools, used at the site of the Project, whether rented from the Construction Manager or other, including installation, repairs and replacements, dismantling, removal, costs of lubrication, transportation and delivery costs thereof, at rental charges consistent with those prevailing in the area.

FIGURE A5.4. (*continued*)

8.2.9 Cost of the premiums for all insurance which the Construction Manager is required to procure by this Agreement or is deemed necessary by the Construction Manager.

8.2.10 Sales, use, gross receipts or similar taxes related to the Project imposed by any governmental authority, and for which the Construction Manager is liable.

8.2.11 Permit fees, licenses, tests, royalties, damages for infringement of patents and costs of defending suits therefor, and deposits lost for causes other than the Construction Manager's negligence. If royalties or losses and damages, including costs of defense, are incurred which arise from a particular design, process, or the product of a particular manufacturer or manufacturers specified by the Owner or Architect/Engineer, and the Construction Manager has no reason to believe there will be infringement of patent rights, such royalties, losses and damages shall be paid by the Owner and not considered as within the Guaranteed Maximum Price.

8.2.12 Losses, expenses or damages to the extent not compensated by insurance or otherwise (including settlement made with the written approval of the Owner).

8.2.13 The cost of corrective work subject, however, to the Guaranteed Maximum Price.

8.2.14 Minor expenses such as telegrams, long-distance telephone calls, telephone service at the site, expressage, and similar petty cash items in connection with the Project.

8.2.15 Cost of removal of all debris.

8.2.16 Cost incurred due to an emergency affecting the safety of persons and property.

8.2.17 Cost of data processing services required in the performance of the services outlined in Article 2.

8.2.18 Legal costs reasonably and properly resulting from prosecution of the Project for the Owner.

8.2.19 All costs directly incurred in the performance of the Project and not included in the Construction Manager's Fee as set forth in Paragraph 7.3.

ARTICLE 9

Changes in the Project

9.1 The Owner, without invalidating this Agreement, may order Changes in the Project within the general scope of this Agreement consisting of additions, deletions or other revisions, the Guaranteed Maximum Price, if established, the Construction Manager's Fee and the Construction Completion Date being adjusted accordingly. All such Changes in the Project shall be authorized by Change Order.

9.1.1 A Change Order is a written order to the Construction Manager signed by the Owner or his authorized agent issued after the execution of this Agreement, authorizing a Change in the Project or the method or manner of performance and/or an adjustment in the Guaranteed Maximum Price, the Construction Manager's Fee, or the Construction Completion Date. Each adjustment in the Guaranteed Maximum Price resulting from a Change Order shall clearly separate the amount attributable to the Cost of the Project and the Construction Manager's Fee.

9.1.2 The increase or decrease in the Guaranteed Maximum Price resulting from a Change in the Project shall be determined in one or more of the following ways:

.1 by mutual acceptance of a lump sum properly itemized and supported by sufficient substantiating data to permit evaluation;

.2 by unit prices stated in the Agreement or subsequently agreed upon;

.3 by cost as defined in Article 8 and a mutually acceptable fixed or percentage fee; or

.4 by the method provided in Subparagraph 9.1.3.

FIGURE A5.4. (*continued*)

9.1.3 If none of the methods set forth in Clauses 9.1.2.1 through 9.1.2.3 is agreed upon, the Construction Manager, provided he receives a written order signed by the Owner, shall promptly proceed with the Work involved. The cost of such Work shall then be determined on the basis of the reasonable expenditures and savings of those performing the Work attributed to the change, including, in the case of an increase in the Guaranteed Maximum Price, a reasonable increase in the Construction Manager's Fee. In such case, and also under Clauses 9.1.2.3 and 9.1.2.4 above, the Construction Manager shall keep and present, in such form as the Owner may prescribe, an itemized accounting together with appropriate supporting data of the increase in the Cost of the Project as outlined in Article 8. The amount of decrease in the Guaranteed Maximum Price to be allowed by the Construction Manager to the Owner for any deletion or change which results in a net decrease in cost will be the amount of the actual net decrease. When both additions and credits are involved in any one change, the increase in Fee shall be figured on the basis of net increase, if any.

9.1.4 If unit prices are stated in the Agreement or subsequently agreed upon, and if the quantities originally contemplated are so changed in a proposed Change Order or as a result of several Change Orders that application of the agreed unit prices to the quantities of Work proposed will cause substantial inequity to the Owner or the Construction Manager, the applicable unit prices and Guaranteed Maximum Price shall be equitably adjusted.

9.1.5 Should concealed conditions encountered in the performance of the Work below the surface of the ground or should concealed or unknown conditions in an existing structure be at variance with the conditions indicated by the Drawings, Specifications, or Owner-furnished information or should unknown physical conditions below the surface of the ground or should concealed or unknown conditions in an existing structure of an unusual nature, differing materially from those ordinarily encountered and generally recognized as inherent in work of the character provided for in this Agreement, be encountered, the Guaranteed Maximum Price and the Construction Completion Date shall be equitably adjusted by Change Order upon claim by either party made within a reasonable time after the first observance of the conditions.

9.2 Claims for Additional Cost or Time

9.2.1 If the Construction Manager wishes to make a claim for an increase in the Guaranteed Maximum Price, an increase in his fee, or an extension in the Construction Completion Date, he shall give the Owner written notice thereof within a reasonable time after the occurrence of the event giving rise to such claim. This notice shall be given by the Construction Manager before proceeding to execute any Work, except in an emergency endangering life or property in which case the Construction Manager shall act, at his discretion, to prevent threatened damage, injury or loss. Claims arising from delay shall be made within a reasonable time after the delay. No such claim shall be valid unless so made. If the Owner and the Construction Manager cannot agree on the amount of the adjustment in the Guaranteed Maximum Price, Construction Manager's Fee or Construction Completion Date, it shall be determined pursuant to the provisions of Article 16. Any change in the Guaranteed Maximum Price, Construction Manager's Fee or Construction Completion Date resulting from such claim shall be authorized by Change Order.

9.3. Minor Changes in the Project

9.3.1 The Architect/Engineer will have authority to order minor Changes in the Project not involving an adjustment in the Guaranteed Maximum Price or an extension of the Construction Completion Date and not inconsistent with the intent of the Drawings and Specifications. Such Changes may be effected by written order and shall be binding on the Owner and the Construction Manager.

9.4 Emergencies

9.4.1 In any emergency affecting the safety of persons or property, the Construction Manager shall act, at his discretion, to prevent threatened damage, injury or loss. Any increase in the Guaranteed Maximum Price or extension of time claimed by the Construction Manager on account of emergency work shall be determined as provided in this Article.

ARTICLE 10

Discounts

All discounts for prompt payment shall accrue to the Owner to the extent the Cost of the Project is paid directly by the

FIGURE A5.4. (*continued*)

Owner or from a fund made available by the Owner to the Construction Manager for such payments. To the extent the Cost of the Project is paid with funds of the Construction Manager, all cash discounts shall accrue to the Construction Manager. All trade discounts, rebates and refunds, and all returns from sale of surplus materials and equipment, shall be credited to the Cost of the Project.

ARTICLE 11

Payments to the Construction Manager

11.1 The Construction Manager shall submit monthly to the Owner a statement, sworn to if required, showing in detail all moneys paid out, costs accumulated or costs incurred on account of the Cost of the Project during the previous month and the amount of the Construction Manager's Fee due as provided in Article 7. Payment by the Owner to the Construction Manager of the statement amount shall be made within ten (10) days after it is submitted.

11.2 . Final payment constituting the unpaid balance of the Cost of the Project and the Construction Manager's Fee shall be due and payable when the Project is delivered to the Owner, ready for beneficial occupancy, or when the Owner occupies the Project, whichever event first occurs, provided that the Project be then substantially completed and this Agreement substantially performed. If there should remain minor items to be completed, the Construction Manager and Architect/Engineer shall list such items and the Construction Manager shall deliver, in writing, his unconditional promise to complete said items within a reasonable time thereafter. The Owner may retain a sum equal to 150% of the estimated cost of completing any unfinished items, provided that said unfinished items are listed separately and the estimated cost of completing any unfinished items likewise listed separately. Thereafter, Owner shall pay to Construction Manager, monthly, the amount retained for incomplete items as each of said items is completed.

11.3 The Construction Manager shall promptly pay all the amounts due Trade Contractors or other persons with whom he has a contract upon receipt of any payment from the Owner, the application for which includes amounts due such Trade Contractor or other persons. Before issuance of final payment, the Construction Manager shall submit satisfactory evidence that all payrolls, materials bills and other indebtedness connected with the Project have been paid or otherwise satisfied.

11.4 If the Owner should fail to pay the Construction Manager within seven (7) days after the time the payment of any amount becomes due, then the Construction Manager may, upon seven (7) additional days' written notice to the Owner and the Architect/Engineer, stop the Project until payment of the amount owing has been received.

11.5 Payments due but unpaid shall bear interest at the rate the Owner is paying on his construction loan or at the legal rate, whichever is higher.

ARTICLE 12

Insurance, Indemnity and Waiver of Subrogation

12.1 Indemnity

12.1.1 The Construction Manager agrees to indemnify and hold the Owner harmless from all claims for bodily injury and property damage (other than the Work itself and other property insured under Paragraph 12.4) that may arise from the Construction Manager's operations under this Agreement.

12.1.2 The Owner shall cause any other contractor who may have a contract with the Owner to perform construction or installation work in the areas where Work will be performed under this Agreement, to agree to indemnify the Owner and the Construction Manager and hold them harmless from all claims for bodily injury and property damage (other than property insured under Paragraph 12.4) that may arise from that contractor's operations. Such provisions shall be in a form satisfactory to the Construction Manager.

FIGURE A5.4. (*continued*)

12.2 Construction Manager's Liability Insurance

12.2.1 The Construction Manager shall purchase and maintain such insurance as will protect him from the claims set forth below which may arise out of or result from the Construction Manager's operations under this Agreement whether such operations be by himself or by any Trade Contractor or by anyone directly or indirectly employed by any of them, or by anyone for whose acts any of them may be liable:

12.2.1.1 Claims under workers' compensation, disability benefit and other similar employee benefit acts which are applicable to the Work to be performed.

12.2.1.2 Claims for damages because of bodily injury, occupational sickness or disease, or death of his employees under any applicable employer's liability law.

12.2.1.3 Claims for damages because of bodily injury, death of any person other than his employees.

12.2.1.4 Claims for damages insured by usual personal injury liability coverage which are sustained (1) by any person as a result of an offense directly or indirectly related to the employment of such person by the Construction Manager or (2) by any other person.

12.2.1.5 Claims for damages, other than to the Work itself, because of injury to or destruction of tangible property, including loss of use therefrom.

12.2.1.6 Claims for damages because of bodily injury or death of any person or property damage arising out of the ownership, maintenance or use of any motor vehicle.

12.2.2 The Construction Manager's Comprehensive General Liability Insurance shall include premises — operations (including explosion, collapse and underground coverage) elevators, independent contractors, completed operations, and blanket contractual liability on all written contracts, all including broad form property damage coverage.

12.2.3 The Construction Manager's Comprehensive General and Automobile Liability Insurance, as required by Subparagraphs 12.2.1 and 12.2.2 shall be written for not less than limits of liability as follows:

a. Comprehensive General Liability
  1. Personal Injury $_____ Each Occurrence
                              $_____ Aggregate
                              (Completed Operations)
  2. Property Damage $_____ Each Occurrence
                              $_____ Aggregate

b. Comprehensive Automobile Liability
  1. Bodily Injury $_____ Each Person
                              $_____ Each Occurrence
  2. Property Damage $_____ Each Occurrence

12.2.4 Comprehensive General Liability Insurance may be arranged under a single policy for the full limits required or by a combination of underlying policies with the balance provided by an Excess or Umbrella Liability policy.

12.2.5 The foregoing policies shall contain a provision that coverages afforded under the policies will not be cancelled or not renewed until at least sixty (60) days' prior written notice has been given to the Owner. Certificates of Insurance showing such coverages to be in Force shall be filed with the Owner prior to commencement of the Work.

12.3 Owner's Liability Insurance

12.3.1 The Owner shall be responsible for purchasing and maintaining his own liability insurance and, at his option, may

---

AGC DOCUMENT NO. 8 • OWNER-CONSTRUCTION MANAGER AGREEMENT JULY 1980

FIGURE A5.4. (*continued*)

purchase and maintain such insurance as will protect him against claims which may arise from operations under this Agreement.

12.4 Insurance to Protect Project

12.4.1 The Owner shall purchase and maintain property insurance in a form acceptable to the Construction Manager upon the entire Project for the full cost of replacement as of the time of any loss. This insurance shall include as named insureds the Owner, the Construction Manager, Trade Contractors and their Trade Subcontractors and shall insure against loss from the perils of Fire, Extended Coverage, and shall include "All Risk" insurance for physical loss or damage including, without duplication of coverage, at least theft, vandalism, malicious mischief, transit, collapse, flood, earthquake, testing, and damage resulting from defective design, workmanship or material. The Owner will increase limits of coverage, if necessary, to reflect estimated replacement cost. The Owner will be responsible for any co-insurance penalties or deductibles. If the Project covers an addition to or is adjacent to an existing building, the Construction Manager, Trade Contractors and their Trade Subcontractors shall be named as additional insureds under the Owner's Property Insurance covering such building and its contents.

12.4.1.1 If the Owner finds it necessary to occupy or use a portion or portions of the Project prior to Substantial Completion thereof, such occupancy shall not commence prior to a time mutually agreed to by the Owner and Construction Manager and to which the insurance company or companies providing the property insurance have consented by endorsement to the policy or policies. This insurance shall not be cancelled or lapsed on account of such partial occupancy. Consent of the Construction Manager and of the insurance company or companies to such occupancy or use shall not be unreasonably withheld.

12.4.2 The Owner shall purchase and maintain such boiler and machinery insurance as may be required or necessary. This insurance shall include the interests of the Owner, the Construction Manager, Trade Contractors and their Trade Subcontractors in the Work.

12.4.3 The Owner shall purchase and maintain such insurance as will protect the Owner and Construction Manager against loss of use of Owner's property due to those perils insured pursuant to Subparagraph 12.4.1. Such policy will provide coverage for expediting expenses of materials, continuing overhead of the Owner and Construction Manager, necessary labor expense including overtime, loss of income by the Owner and other determined exposures. Exposures of the Owner and the Construction Manager shall be determined by mutual agreement and separate limits of coverage fixed for each item.

12.4.4 The Owner shall file a copy of all policies with the Construction Manager before an exposure to loss may occur. Copies of any subsequent endorsements will be furnished to the Construction Manager. The Construction Manager will be given sixty (60) days notice of cancellation, non-renewal, or any endorsements restricting or reducing coverage. If the Owner does not intend to purchase such insurance, he shall inform the Construction Manager in writing prior to the commencement of the Work. The Construction Manager may then effect insurance which will protect the interest of himself, the Trade Contractors and their Trade Subcontractors in the Project, the cost of which shall be a Cost of the Project pursuant to Article 8, and the Guaranteed Maximum Price shall be increased by Change Order. If the Construction Manager is damaged by failure of the Owner to purchase or maintain such insurance or to so notify the Construction Manager, the Owner shall bear all reasonable costs properly attributable thereto.

12.5 Property Insurance Loss Adjustment

12.5.1 Any insured loss shall be adjusted with the Owner and the Construction Manager and made payable to the Owner and Construction Manager as trustees for the insureds, as their interests may appear, subject to any applicable mortgagee clause.

12.5.2 Upon the occurrence of an insured loss, monies received will be deposited in a separate account and the trustees shall make distribution in accordance with the agreement of the parties in interest, or in the absence of such agreement, in accordance with an arbitration award pursuant to Article 16. If the trustees are unable to agree on the settlement of the loss, such dispute shall also be submitted to arbitration pursuant to Article 16.

12.6 Waiver of Subrogation

12.6.1 The Owner and Construction Manager waive all rights against each other, the Architect/Engineer, Trade Contractors, and their Trade Subcontractors for damages caused by perils covered by insurance provided under Paragraph 12.4, except such rights as they may have to the proceeds of such insurance held by the Owner and Construction Manager as trustees. The Construction Manager shall require similar waivers from all Trade Contractors and their Trade Subcontractors.

FIGURE A5.4. (*continued*)

12.6.2 The Owner and Construction Manager waive all rights against each other and the Architect Engineer, Trade Contractors and their Trade Subcontractors for loss or damage to any equipment used in connection with the Project and covered by any property insurance. The Construction Manager shall require similar waivers from all Trade Contractors and their Trade Subcontractors.

12.6.3 The Owner waives subrogation against the Construction Manager, Architect/Engineer, Trade Contractors, and their Trade Subcontractors on all property and consequential loss policies carried by the Owner on adjacent properties and under property and consequential loss policies purchased for the Project after its completion.

12.6.4 If the policies of insurance referred to in this Paragraph require an endorsement to provide for continued coverage where there is a waiver of subrogation, the owners of such policies will cause them to be so endorsed.

ARTICLE 13

Termination of the Agreement and Owner's Right to Perform Construction Manager's Obligations

13.1 Termination by the Construction Manager

13.1.1 If the Project, in whole or substantial part, is stopped for a period of thirty days under an order of any court or other public authority having jurisdiction, or as a result of an act of government, such as a declaration of a national emergency making materials unavailable, through no act or fault of the Construction Manager, or if the Project should be stopped for a period of thirty days by the Construction Manager for the Owner's failure to make payment thereon, then the Construction Manager may, upon seven days' written notice to the Owner and the Architect/Engineer, terminate this Agreement and recover from the Owner payment for all work executed, the Construction Manager's Fee earned to date, and for any proven loss sustained upon any materials, equipment, tools, construction equipment and machinery, cancellation charges on existing obligations of the Construction Manager, and a reasonable profit.

13.2 Owner's Right to Perform Construction Manager's Obligations and Termination by the Owner for Cause

13.2.1 If the Construction Manager fails to perform any of his obligations under this Agreement including any obligation he assumes to perform Work with his own forces, the Owner may, after seven days' written notice during which period the Construction Manager fails to perform such obligation, make good such deficiencies. The Guaranteed Maximum Price, if any, shall be reduced by the cost to the Owner of making good such deficiencies.

13.2.2 If the Construction Manager is adjudged a bankrupt, or if he makes a general assignment for the benefit of his creditors, or if a receiver is appointed on account of his insolvency, or if he persistently or repeatedly refuses or fails, except in cases for which extension of time is provided, to supply enough properly skilled workmen or proper materials, or if he fails to make proper payment to Trade Contractors or for materials or labor, or persistently disregards laws, ordinances, rules, regulations or orders of any public authority having jurisdiction, or otherwise is guilty of a substantial violation of a provision of the Agreement, then the Owner may, without prejudice to any right or remedy and after giving the Construction Manager and his surety, if any, seven days' written notice, during which period the Construction Manager fails to cure the violation, terminate the employment of the Construction Manager and take possession of the site and of all materials, equipment, tools, construction equipment and machinery thereon owned by the Construction Manager and may finish the Project by whatever reasonable method he may deem expedient. In such case, the Construction Manager shall not be entitled to receive any further payment until the Project is finished nor shall he be relieved from his obligations assumed under Article 6.

13.3 Termination by Owner Without Cause

13.3.1 If the Owner terminates this Agreement other than pursuant to Subparagraph 13.2.2 or Subparagraph 13.3.2, he shall reimburse the Construction Manager for any unpaid Cost of the Project due him under Article 8, plus (1) the unpaid balance of the Fee computed upon the Cost of the Project to the date of termination at the rate of the percentage named in Subparagraph 7.2.1 or if the Construction Manager's Fee be stated as a fixed sum, such an amount as will increase the payment on account of his fee to a sum which bears the same ratio to the said fixed sum as the Cost of the Project at the time of termination bears to the adjusted Guaranteed Maximum Price, if any, otherwise to a reasonable estimated Cost of the Project when completed. The Owner shall also pay to the Construction Manager fair compensation, either by purchase or rental at the

FIGURE A5.4. *(continued)*

election of the Owner, for any equipment retained. In case of such termination of the Agreement the Owner shall further assume and become liable for obligations, commitments and unsettled claims that the Construction Manager has previously undertaken or incurred in good faith in connection with said Project. The Construction Manager shall, as a condition of receiving the payments referred to in this Article 13, execute and deliver all such papers and take all such steps, including the legal assignment of his contractual rights, as the Owner may require for the purpose of fully vesting in him the rights and benefits of the Construction Manager under such obligations or commitments.

13.3.2 After the completion of the Design Phase, if the final cost estimates make the Project no longer feasible from the standpoint of the Owner, the Owner may terminate this Agreement and pay the Construction Manager his Fee in accordance with Subparagraph 7.1.1 plus any costs incurred pursuant to Article 9.

ARTICLE 14

Assignment and Governing Law

14.1 Neither the Owner nor the Construction Manager shall assign his interest in this Agreement without the written consent of the other except as to the assignment of proceeds.

14.2 This Agreement shall be governed by the law of the place where the Project is located.

ARTICLE 15

Miscellaneous Provisions

15.1 It is expressly understood that the Owner shall be directly retaining the services of an Architect/Engineer.

ARTICLE 16

Arbitration

16.1 All claims, disputes and other matters in questions arising out of, or relating to, this Agreement or the breach thereof, except with respect to the Architect/Engineer's decision on matters relating to artistic effect, and except for claims which have been waived by the making or acceptance of final payment shall be decided by arbitration in accordance with the Construction Industry Arbitration Rules of the American Arbitration Association then obtaining unless the parties mutually agree otherwise. This Agreement to arbitrate shall be specifically enforceable under the prevailing arbitration law.

16.2 Notice of the demand for arbitration shall be filed in writing with the other party to this Agreement and with the American Arbitration Association. The demand for arbitration shall be made within a reasonable time after the claim, dispute or other matter in question has arisen, and in no event shall it be made after the date when institution of legal or equitable proceedings based on such claim, dispute or other matter in question would be barred by the applicable statute of limitations.

16.3 The award rendered by the arbitrators shall be final and judgment may be entered upon it in accordance with applicable law in any court having jurisdiction thereof.

16.4 Unless otherwise agreed in writing, the Construction Manager shall carry on the Work and maintain the Contract Completion Date during any arbitration proceedings, and the Owner shall continue to make payments in accordance with this Agreement.

16.5 All claims which are related to or dependent upon each other, shall be heard by the same arbitrator or arbitrators even though the parties are not the same unless a specific contract prohibits such consolidation.

FIGURE A5.4. (*continued*)

This Agreement executed the day and year first written above.

ATTEST: OWNER:

ATTEST: CONSTRUCTION MANAGER:

AGC DOCUMENT NO. 8 • OWNER-CONSTRUCTION MANAGER AGREEMENT   APRIL 1980

FIGURE A5.4.   *(continued)*

THE ASSOCIATED GENERAL CONTRACTORS

AMENDMENT TO OWNER-CONSTRUCTION MANAGER CONTRACT

Pursuant to Article 6 of the original Agreement, AGC Form No. 8, dated _____

between _____ (Owner)

and _____ (the Construction Manager),

for _____ (the Project),

the Owner desires to fix a Guaranteed Maximum Price for the Project and the Construction Manager agrees that the design, plans and specifications are sufficiently complete for such purpose. Therefore, the Owner and Construction Manager agree as set forth below.

> AIA copyrighted material has been reproduced with permission of the American Institute of Architects under permission number 81067. Further reproduction is prohibited.
>
> Because AIA Documents are revised from time to time, users should ascertain from the AIA the current edition(s) of the Document(s) reproduced herein.

ARTICLE I

Guaranteed Maximum Price

The Construction Manager's Guaranteed Maximum Price for the Project, including the Cost of the Work as defined in Article 8 and the Construction Manager's Fee as defined in Article 7 is _____ Dollars ($ _____). This price is for the performance of the Work in accordance with the documents listed and attached to this Amendment and marked Amendment Exhibit A.

(*OPTIONAL SAVINGS CLAUSE*) It is further agreed that if, upon completion of the work, the actual cost of the work plus the Construction Manager's Fee is less than the Guaranteed Maximum Price as set forth herein and as adjusted by approved change orders that the Owner agrees to pay to the Construction Manager an amount equal to _____% of such savings, as additional compensation.

AGC DOCUMENT NO. 8A • AMENDMENT TO OWNER-CONSTRUCTION MANAGER CONTRACT • JUNE 1977
©ASSOCIATED GENERAL CONTRACTORS OF AMERICA 1977

FIGURE A5.5. Associated General Contractors (AGC) Document No. 8A

ARTICLE II

Time Schedule

The Construction Completion date established by this Amendment is:

OWNER:

ATTEST:

By: _____

_____

Date: _____

CONSTRUCTION MANAGER:

_____

ATTEST:

By: _____

_____

Date: _____

AGC DOCUMENT NO. 8A • AMENDMENT TO OWNER-CONSTRUCTION MANAGER CONTRACT • JUNE 1977
©ASSOCIATED GENERAL CONTRACTORS OF AMERICA 1977

FIGURE A5.5. *(continued)*

THE ASSOCIATED GENERAL CONTRACTORS

GENERAL CONDITIONS FOR TRADE CONTRACTORS UNDER CONSTRUCTION MANAGEMENT AGREEMENTS

INSTRUCTIONS FOR CONSTRUCTION MANAGER

1. These conditions primarily govern the obligations of the Trade Contractors and in addition establish the general procedures for the administration of construction. They have been drafted to cover Trade Contracts with either the Owner or the Construction Manager.

2. In all cases your attorney should be consulted to advise you on their use and any modifications.

3. Nothing contained herein is intended to conflict with local, state or federal laws or regulations.

4. It is recommended all insurance matters be reviewed with your insurance consultant and carrier such as implications of errors and omission liability, completed operations, and waiver of subrogation.

5. Each article should be reviewed by the Construction Manager as to the applicability to a given project and contractual conditions.

6. Special conditions and terms for the project or the Trade Contractor Agreements should cover the following:
   — trade contractor retainages
   — payment schedules
   — insurance limits
   — owner's protective insurance if required of trade contractors
   — builder's risk deductible, if any.

   > AIA copyrighted material has been reproduced with permission of the American Institute of Architects under permission number 81067. Further reproduction is prohibited.
   > Because AIA Documents are revised from time to time, users should ascertain from the AIA the current edition(s) of the Document(s) reproduced herein.

7. If the Owner does not provide Builder's Risk Insurance, Paragraph 12.2 will need to be modified.

Certain provisions of this document have been derived, with modifications, from the following document published by The American Institute of Architects: AIA Document A201, General Conditions, © 1976, by The American Institute of Architects. Usage made of AIA language, with the permission of AIA, does not imply AIA endorsement or approval of this document. Further reproduction of copyrighted AIA materials without separate written permission from AIA is prohibited.

AGC DOCUMENT NO. 8b • GENERAL CONDITIONS FOR TRADE CONTRACTORS UNDER CONSTRUCTION MANAGEMENT AGREEMENTS • JULY 1980
© 1980 Associated General Contractors of America

FIGURE A5.6. Associated General Contractors (AGC) Document No. 8b

THE ASSOCIATED GENERAL CONTRACTORS

GENERAL CONDITIONS FOR TRADE CONTRACTORS UNDER CONSTRUCTION MANAGEMENT AGREEMENTS

TABLE OF CONTENTS

ARTICLES		PAGE
1	Contract Documents	1
2	Owner	2
3	Architect/Engineer	2
4	Construction Manager	3
5	Trade Contractors	4
6	Trade Subcontractors	8
7	Separate Trade Contracts	9
8	Miscellaneous Provisions	10
9	Time	11
10	Payments and Completion	12
11	Protection of Persons and Property	15
12	Insurance	16
13	Changes in the Work	18
14	Uncovering and Correction of Work	19
15	Termination of the Contract	20

Certain provisions of this document have been derived, with modifications, from the following document published by The American Institute of Architects: AIA Document A201, General Conditions, ©1976, by The American Institute of Architects. Usage made of AIA language, with the permission of AIA, does not imply AIA endorsement or approval of this document. Further reproduction of copyrighted AIA materials without separate written permission from AIA is prohibited.

AGC DOCUMENT NO. 8b • GENERAL CONDITIONS FOR TRADE CONTRACTORS UNDER CONSTRUCTION MANAGEMENT AGREEMENTS • JULY 1980
©1980 Associated General Contractors of America

FIGURE A5.6. *(continued)*

ARTICLE 1

CONTRACT DOCUMENTS

1.1 DEFINITIONS

1.1.1 THE CONTRACT DOCUMENTS

The Contract Documents consist of the Agreement between the Owner or Construction Manager, as the case may be, and the Trade Contractor, the Conditions of the Contract (General, Supplementary and other Conditions), the Drawings (and criteria if the drawings are not complete), the Specifications, all Addenda issued prior to execution of the Contract, and all Modifications issued after the execution of the contract. A modification is (1) a written amendment to the Contract signed by both parties, (2) a Change Order, (3) a written interpretation issued by the Architect/Engineer pursuant to Subparagraph 3.2.2, or (4) a written order for a minor change in the Work issued on the Owner's behalf pursuant to Paragraph 13.4. The Contract Documents do not include Bidding or Proposal Documents such as the Advertisement or Invitation To Bid, Requests for Proposals, sample forms, Trade Contractors Bid or Proposal, or portions of Addenda relative to any of these, or any other documents other than those set forth in this subparagraph unless specifically set forth in the Agreement with the Trade Contractor. In the event of an inconsistency between the Agreement and the other Contract Documents, the provisions of the Agreement will control.

1.1.2 THE CONTRACT

The Contract Documents form the Contract with the Trade Contractor. This Contract represents the entire and integrated agreement and supersedes all prior negotiations, representations, or agreements, either written or oral. The Contract may be amended or modified only by a Modification as defined in Subparagraph 1.1.1.

1.1.3 THE WORK

The Work comprises the completed construction performed by the Construction Manager with his own forces or required by a Trade Contractor's contract and includes all labor necessary to produce such construction required of the Construction Manager or a Trade Contractor, and all materials and equipment incorporated or to be incorporated in such construction.

1.1.4 THE PROJECT

The Project is the total construction to be performed under the Agreement between the Owner and Construction Manager of which the Work is a part.

1.2 EXECUTION, CORRELATION AND INTENT

1.2.1 By executing his Agreement, each Trade Contractor represents that he has visited the site, familiarized himself with the local conditions under which the Work is to be performed and correlated his observations with the requirements of the Contract Documents.

1.2.2 The intent of the Contract Documents is to include all items necessary for the proper execution and completion of the Work. The Contract Documents are complementary, and what is required by any one shall be as binding as if required by all. Work not covered in the Contract Documents will not be required unless it is consistent therewith and is reasonably inferable therefrom as being necessary to produce the intended results. Words and abbreviations in the Contract Documents which have well-known technical or trade meanings are used in accordance with such recognized meanings.

1.2.3 The organization of the Specifications into divisions, sections and articles, and the arrangements of Drawings shall not control the Construction Manager in dividing the Work among Trade Contractors or in establishing the extent of Work to be performed by any trade.

1.3 OWNERSHIP AND USE OF DOCUMENTS

1.3.1 Unless otherwise provided in the Contract Documents, the Trade Contractor will be furnished, free of charge, all copies of Drawings and Specifications reasonably necessary for the execution of the Work.

FIGURE A5.6.   (*continued*)

1.3.2 All Drawings, Specifications and copies thereof furnished by the Architect/Engineer are and shall remain his property. They are to be ued only with respect to this Project and are not to be used on any other project. With the exception of one contract set for each party, such documents are to be returned or suitably accounted for to the Architect/Engineer on request at the completion of the Work. Submission or distribution to meet official regulatory requirements or for other purposes in connection with the Project is not to be construed as publication in derogation of the Architect/Engineer's common law copyright or other reserved rights.

ARTICLE 2

OWNER

2.1 DEFINITION

2.1.1 The Owner is the person or entity identified as such in the Agreement between the Owner and Construction Manager and is referred to throughout the Contract Documents as if singular in number and masculine in gender. The term Owner means the Owner or his authorized representative.

2.2 INFORMATION AND SERVICES FURNISHED BY THE OWNER

2.2.1 The Owner will furnish all surveys describing the physical characteristics, legal limitations and utility locations for the site of the Project, and a legal description of the site.

2.2.2 Except as provided in Subparagraph 5.7.1 the Owner will secure and pay for necessary approvals, easements, assessments and charges required for the construction, use, or occupancy of permanent structures or for permanent changes in existing facilities.

2.2.3 Information or services under the Owner's control will be furnished by the Owner with reasonable promptness to avoid delay in the orderly progress of the Work.

2.2.4 The Owner shall forward all instructions to the Trade Contractors through the Construction Manager even when the Owner has direct contracts with Trade Contractors.

ARTICLE 3

ARCHITECT/ENGINEER

3.1 DEFINITION

3.1.1 The Architect/Engineer is the person lawfully licensed to practice architecture or engineering or an entity lawfully practicing architecture or engineering and identified as such in the Agreement between the Owner and Construction Manager and is referred to throughout the Contract Documents as if singular in number and masculine in gender. The term Architect/Engineer means the Architect/Engineer or his authorized representative.

3.1.2 Nothing contained in the Contract Documents shall create any contractual relationship between the Architect/Engineer and any Trade Contractor.

3.2 ARCHITECT/ENGINEER'S DUTIES DURING CONSTRUCTION

3.2.1 The Architect/Engineer shall at all times have access to the Work wherever it is in preparation and progress. When directed by the Construction Manager, the Trade Contractor shall provide facilities for such access so the Architect/Engineer may perform his functions under the Contract Documents.

3.2.2 The Architect/Engineer will be the interpreter of the requirements of the Drawings and Specifications. The Architect/Engineer will, within a reasonable time, render such interpretations as are necessary for the proper execution of the progress of the Work.

FIGURE A5.6. (*continued*)

3.2.3 All interpretations of the Architect/Engineer shall be consistent with the intent of and reasonably inferable from the Contract Documents and will be in writing or in the form of drawings. All requests for interpretations shall be directed through the Construction Manager. The Architect/Engineer shall not be liable to the Trade Contractor for the result of any interpretation or decision rendered in good faith in such capacity.

3.2.4 The Architect/Engineer's decisions in matters relating to artistic effect will be final if consistent with the intent of the Contract Documents.

3.2.5 The Architect/Engineer will have authority to reject Work which does not conform to the Contract Documents. Whenever, in his opinion, he considers it necessary or advisable for the implementation of the intent of the Contract Documents, he will have authority to require special inspection or testing of the Work in accordance with Subparagraph 8.7.2 whether or not such Work be then fabricated, installed or completed. However, neither the Architect/Engineer's authority to act under this Subparagraph 3.2.5, nor any decision made by him in good faith either to exercise or not to exercise such authority, shall give rise to any duty or responsibility of the Architect/Engineer to the Trade Contractor, any Trade Subcontractor, any of their agents or employees, or any other person performing any of the Work.

3.2.6 The Architect/Engineer will review and approve or take other appropriate action upon Trade Contractor's submittals such as Shop Drawings, Product Data and Samples, but only for conformance with the design concept of the Work and with the information given in the Contract Documents. Such action shall be taken with reasonable promptness so as to cause no delay. The Architect/Engineer's approval of a specific item shall not indicate approval of an assembly of which the item is a component.

3.2.7 The Architect/Engineer along with the Construction Manager will conduct inspections to determine the dates of Substantial Completion and final completion, will receive and review written warranties and related documents required by the Contract and assembled by the Trade Contractor.

3.2.8 The Architect/Engineer will communicate with the Trade Contractors through the Construction Manager.

ARTICLE 4

CONSTRUCTION MANAGER

4.1 DEFINITION

4.1.1 The Construction Manager is the person or entity who has entered into an agreement with the Owner to serve as Construction Manager and is referred to throughout the Contract Documents as if singular in number and masculine in gender. The term Construction Manager means the Construction Manager acting through his authorized representative.

4.1.2 Whether the Trade Contracts are between the Owner and Trade Contractors, or the Construction Manager and Trade Contractors, it is the intent of these General Conditions to allow the Construction Manager to direct and schedule the performance of all Work and the Trade Contractors are expected to follow all such directions and schedules.

4.2 ADMINISTRATION OF THE CONTRACT

4.2.1 The Construction Manager will provide, as the Owner's authorized representative, the general administration of the Project as herein described.

4.2.2 The Construction Manager will be the Owner's construction representative during construction until final payment and shall have the responsibility to supervise and coordinate the work of all Trade Contractors.

4.2.3 The Construction Manager shall prepare and update all Construction Schedules and shall direct the Work with respect to such schedules.

4.2.4 The Construction Manager shall have the authority to reject Work which does not conform to the Contract Documents and to require any Special Inspection and Testing in accordance with Subparagraph 8.7.2.

4.2.5 The Construction Manager will prepare and issue Change Orders to the Trade Contractors in accordance with Article 13.

AGC DOCUMENT NO. 8b • GENERAL CONDITIONS FOR TRADE CONTRACTORS UNDER CONSTRUCTION MANAGEMENT AGREEMENTS • JULY 1980

FIGURE A5.6. (*continued*)

4.2.6 The Construction Manager along with the Architect/Engineer will conduct inspections to determine the dates of Substantial Completion and final completion, and will receive and review written warranties and related documents required by the Contract and assembled by the Trade Contractor.

4.2.7 Nothing contained in the Contract Documents between a Trade Contractor and the Owner shall create any contractual relationship between the Construction Manager and any Trade Contractor.

4.3 OWNER'S AND CONSTRUCTION MANAGER'S RIGHT TO STOP WORK

4.3.1 If the Trade Contractor fails to correct defective Work as required by Paragraph 14.2 or persistently fails to carry out the Work in accordance with the Contract Documents, the Construction Manager or the Owner through the Construction Manager may order the Trade Contractor to stop the Work, or any portion thereof, until the cause for such order has been eliminated.

4.3.2 If the Trade Contractor defaults or neglects to carry out the Work in accordance with the Contract Documents and fails within seven days after receipt of written notice from the Construction Manager to commence and continue correction of such default or neglect with diligence and promptness, the Construction Manager may, by written notice, and without prejudice to any other remedy he or the Owner may have, make good such deficiencies. In such case an appropriate Change Order shall be issued deducting from the payments then or thereafter due the Trade Contractor the cost of correcting such deficiencies, including compensation for the Architect/Engineer's and Construction Manager's additional services made necessary by such default, neglect or failure.

ARTICLE 5

TRADE CONTRACTORS

5.1 DEFINITION

5.1.1 A Trade Contractor is the person or entity identified as such in the Agreement between the Owner or Construction Manager and a Trade Contractor and is referred to throughout the Contract Document as if singular in number and masculine in gender. The term Trade Contractor means the Trade Contractor or his authorized representative.

5.1.2 The Agreements with the Trade Contractors may either be with the Owner or with the Construction Manager. These conditions in several instances make reference to obligations and rights of the "Owner or Construction Manager" to cover both possibilities. Such references are only to cover either possibility and such use does not create a joint obligation on the Owner and Construction Manager to the Trade Contractor. The contract obligation with the Trade Contractor is solely with the person or entity with whom he has his Agreement.

5.1.3 If the Trade Contracts are with the Construction Manager, the Trade Contractor assumes toward the Construction Manager all the obligations and responsibilities which the Construction Manager assumes toward the Owner under the Agreement between the Owner and the Construction Manager. A copy of the pertinent parts of this Agreement will be made available on request.

5.2 REVIEW OF CONTRACT DOCUMENTS

5.2.1 The Trade Contractor shall carefully study and compare the Contract Documents and shall at once report to the Construction Manager any error, inconsistency or omission he may or reasonably should discover. The Trade Contractor shall not be liable to the Owner or the Architect/Engineer or the Construction Manager for any damage resulting from any such errors, inconsistencies or omissions.

5.3 SUPERVISION AND CONSTRUCTION PROCEDURES

5.3.1 The Trade Contractor shall supervise and direct the Work, using his best skill and attention. He shall be solely responsible for all construction means, methods, techniques, sequences and procedures and for coordinating all portions of the Work under the Contract subject to the overall coordination of the Construction Manager.

FIGURE A5.6. (*continued*)

5.3.2 The Trade Contractor shall be responsible to the Owner and the Construction Manager for the acts and omissions of his employees and all his Trade Subcontractors and their agents and employees and other persons performing any of the Work under a contract with the Trade Contractor.

5.3.3 Neither observations nor inspections, tests or approvals by persons other than the Trade Contractor shall relieve the Trade Contractor from his obligations to perform the Work in accordance with the Contract Documents.

5.4 LABOR AND MATERIALS

5.4.1 Unless otherwise specifically provided in the Contract Documents, the Trade Contractor shall provide and pay for all labor, materials, equipment, tools, construction equipment and machinery, transportation, and other facilities and services necessary for the proper execution and completion of the Work.

5.4.2 The Trade Contractor shall at all times enforce strict discipline and good order among his employees and shall not employ on the Work any unfit person or anyone not skilled in the task assigned to him.

5.5 WARRANTY

5.5.1 The Trade Contractor warrants to the Owner and the Construction Manager that all materials and equipment furnished under this Contract will be new unless otherwise specified, and that all Work will be of good quality, free from faults and defects and in conformance with the Contract Documents. All Work not so conforming to these requirements, including substitutions not properly approved and authorized, may be considered defective. If required by the Construction Manager, the Trade Contractor shall furnish satisfactory evidence as to the kind and quality of materials and equipment. This warranty is not limited by the provisions of Paragraph 14.2.

5.6 TAXES

5.6.1 The Trade Contractor shall pay all sales, consumer, use and other similar taxes for the Work or portions thereof provided by the Trade Contractor which are legally enacted at the time bids or proposals are received, whether or not yet effective.

5.7 PERMITS, FEES AND NOTICES

5.7.1 Unless otherwise provided in the Contract Documents, the Trade Contractor shall secure and pay for all permits, governmental fees, licenses and inspections necessary for the proper execution and completion of his Work, which are customarily secured after execution of the contract and which are legally required at the time bids or proposals are received.

5.7.2 The Trade Contractor shall give all notices and comply with all laws, ordinances, rules, regulations and orders of any public authority bearing on the performance of the Work.

5.7.3 Unless otherwise provided in the Contract Documents, it is not the responsibility of the Trade Contractor to make certain that the Contract Documents are in accordance with applicable laws, statutes, building codes and regulations. If the Trade Contractor observes that any of the Contract Documents are at variance therewith in any respect, he shall promptly notify the Construction Manager in writing, and any necessary changes shall be by appropriate Modification.

5.7.4 If the Trade Contractor performs any Work knowing it to be contrary to such laws, ordinances, rules and regulations, and without such notice to the Construction Manager, he shall assume full responsibility therefor and shall bear all costs attributable thereto.

5.8 ALLOWANCES

5.8.1 The Trade Contractor shall include in the Contract Sum as defined in 10.1.1 all allowances stated in the Contract Documents. Items covered by these allowances shall be supplied for such amounts and by such persons as the Construction Manager may direct, but the Trade Contractor will not be required to employ persons against whom he makes a reasonable objection.

5.8.2 Unless otherwise provided in the Contract Documents:

AGC DOCUMENT NO. 8b • GENERAL CONDITIONS FOR TRADE CONTRACTORS UNDER CONSTRUCTION MANAGEMENT AGREEMENTS • JULY 1980

FIGURE A5.6. (*continued*)

.1 These allowances shall cover the cost to the Trade Contractor, less applicable trade discount, of the materials and equipment required by the allowance delivered at the site, and all applicable taxes;

.2 The Trade Contractor's costs for unloading and handling on the site, labor, installation costs, overhead, profit and other expenses contemplated for the original allowance shall be included in the Contract Sum and not in the allowance;

.3 Whenever the cost is more than or less than the allowance, the Contract Sum shall be adjusted accordingly by Change Order, the amount of which will recognize changes, if any, in handling costs on the site, labor, installation costs, overhead, profit and other expenses.

5.9 SUPERINTENDENT

5.9.1 The Trade Contractor shall employ a competent superintendent and necessary assistants who shall be in attendance at the Project site during the progress of the Work. The superintendent shall be satisfactory to the Construction Manager, and shall not be changed except with the consent of the Construction Manager, unless the superintendent proves to be unsatisfactory to the Trade Contractor or ceases to be in his employ. The superintendent shall represent the Trade Contractor and all communications given to the superintendent shall be as binding as if given to the Trade Contractor. Important communications shall be confirmed in writing. Other communications shall be so confirmed on written request in each case.

5.10 PROGRESS SCHEDULE

5.10.1 The Trade Contractor, immediately after being awarded the Contract, shall prepare and submit for the Construction Manager's information an estimated progress schedule for the Work. The progress schedule shall be related to the entire Project to the extent required by the Contract Documents and shall provide for expeditious and practicable execution of the Work. This schedule shall indicate the dates for the starting and completion of the various stages of construction, shall be revised as required by the conditions of the Work, and shall be subject to the Construction Manager's approval.

5.11 DRAWINGS AND SPECIFICATIONS AT THE SITE

5.11.1 The Trade Contractor shall maintain at the site for the Construction Manager and Architect/Engineer two copies of all Drawings, Specifications, Addenda, Change Orders and other Modifications, in good order and marked currently to record all changes made during construction. These Drawings, marked to record all changes during construction, and approved Shop Drawings, Product Data and Samples shall be delivered to the Construction Manager for the Owner upon completion of the Work.

5.12 SHOP DRAWINGS, PRODUCT DATA AND SAMPLES

5.12.1 Shop Drawings are drawings, diagrams, schedules and other data especially prepared for the Work by the Trade Contractor or any Trade Subcontractor, manufacturer, supplier or distributor to illustrate some portion of the Work.

5.12.2 Product Data are illustrations, standard schedules, performance charts, instructions, brochures, diagrams and other information furnished by the Trade Contractor to illustrate a material, product or system for some portion of the Work.

5.12.3 Samples are physical examples which illustrate materials, equipment or workmanship and establish standards by which the Work will be judged.

5.12.4 The Trade Contractor shall review, approve and submit through the Construction Manager with reasonable promptness and in such sequence as to cause no delay in the Work or in the work of any separate contractor, all Shop Drawings, Product Data and Samples required by the Contract Documents.

5.12.5 By approving and submitting Shop Drawings, Product Data and Samples, the Trade Contractor represents that he has determined and verified all materials, field measurements, and field construction criteria related thereto, or will do so, and that he has checked and coordinated the information contained within such submittals with the requirements of the Work and of the Contract Documents.

FIGURE A5.6. (*continued*)

5.12.6 The Construction Manager, if he finds such submittals to be in order, will forward them to the Architect/Engineer. If the Construction Manager finds them not to be complete or in proper form, he may return them to the Trade Contractor for correction or completion.

5.12.7 The Trade Contractor shall not be relieved of responsibility for any deviation from the requirements of the Contract Documents by the Construction Manager's forwarding them to the Architect/Engineer, or by the Architect/Engineer's approval of Shop Drawings, Product Data or Samples under Subparagraph 3.2.6 unless the Trade Contractor has specifically informed the Architect/Engineer and Construction Manager in writing of such deviation at the time of submission and the Architect/Engineer has given written approval to the specific deviation: The Trade Contractor shall not be relieved from responsibility for errors or omissions in the Shop Drawings, Product Data or Samples by the Construction Manager's forwarding or the Architect/Engineer's approval thereof.

5.12.8 The Trade Contractor shall direct specific attention, in writing or on resubmitted Shop Drawings, Product Data or Samples, to revisions other than those requested by the Architect/Engineer or Construction Manager on previous submittals.

5.12.9 No portion of the Work requiring submission of a Shop Drawing, Product Data or Sample shall be commenced until the submittal has been approved by the Architect/Engineer. All such portions of the Work shall be in accordance with approved submittals.

5.13 USE OF SITE

5.13.1 The Trade Contractor shall confine operations at the site to areas designated by the Construction Manager, permitted by law, ordinances, permits and the Contract Documents and shall not unreasonably encumber the site with any materials or equipment.

5.14 CUTTING AND PATCHING OF WORK

5.14.1 The Trade Contractor shall be responsible for all cutting, fitting or patching that may be required to complete the Work or to make its several parts fit together properly. He shall provide protection of existing Work as required.

5.14.2 The Trade Contractor shall not damage or endanger any portion of the Work or the work of the Construction Manager or any separate contractors by cutting, patching or otherwise altering any work, or by excavation. The Trade Contractor shall not cut or otherwise alter the work of the Construction Manager or any separate contractor except with the written consent of the Construction Manager and of such separate contractor. The Trade Contractor shall not unreasonably withhold from the Construction Manager or any separate contractor his consent to cutting or otherwise altering the Work.

5.15 CLEANING UP

5.15.1 The Trade Contractor at all times shall keep the premises free from accumulation of waste materials or rubbish caused by his operations. At the completion of the Work he shall remove all his waste materials and rubbish from and about the Project as well as all his tools, construction equipment, machinery and surplus materials.

5.15.2 If the Trade Contractor fails to clean up, the Construction manager may do so and the cost thereof shall be charged to the Trade Contractor.

5.16 COMMUNICATIONS

5.16.1 The Trade Contractor shall forward all communications to the Owner and Architect/Engineer through the Construction Manager.

5.17 ROYALTIES AND PATENTS

5.17.1 The Trade Contractor shall pay all royalties and license fees. He shall defend all suits or claims for infringement of any patent rights and shall save the Owner and Construction Manager harmless from loss on account thereof, except that the Owner shall be responsible for all such loss when a particular design, process or the product of a particular manufacturer or manufacturers is specified, but if the Trade Contractor has reason to believe that the design, process or product specified is an infringement of a patent, he shall be responsible for such loss unless he promptly gives such information to the Construction Manager.

FIGURE A5.6. (*continued*)

5.18 INDEMNIFICATION

5.18.1 To the fullest extent permitted by law, the Trade Contractor shall indemnify and hold harmless the Owner, the Construction Manager and the Architect/Engineer and their agents and employees from and against all claims, damages, losses and expenses, including but not limited to attorneys' fees, arising out of or resulting from the performance of the Work, provided that any such claim, damage, loss or expense (1) is attributable to bodily injury, sickness, disease or death, or to injury to or destruction of tangible property (other than the Work itself) including the loss of use resulting therefrom, and (2) is caused in whole or in part by any negligent act or omission of the Trade Contractor, any Trade Subcontractor, anyone directly or indirectly employed by any of them or anyone for whose acts any of them may be liable, regardless of whether or not it is caused in part by a party indemnified hereunder. Such obligation shall not be construed to negate, abridge or otherwise reduce any other right or obligation of indemnity which would otherwise exist as to any party or person described in this Paragraph 5.18.

5.18.2 In any and all claims against the Owner, the Construction Manager or the Architect/Engineer or any of their agents or employees by any employee of the Trade Contractor, any Trade Subcontractor, anyone directly or indirectly employed by any of them or anyone for whose acts any of them may be liable, the indemnification obligation under this Paragraph 5.18 shall not be limited in any way by any limitation on the amount or type of damages, compensation or benefits payable by or for the Trade Contractor or any Trade Subcontractor under workers' or workmen's compensation acts, disability benefit acts or other employee benefit acts.

5.18.3 The obligations of the Trade Contractor under this Paragraph 5.18 shall not extend to the liability of the Architect/Engineer, his agents or employees arising out of (1) the preparation or approval of maps, drawings, opinions, reports, surveys, designs or specifications, or (2) the giving of or the failure to give directions or instruction by the Architect/Engineer, his agents or employees provided such giving or failure to give is the primary cause of the injury or damage.

ARTICLE 6

TRADE SUBCONTRACTORS

6.1 DEFINITION

6.1.1 A Trade Subcontractor is a person or entity who has a direct contract with a Trade Contractor to perform any of the Work at the site. The term Trade Subcontractor is referred to throughout the Contract Documents as if singular in number and masculine in gender and means a Trade Subcontractor or his authorized representative.

6.1.2 A Trade Subsubcontractor is a person or entity who has a direct or indirect contract with a Trade Subcontractor to perform any of the Work at the site. The term Trade Subsubcontractor is referred to throughout the Contract Documents as if singular in number and masculine in gender and means a Trade Subsubcontractor or an authorized representative thereof.

6.2 AWARD OF TRADE SUBCONTRACTS AND OTHER CONTRACTS FOR PORTIONS OF THE WORK

6.2.1 Unless otherwise required by the Contract Documents or in the Bidding or Proposal Documents, the Trade Contractor shall furnish to the Construction Manager in writing, for acceptance by the Owner and the Construction Manager in writing, the names of the persons or entities (including those who are to furnish materials or equipment fabricated to a special design) proposed for each of the principal portions of the Work. The Construction Manager will promptly reply to the Trade Contractor in writing if either the Owner or the Construction Manager, after due investigation, has reasonable objection to any such proposed person or entity. Failure of the Owner or Construction Manager to reply promptly shall constitute notice of no reasonable objection.

6.2.2 The Trade Contractor shall not contract with any such proposed person or entity to whom the Owner or the Construction Manager has made reasonable objection under the provisions of Subparagraph 6.2.1. The Trade Contractor shall not be required to contract with anyone to whom he has a reasonable objection.

6.2.3 If the Owner or Construction Manager refuses to accept any person or entity on a list submitted by the Trade Contractor in response to the requirements of the Contract Documents, the Trade Contractor shall submit an acceptable substitute; however, no increase in the Contract Sum shall be allowed for any such substitution.

AGC DOCUMENT NO. 8b • GENERAL CONDITIONS FOR TRADE CONTRACTORS UNDER CONSTRUCTION MANAGEMENT AGREEMENTS • JULY 1980

FIGURE A5.6. (*continued*)

6.2.4 The Trade Contractor shall me; however, no increase in the Contract Sum shall be allowed for any such substitution.

6.2.4 The Trade Contractor shall make no substitution for any Trade Subcontractor, person or entity previously selected if the Owner or Construction Manager makes reasonable objection to such substitution.

6.3 TRADE SUBCONTRACTUAL RELATIONS

6.3.1 By an appropriate agreement, written where legally required for validity, the Trade Contractor shall require each Trade Subcontractor, to the extent of the work to be performed by the Trade Subcontrator, to be bound to the Trade Contractor by the terms of the Contract Documents, and to assume toward the Trade Contractor all the obligations and responsibilities which the Trade Contractor, by these Documents, assumes toward the Owner, the Construction Manager, or the Architect/Engineer. Said agreement shall preserve and protect the rights of the Owner, the Construction Manager and the Architect/Engineer under the Contract Documents with respect to the Work to be performed by the Trade Subcontractor so that the subcontracting thereof will not prejudice such rights, and shall allow to the Trade Subcontractor, unless specifically provided otherwise in the Trade Contractor-Trade Subcontractor agreement, the benefit of all rights, remedies and redress against the Trade Contractor that the Trade Contractor, by these Documents, has against the Owner or Construction Manager. Where appropriate, the Trade Contractor shall require each Trade Subcontractor to enter into similar agreements with his Trade Subsubcontractors. The Trade Contractor shall make available to each proposed Trade Subcontractor, prior to the execution of the Trade Subcontract, copies of the Contract Documents to which the Trade Subcontractor will be bound by this Paragraph 6.3, and shall identify to the Trade Subcontractor any terms and conditions of the proposed Trade Subcontract which may be at variance with the Contract Documents. Each Trade Subcontractor shall similarly make copies of such Documents available to his Trade Subsubcontractors.

ARTICLE 7

SEPARATE TRADE CONTRACTS

7.1 MUTUAL RESPONSIBILITY OF TRADE CONTRACTORS

7.1.1 The Trade Contractor shall afford the Construction Manager and other trade contractors reasonable opportunity for the introduction and storage of their materials and equipment and the execution of their work, and shall connect and coordinate his Work with others under the general direction of the Construction Manager.

7.1.2 If any part of the Trade Contractor's Work depends, for proper execution or results, upon the work of the Construction Manager or any separate trade contractor, the Trade Contractor shall, prior to proceeding with the Work, promptly report to the Construction Manager any apparent discrepancies or defects in such work that render it unsuitable for such proper execution and results. Failure of the Trade Contractor so to report shall constitute an acceptance of the other trade contractor's or Construction Manager's work as fit and proper to receive his Work, except as to defects which may subsequently become apparent in such work by others.

7.1.3 Any costs caused by defective or ill-timed work shall be borne by the party responsible thereof.

7.1.4 Should the Trade Contractor wrongfully cause damage to the work or property of the Owner or to other work on the site, the Trade Contractor shall promptly remedy such damage as provided in Subparagraph 11.2.5.

7.1.5 Should the Trade Contractor wrongfully cause damage to the work or property of any separate trade contractor or other contractor, the Trade Contractor shall, upon due notice, promptly attempt to settle with the separate trade contractor or other contractor by agreement, or otherwise resolve the dispute. If such separate trade contractor or other contractor sues the Owner or the Construction Manager or initiates an arbitration proceeding against the Owner or Construction Manager on account of any damage alleged to have been caused by the Trade Contractor, the Owner or Construction Manager shall notify the Trade Contractor who shall defend such proceedings at the Trade Contractor's expense, and if any judgment or award against the Owner or Construction Manager arises therefrom, the Trade Contractor shall pay or satisfy it and shall reimburse the Owner or Construction Manager for all attorney's fees and court or arbitration costs which the Owner or Construction Manager has incurred.

AGC DOCUMENT NO. 8b • GENERAL CONDITIONS FOR TRADE CONTRACTORS UNDER CONSTRUCTION MANAGEMENT AGREEMENTS • JULY 1980

FIGURE A5.6. (*continued*)

7.2 CONSTRUCTION MANAGER'S RIGHT TO CLEAN UP

7.2.1 If a dispute arises between the separate Trade Contractors as to their responsibility for cleaning up as required by Paragraph 5.15, the Construction Manager may clean up and charge the cost thereof to the Trade Contractors responsible therefor as the Construction Manager shall determine to be just.

ARTICLE 8
MISCELLANEOUS PROVISIONS

8.1 GOVERNING LAW

8.1.1 The Contract shall be governed by the law of the place where the Project is located.

8.2 SUCCESSORS AND ASSIGNS

8.2.1 The Owner or Construction Manager (as the case may be) and the Trade Contractor each binds himself, his partners, successors, assigns and legal representatives to the other party hereto and to the partners, successors, assigns and legal representatives of such other party in respect to all covenants, agreements and obligations contained in the Contract Documents. Neither party to the Contract shall assign the Contract or sublet it as a whole without the written consent of the other.

8.3 WRITTEN NOTICE

8.3.1 Written notice shall be deemed to have been duly served if delivered in person to the individual or member of the firm or entity or to an officer of the corporation for whom it was intended, or if delivered at or sent by registered or certified mail to the last business address known to him who gives the notice.

8.4 CLAIMS FOR DAMAGES

8.4.1 Should either party to the Trade Contract suffer injury or damage to person or property because of any act or omission of the other party or of any of his employees, agents or others for whose acts he is legally liable, claim shall be made in writing to such other party within a reasonable time after the first observance of such injury or damage.

8.5 PERFORMANCE BOND AND LABOR AND MATERIAL PAYMENT BOND

8.5.1 The Owner or Construction Manager shall have the right to require the Trade Contractor to furnish bonds in a form and with a corporate surety acceptable to the Construction Manager covering the faithful performance of the Contract and the payment of all obligations arising thereunder if and as required in the Bidding or Proposal Documents or in the Contract Documents.

8.6 RIGHTS AND REMEDIES

8.6.1 The duties and obligations imposed by the Contract Documents and the rights and remedies available thereunder shall be in addition to and not a limitation of any duties, obligations, rights and remedies otherwise imposed or available by law.

8.6.2 No action or failure to act by the Construction Manager, Architect/Engineer or Trade Contractor shall constitute a waiver of any right or duty afforded any of them under the Contract Documents, nor shall any such action or failure to act constitute an approval of or acquiescence in any breach thereunder, except as may be specifically agreed in writing.

8.7 TESTS

8.7.1 If the Contract Documents, laws, ordinances, rules, regulations or orders of any public authority having jurisdiction require any portion of the Work to be inspected, tested or approved, the Trade Contractor shall give the Construction Manager timely notice of its readiness so the Architect/Engineer and Construction Manager may observe such inspection, testing or approval. The Trade Contractor shall bear all costs of such inspections, tests or approvals unless otherwise provided.

AGC DOCUMENT NO. 8b • GENERAL CONDITIONS FOR TRADE CONTRACTORS UNDER CONSTRUCTION MANAGEMENT AGREEMENTS • JULY 1980

FIGURE A5.6. (*continued*)

8.7.2 If the Architect/Engineer or Construction Manager determines that any Work requires special inspection, testing or approval which Subparagraph 8.7.1 does not include, he will, through the Construction Manager, instruct the Trade Contractor to order such special inspection, testing or approval and the Trade Contractor shall give notice as in Subparagraph 8.7.1. If such special inspection or testing reveals a failure of the Work to comply with the requirements of the Contract Documents, the Trade Contractor shall bear all costs thereof, including compensation for the Architect/Engineer's and Construction Manager's additional services made necessary by such failure. If the Work complies, the Owner or Construction Manager (as the case may be) shall bear such costs and an appropriate Change Order shall be issued.

8.7.3 Required certificates of inspection, testing or approval shall be secured by the Trade Contractor and promptly delivered by him through the Construction Manager to the Architect/Engineer.

8.7.4 If the Architect/Engineer or Construction Manager is to observe the inspections, tests or approvals required by the Contract Documents, he will do so promptly and, where practicable, at the source of supply.

8.8 INTEREST

8.8.1 Payments due and unpaid under the Contract Documents shall bear interest from the date payment is due at such rate upon which the parties may agree in writing or, in the absence thereof, at the legal rate prevailing at the place of the Project.

8.9 ARBITRATION

8.9.1 All claims, disputes and other matters in question arising out of, or relating to this Contract or the breach thereof, except as set forth in Subparagraph 3.2.4 with respect to the Architect/Engineer's decisions on matters relating to artistic effect, and except for claims which have been waived by the making or acceptance of final payment provided by Subparagraphs 10.8.4. and 10.8.5, shall be decided by arbitration in accordance with the Construction Industry Arbitration Rules of the American Arbitration Association then obtaining unless the parties mutually agree otherwise. This agreement to arbitrate shall be specifically enforceable under the prevailing arbitration law. The award rendered by the arbitrators shall be final, and judgment may be entered upon it in accordance with applicable law in any court having jurisdiction thereof.

8.9.2 Notice of the demand for arbitration shall be filed in writing with the other party to the Contract and with the American Arbitration Association. The demand for arbitration shall be made within a reasonable time after the claim, dispute or other matter in question has arisen, and in no event shall it be made after the date when institution of legal or equitable proceedings based on such claim, dispute or other matter in question would be barred by the applicable statute of limitations.

8.9.3 The Trade Contractor shall carry on the Work and maintain the progress schedule during any arbitration proceedings, unless otherwise agreed by him and the Construction Manager in writing.

8.9.4 All claims which are related to or dependent upon each other shall be heard by the same arbitrator or arbitrators even though the parties are not the same unless a specific contract prohibits such consolidation.

ARTICLE 9

TIME

9.1 DEFINITIONS

9.1.1 Unless otherwise provided, the Contract Time is the period of time allotted in the Contract Documents for the Substantial Completion of the Work as defined in Subparagraph 9.1.3 including authorized adjustments thereto.

9.1.2 The date of commencement of the Work is the date established in a notice to proceed. If there is no notice to proceed, it shall be the date of the Trade Contractor Agreement or such other date as may be established therein.

9.1.3 The Date of Substantial Completion of the Work or designated portion thereof is the Date certified by the Architect/Engineer when construction is sufficiently complete, in accordance with the Contract Documents, so the Owner can occupy or utilize the Work or designated portion thereof for the use for which it is intended.

9.1.4 The term day as used in the Contract Documents shall mean calendar day unless otherwise specifically designated.

AGC DOCUMENT NO. 8b • GENERAL CONDITIONS FOR TRADE CONTRACTORS UNDER CONSTRUCTION MANAGEMENT AGREEMENTS • JULY 1980

FIGURE A5.6. (*continued*)

9.2 PROGRESS AND COMPLETION

9.2.1 All time limits stated in the Contract Documents are of the essence of the Contract.

9.2.2 The Trade Contractor shall begin the Work on the date of commencement as defined in Subparagraph 9.1.2. He shall carry the Work forward expeditiously with adequate forces and shall achieve Substantial Completion within the Contract Time.

9.3 DELAYS AND EXTENSIONS OF TIME

9.3.1 If the Trade Contractor is delayed at any time in the progress of the Work by any act or neglect of the Owner, Construction Manager, or the Architect/Engineer, or by any employee of either, or by any separate contractor employed by the Owner, or by changes ordered in the Work, or by labor disputes, fire, unusual delay in transportation, adverse weather conditions not reasonably anticipatable, unavoidable casualties or any causes beyond the Trade Contractor's control, or by delay authorized by the Owner or Construction Manager pending arbitration, or by any other cause which the Construction Manager determines may justify the delay, then the Contract Time shall be extended by Change Order for such reasonable time as the Construction Manager may determine.

9.3.2 Any claim for extension of time shall be made in writing to the Construction Manager not more than twenty (20) days after the commencement of the delay; otherwise, it shall be waived. In the case of a continuing delay only one claim is necessary. The Trade Contractor shall provide an estimate of the probable effect of such delay on the progress of the Work.

9.3.3 If no agreement is made stating the dates upon which interpretations as set forth in Subparagraph 3.2.2 shall be furnished, then no claim for delay shall be allowed on account of failure to furnish such interpretations until fifteen days after written request is made for them, and not then unless such claim is reasonable.

9.3.4 It shall be recognized by the Trade Contractor that he may reasonably anticipate that as the job progresses, the Construction Manager will be making changes in and updating Construction Schedules pursuant to the authority given him in Subparagraph 4.2.3. Therefore, no claim for an increase in the Contract Sum for either acceleration or delay will be allowed for extensions of time pursuant to this Paragraph 9.3 or for other changes in the Construction Schedules which are of the type ordinarily experienced in projects of similar size and complexity.

9.3.5 This Paragraph 9.3 does not exclude the recovery of damages for delay by either party under other provisions of the Contract Documents.

ARTICLE 10

PAYMENTS AND COMPLETION

10.1 CONTRACT SUM

10.1.1 The Contract Sum is stated in the Agreement between the Owner or Construction Manager and the Trade Contractor including adjustments thereto and is the total amount payable to the Trade Contractor for the performance of the Work under the Contract Documents.

10.2 SCHEDULE OF VALUES

10.2.1 Before the first Application for Payment, the Trade Contractor shall submit to the Construction Manager a schedule of values allocated to the various portions of the Work prepared in such form and supported by such data to substantiate its accuracy as the Construction Manager may require. This schedule, unless objected to by the Construction Manager, shall be used only as a basis for the Trade Contractor's Application for Payment.

10.3 APPLICATIONS FOR PAYMENT

10.3.1 At least ten days before the date for each progress payment established in the Trade Contractor's Agreement, the Trade Contractor shall submit to the Construction Manager an itemized Application for Payment, notarized if required, supported by such data substantiating the Trade Contractor's right to payment as the Owner or the Construction Manager may require, and reflecting retainage, if any, as provided elsewhere in the Contract Documents.

AGC DOCUMENT NO. 8b • GENERAL CONDITIONS FOR TRADE CONTRACTORS UNDER CONSTRUCTION MANAGEMENT AGREEMENTS • JULY 1980

FIGURE A5.6. (*continued*)

10.3.2 Unless otherwise provided in the Contract Documents, payments will be made on account of materials or equipment not incorporated in the Work but delivered and suitably stored at the site and, if approved in advance by the Construction Manager, payments may similarly be made for materials or equipment stored at some other location agreed upon in writing. Payments made for materials or equipment stored on or off the site shall be conditioned upon submission by the Trade Contractor of bills of sale or such other procedures satisfactory to the Construction Manager to establish the Owner's title to such materials or equipment or otherwise protect the Owner's interest, including applicable insurance and transportation to the site for those materials and equipment stored off the site.

10.3.3 The Trade Contractor warrants that title to all Work, materials and equipment covered by an Application for Payment will pass to the Owner either by incorporation in the construction or upon the receipt of payment by the Trade Contractor, whichever occurs first, free and clear of all liens, claims, security interests or encumbrances, hereinafter referred to in this Article 10 as "liens;" and that no Work, materials or equipment covered by an Application for Payment will have been acquired by the Trade Contractor, or by any other person performing his Work at the site or furnishing materials and equipment for his Work, subject to an agreement under which an interest therein or an encumbrance thereon is retained by the seller or otherwise imposed by the Trade Contractor or such other person. All Trade Subcontractors and Trade Subsubcontractors agree that title will so pass upon their receipt of payment from the Trade Contractor.

10.4 PROGRESS PAYMENTS

10.4.1 If the Trade Contractor has made Application for Payment as above, the Construction Manager will, with reasonable promptness but not more than seven days after the receipt of the Application, review and process such Application for payment in accordance with the Contract.

10.4.2 No approval of an application for a progress payment, nor any progress payment, nor any partial or entire use or occupancy of the Project by the Owner, shall constitute an acceptance of any Work not in accordance with the Contract Documents.

10.4.3 The Trade Contractor shall promptly pay each Trade Subcontractor upon receipt of payment out of the amount paid to the Trade Contractor on account of such Trade Subcontractor's Work, the amount to which said Trade Subcontractor is entitled, reflecting the percentage actually retained, if any, from payments to the Trade Contractor on account of such Trade Subcontractor's Work. The Trade Contractor shall, by an appropriate agreement with each Trade Subcontractor, also require each Trade Subcontractor to make payments to his Trade Subsubcontractors in a similar manner.

10.5 PAYMENTS WITHHELD

10.5.1 The Construction Manager may decline to approve an Application for Payment if in his opinion the Application is not adequately supported. If the Trade Contractor and Construction Manager cannot agree on a revised amount, the Construction Manager shall process the Application for the amount he deems appropriate. The Construction Manager may also decline to approve any Applications for Payment or, because of subsequently discovered evidence or subsequent inspections, he may nullify in whole or in part any approval previously made to such extent as may be necessary in his opinion because of:

.1 defective work not remedied;

.2 third party claims filed or reasonable evidence indicating probable filing of such claims;

.3 failure of the Trade Contractor to make payments properly to Trade Subcontractors or for labor, materials or equipment;

.4 reasonable evidence that the Work cannot be completed for the unpaid balance of the Contract Sum;

.5 damage to the Construction Manager, the Owner, or another contractor working at the Project;

.6 reasonable evidence that the Work will not be completed within the Contract Time; or

.7 persistent failure to carry out the Work in accordance with the Contract Documents.

10.5.2 When the above grounds in Subparagraph 10.5.1 are removed, payment shall be made for amounts withheld because of them.

FIGURE A5.6. (*continued*)

10.6 FAILURE OF PAYMENT

10.6.1 If the Trade Contractor is not paid within seven days after any amount is approved for payment by the Construction Manager and has become due and payable, then the Trade Contractor may, upon seven additional days' written notice to the Owner and Construction Manager, stop the Work until payment of the amount owing has been received. The Contract Sum shall be increased by the amount of the Trade Contractor's reasonable costs of shutdown, delay and start up, which shall be effected by appropriate Change Order in accordance with Paragraph 13.3.

10.7 SUBSTANTIAL COMPLETION

10.7.1 When the Trade Contractor considers that the Work, or a designated portion thereof which is acceptable to the Owner, is substantially complete as defined in Subparagraph 9.1.3, the Trade Contractor shall prepare for submission to the Construction Manager a list of items to be completed or corrected. The failure to include any items on such list does not alter the responsibility of the Trade Contractor to complete all Work in accordance with the Contract Documents. When the Construction Manager and Architect/Engineer on the basis of inspection determine that the Work or designated portion thereof is substantially complete, the Architect/Engineer will then prepare a Certificate of Substantial Completion which shall establish the Date of Substantial Completion, shall state the responsibilities of the Owner, the Construction Manager and the Trade Contractor for security, maintenance, heat, utilities, damage to the Work, and insurance, and shall fix the time within which the Trade Contractor shall complete the items listed therein. Warranties required by the Contract Documents shall commence on the Date of Substantial Completion of the Work or designated portion thereof unless otherwise provided in the Certificate of Substantial Completion. The Certificate of Substantial Completion shall be submitted to the Owner, the Construction Manager and the Trade Contractor for their written acceptance of the responsibilities assigned to them in such Certificate.

10.8 FINAL COMPLETION AND FINAL PAYMENT

10.8.1 Upon receipt of written notice that the Work is ready for final inspection and acceptance and upon receipt of a final Application for Payment, the Architect/Engineer and the Construction Manager will promptly make such inspection and, when they find the Work acceptable under the Contract Documents and the Contract fully performed, the Construction Manager will promptly approve final payment.

10.8.2 Neither the final payment nor the remaining retained percentage shall become due until the Trade Contractor submits to the Construction Manager (1) an affidavit that all payrolls, bills for materials and equipment, and other indebtedness connected with the Work for which the Owner or his property might in any way be responsible, have been paid or otherwise satisfied, (2) consent of surety, if any, to final payment, and (3) if required by the Owner, other data establishing payment or satisfaction of all such obligations, such as receipts, releases and waivers of liens arising out of the Contract, to the extent and in such form as may be designated by the Owner. If any Trade Subcontractor refuses to furnish a release or waiver required by the Owner or Construction Manager, the Trade Contractor may furnish a bond satisfactory to the Owner and Construction Manager to indemnify them against any such lien. If any such lien remains unsatisfied after all payments are made, the Trade Contractor shall refund to the Owner or Construction Manager all moneys that the latter may be compelled to pay in discharging such lien, including all costs and reasonable attorneys' fees.

10.8.3 If, after Substantial Completion of the Work, final completion thereof is materially delayed through no fault of the Trade Contractor or by the issuance of Change Orders affecting final completion, and the Construction Manager so confirms, the Owner or Construction Manager shall, upon certification by the Construction Manager, and without terminating the Contract, make payment of the balance due for that portion of the Work fully completed and accepted. If the remaining balance for Work not fully completed or corrected is less than the retainage stipulated in the Contract Documents, and if bonds have been furnished as provided in Paragraph 8.5, the written consent of the surety to the payment of the balance due for that portion of the Work fully completed and accepted shall be submitted by the Trade Contractor to the Construction Manager prior to such payment. Such payment shall be made under the terms and conditions governing final payment, except that it shall not constitute a waiver of claims.

10.8.4 The making of final payment shall constitute a waiver of all claims by the Owner or Construction Manager except those arising from:

    .1   unsettled liens;

    .2   faulty or defective Work appearing after Substantial Completion;

FIGURE A5.6. (*continued*)

.3 failure of the Work to comply with the requirements of the Contract Documents; or

.4 terms of any special warranties required by the Contract Documents.

10.8.5 The acceptance of final payment shall constitute a waiver of all claims by the Trade Contractor except those previously made in writing and identified by the Trade Contractor as unsettled at the time of the Final Application for Payment.

ARTICLE 11

PROTECTION OF PERSONS AND PROPERTY

11.1 SAFETY PRECAUTIONS AND PROGRAMS

11.1.1 The Trade Contractor shall be responsible for initiating, maintaining and supervising all safety precautions and programs in connection with the Work.

11.1.2 If the Trade Contractor fails to maintain the safety precautions required by law or directed by the Construction Manager, the Construction Manager may take such steps as necessary and charge the Trade Contractor therefor.

11.1.3 The failure of the Construction Manager to take any such action shall not relieve the Trade Contractor of his obligations in Subparagraph 11.1.1.

11.2 SAFETY OF PERSONS AND PROPERTY

11.2.1 The Trade Contractor shall take all reasonable precautions for the safety of, and shall provide all reasonable protection to prevent damage, injury or loss to:

.1 all employees on the Work and all other persons who may be affected thereby;

.2 all the Work and all materials and equipment to be incorporated therein, whether in storage on or off the site, under the care, custody or control of the Trade Contractor or any of his Trade Subcontractors or Trade Sub-subcontractors; and

.3 other property at the site or adjacent thereto, including trees, shrubs, lawns, walks, pavements, roadways, structures and utilities not designated for removal, relocation or replacement in the course of construction.

11.2.2 The Trade Contractor shall give all notices and comply with all applicable laws, ordinances, rules, regulations and lawful orders of any public authority bearing on the safety of persons or property or their protection from damage, injury or loss.

11.2.3 The Trade Contractor shall erect and maintain, as required by existing conditions and progress of the Work, all reasonable safeguards for safety and protection, including posting danger signs and other warnings against hazards, promulgating safety regulations and notifying owners and users of adjacent utilities. If the Trade Contractor fails to so comply he shall, at the direction of the Construction Manager, remove all forces from the Project without cost or loss to the Owner or Construction Manager, until he is in compliance.

11.2.4 When the use or storage of explosives or other hazardous materials or equipment is necessary for the execution of the Work, the Trade Contractor shall exercise the utmost care and shall carry on such activities under the supervision of properly qualified personnel.

11.2.5 The Trade Contractor shall promptly remedy all damage or loss (other than damage or loss insured under Paragraph 12.2) to any property referred to in Clauses 11.2.1.2 and 11.2.1.3 caused in whole or in part by the Trade Contractor, his Trade Subcontractors, his Trade Subsubcontractors, or anyone directly or indirectly employed by any of them, or by anyone for whose acts any of them may be liable and for which the Trade Contractor is responsible under Clauses 11.2.1.2 and 11.2.1.3, except damage or loss attributable to the acts or omissions of the Owner or Architect/Engineer or anyone directly or indirectly employed by either of them or by anyone for whose acts either of them may be liable, and not attributable to the fault or negligence of the Trade Contractor. The foregoing obligations of the Trade Contractor are in addition to his obligations under Paragraph 5.18.

AGC DOCUMENT NO. 8b • GENERAL CONDITIONS FOR TRADE CONTRACTORS UNDER CONSTRUCTION MANAGEMENT AGREEMENTS • JULY 1980

FIGURE A5.6. *(continued)*

11.2.6 The Trade Contractor shall designate a responsible member of his organization at the site whose duty shall be the prevention of accidents. This person shall be the Trade Contractor's superintendent unless otherwise designated by the Trade Contractor in writing to the Construction Manager.

11.2.7 The Trade Contractor shall not load or permit any part of the Work to be loaded so as to endanger its safety.

11.3 EMERGENCIES

11.3.1 In any emergency affecting the safety of persons or property, the Trade Contractor shall act, at his discretion, to prevent threatened damage, injury or loss. Any additional compensation or extension of time claimed by the Trade Contractor on account of emergency work shall be determined as provided in Article 13 for Changes in the Work.

ARTICLE 12

INSURANCE

12.1 TRADE CONTRACTOR'S LIABILITY INSURANCE

12.1.1 The Trade Contractor shall purchase and maintain such insurance as will protect him from claims set forth below which may arise out of or result from the Trade Contractor's operations under the Contract, whether such operations be by himself or by any of his Trade Subcontractors or by anyone directly or indirectly employed by any of them, or by anyone for whose acts any of them may be liable:

.1 claims under workers' or workmen's compensation, disability benefit and other similar employee benefit acts which are applicable to the Work to be performed including the "Broad Form All States" Endorsement;

.2 claims for damages because of bodily injury, occupational sickness or disease, or death of his employees under any employers liability law including, if applicable, those required under maritime or admiralty law for wages, maintenance, and cure;

.3 claims for damages because of bodily injury, sickness or disease, or death of any person other than his employees;

.4 claims for damages insured by usual personal injury liability coverage which are sustained (1) by any person as a result of an offense directly or indirectly related to the employment of such person by the Trade Contractor, or (2) by any other person;

.5 claims for damages other than to the Work itself because of injury to or destruction of tangible property, including loss of use resulting therefrom; and

.6 claims for damages because of bodily injury or death of any person or property damage arising out of the ownership, maintenance or use of any motor vehicle.

12.1.2 The insurance required by Subparagraph 12.1.1 shall be written for not less than any limits of liability specified in the Contract Documents, or required by law, whichever is greater.

12.1.3 The insurance required by Subparagraph 12.1.1 shall include premises-operations (including explosion, collapse and underground coverage), elevators, independent contractors, products and/or completed operations, and contractual liability insurance (on a "blanket basis" designating all written contracts), all including broad form property damage coverage. Liability insurance may be arranged under Comprehensive General Liability policies for the full limits required or by a combination of underlying policies for lesser limits with the remaining limits provided by an Excess or Umbrella Liability Policy.

12.1.4 The foregoing policies shall contain a provision that coverages afforded under the policies will not be cancelled until at least sixty days' prior written notice has been given to the Construction Manager. Certificates of Insurance acceptable to the Construction Manager shall be filed with the Construction Manager prior to commencement of the Work. Upon request, the Trade Contractor shall allow the Construction Manager to examine the actual policies.

AGC DOCUMENT NO. 8b • GENERAL CONDITIONS FOR TRADE CONTRACTORS UNDER CONSTRUCTION MANAGEMENT AGREEMENTS • JULY 1980

FIGURE A5.6. *(continued)*

12.2 PROPERTY INSURANCE AND WAIVER OF SUBROGATION

12.2.1 Unless otherwise provided, the Owner will purchase and maintain property insurance upon the entire Work at the site to the full insurable value thereof. This insurance shall include the interests of the Owner, the Construction Manager, the Trade Contractors, and Trade Subcontractors in the Work and shall insure against the perils of fire and extended coverage, and shall include "all risk" insurance for physical loss or damage.

12.2.2 The Owner will effect and maintain such boiler and machinery insurance as may be necessary and/or required by law. This insurance shall include the interest of the Owner, the Construction Manger, the Trade Contractors, and Trade Subcontractors in the Work.

12.2.3 Any loss insured under Paragraph 12.2 is to be adjusted with the Owner and Construction Manager and made payable to the Owner and Construction Manager as trustees for the insureds, as their interests may appear, subject to the requirements of any applicable mortgagee clause.

12.2.4 The Owner, the Construction Manager, the Architect/Engineer, the Trade Contractors, and the Trade Subcontractors waive all rights against each other and any other contractor or subcontractor engaged in the Project for damages caused by fire or other perils to the extent covered by insurance provided under Paragraph 12.2, or any other property or consequential loss insurance applicable to the Project, equipment used in the Project, or adjacent structures, except such rights as they may have to the proceeds of such insurance. If any policy of insurance requires an endorsement to maintain coverage with such waivers, the owner of such policy will cause the policy to be so endorsed. The Owner will require, by appropriate agreement, written where legally required for validity, similar waivers in favor of the Trade Contractors and Trade Subcontractors by any separate contractor and his subcontractors.

12.2.5 The Owner and Construction Manager shall deposit in a separate account any money received as trustees, and shall distribute it in accordance with such agreement as the parties in interest may reach, or in accordance with an award by arbitration in which case the procedure shall be as provided in Paragraph 8.9. If after such loss no other special agreement is made, replacement of damaged Work shall be covered by an appropriate Change Order.

12.2.6 The Owner and Construction Manger as trustees shall have power to adjust and settle any loss with the insurers unless one of the parties in interest shall object in writing within five days after the occurrence of loss to the Owner's and Construction Manager's exercise of this power, and if such objection be made, arbitrators shall be chosen as provided in Paragraph 8.9. The Owner and Construction Manager as trustees shall, in that case, make settlement with the insurers in accordance with the directions of such arbitrators. If distribution of the insurance proceeds by arbitration is required, the arbitrators will direct such distribution.

12.2.7 If the Owner finds it necessary to occupy or use a portion or portions of the Work prior to Substantial Completion thereof, such occupancy shall not commence prior to a time mutually agreed to by the Owner and Construction Manager and to which the insurance company or companies providing the property insurance have consented by endorsement to the policy or policies. This insurance shall not be cancelled or lapsed on account of such partial occupancy.

ARTICLE 13

CHANGES IN THE WORK

13.1 CHANGE ORDERS

13.1.1 A Change Order is a written order to the Trade Contractor signed by the Owner or Construction Manager, as the case may be, issued after the execution of the Contract, authorizing a Change in the Work or an adjustment in the Contract Sum or the Contract Time. The Contract Sum and the Contract Time may be changed only by Change Order. A Change Order signed by the Trade Contractor indicates his agreement therewith, including the adjustment in the Contract Sum or the Contract Time.

13.1.2 The Owner or Construction Manager, without invalidating the Contract, may order Changes in the Work within the general scope of the Contract consisting of additions, deletions or other revisions, the Contract Sum and the Contract Time being adjusted accordingly. All such changes in the Work shall be authorized by Change Order, and shall be performed under the applicable conditions of the Contract Documents.

FIGURE A5.6. (*continued*)

13.1.3 The cost or credit to the Owner or Construction Manager resulting from a Change in the Work shall be determined in one or more of the following ways:

.1 by mutual acceptance of a lump sum properly itemized and supported by sufficient substantiating data to permit evaluation; or

.2 by unit prices stated in the Contract Documents or subsequently agreed upon; or

.3 by cost to be determined in a manner agreed upon by the parties and a mutually acceptable fixed or percentage fee; or

.4 by the method provided in Subparagraph 13.1.4.

13.1.4 If none of the methods set forth in Clauses 13.1.3.1, 13.1.3.2 or 13.1.3.3 is agreed upon, the Trade Contractor, provided he receives a written order signed by the Owner or the Construction Manager, shall promptly proceed with the Work involved. The cost of such Work shall be determined by the Construction Manager on the basis of the reasonable expenditures and savings of those performing the Work attributable to the change, including, in the case of an increase in the Contract Sum, a reasonable allowance for overhead and profit. In such case, and also under Clauses 13.1.3.3 and 13.1.3.4 above, the Trade Contractor shall keep and present, in such form as the Construction Manager may prescribe, an itemized accounting together with appropriate supporting data for inclusion in a Change Order. Unless otherwise provided in the Contract Documents, cost shall be limited to the following: cost of materials; including sales tax and cost of delivery; cost of labor, including social security, old age and unemployment insurance, and fringe benefits required by agreement or custom; workers' or workmen's compensation insurance; bond premiums; rental value of equipment and machinery; and the additional costs of supervision and field office personnel directly attributable to the change. Pending final determination of cost, payments on account shall be made as determined by the Construction Manager. The amount of credit to be allowed by the Trade Contractor for any deletion or change which results in a net decrease in the Contract Sum will be the amount of the actual net cost as confirmed by the Construction Manager. When both additions and credits covering related Work or substitutions are involved in any one change, the allowance for overhead and profit shall be figured on the basis of the net increase, if any, with respect to that change.

13.1.5 If unit prices are stated in the Contract Documents or subsequently agreed upon, and if the quantities originally contemplated are so changed in a proposed Change Order that application of the agreed unit prices to the quantities of Work proposed will cause substantial inequity to the Owner, the Construction Manager, or the Trade Contractor, the applicable unit prices shall be equitably adjusted.

13.2 CONCEALED CONDITIONS

13.2.1 Should concealed conditions encountered in the performance of the Work below the surface of the ground or should concealed or unknown conditions in an existing structure be at variance with the conditions indicated by the Contract Documents, or should unknown physical conditions below the surface of the ground or should concealed or unknown conditions in an existing structure of an unusual nature, differing materially from those ordinarily encountered and generally recognized as inherent in work of the character provided for in this Contract, be encountered, the Contract Sum shall be equitably adjusted by Change Order upon claim by either party made within twenty days after the first observance of the conditions.

13.3 CLAIMS FOR ADDITIONAL COST

13.3.1 If the Trade Contractor wishes to make a claim for an increase in the Contract Sum, he shall give the Construction Manager written notice thereof within twenty days after the occurrence of the event giving rise to such claim. This notice shall be given by the Trade Contractor before proceeding to execute the Work, except in an emergency endangering life or property in which case the Trade Contractor shall proceed in accordance with Paragraph 11.3. No such claim shall be valid unless so made. Any change in the Contract Sum resulting from such claim shall be authorized by Change Order.

13.3.2 If the Trade Contractor claims that additional cost is involved because of, but not limited to, (1) any written interpretation issued pursuant to Subparagraph 3.2.2, (2) any order by the Owner or Construction Manager to stop the Work pursuant to Paragraph 4.3 where the Trade Contractor was not at fault, or (3) any written order for a minor change in the Work issued pursuant to Paragraph 13.4, the Trade Contractor shall make such claim as provided in Subparagraph 13.3.1.

FIGURE A5.6. (*continued*)

13.4 MINOR CHANGES IN THE WORK

13.4.1 The Architect/Engineer will have authority to order through the Construction Manager minor changes in the Work not involving an adjustment in the Contract Sum or an extension of the Contract Time and not inconsistent with the intent of the Contract Documents. Such changes shall be effected by written order and such changes shall be binding on the Owner, the Construction Manager, and the Trade Contractor. The Trade Contractor shall carry out such written orders promptly.

ARTICLE 14

UNCOVERING AND CORRECTION OF WORK

14.1 UNCOVERING OF WORK

14.1.1 If any portion of the Work should be covered contrary to the request of the Construction Manager or Architect/Engineer, or to requirements specifically expressed in the Contract Documents, it must, if required in writing by the Construction Manager, be uncovered for their observation and replaced, at the Trade Contractor's expense.

14.1.2 If any other portion of the Work has been covered which neither the Construction Manager nor the Architect/Engineer has specifically requested to observe prior to being covered, the Architect/Engineer or Construction Manager may request to see such Work and it shall be uncovered by the Trade Contractor. If such Work be found in accordance with the Contract Documents, the cost of uncovering and replacement shall, by appropriate Change Order, be charged to the Owner or Construction Manager, as the case may be. If such Work be found not in accordance with the Contract Documents, the Trade Contractor shall pay such costs unless it be found that this condition was caused by a separate trade contractor employed as provided in Article 7, and in that event the separate trade contractor shall be responsible for the payment of such costs.

14.2 CORRECTION OF WORK

14.2.1 The Trade Contractor shall promptly correct all Work rejected by the Architect/Engineer or the Construction Manager as defective or as failing to conform to the Contract Documents whether observed before or after Substantial Completion and whether or not fabricated, installed or completed. The Trade Contractor shall bear all costs of correcting such rejected Work, including compensation for the Architect/Engineer's and/or Construction Manager's additional services made necessary thereby.

14.2.2 If, within one year after the Date of Substantial Completion of Work or designated portion thereof, or within one year after acceptance by the Owner of designated equipment or within such longer period of time as may be prescribed by law or by the terms of any applicable special warranty required by the Contract Documents, any of the Work is found to be defective or not in accordance with the Contract Documents, the Trade Contractor shall correct it promptly after receipt of a written notice from the Owner or Construction Manager to do so unless the Owner or Construction Manager has previously given the Trade Contractor a written acceptance of such condition. This obligation shall survive the termination of the Contract. The Owner or Construction Manager shall give such notice promptly after discovery of the condition.

14.2.3 The Trade Contractor shall remove from the site all portions of the Work which are defective or non-conforming and which have not been corrected under Subparagraphs 5.5.1, 14.2.1 and 14.2.2, unless removal has been wavied by the Owner.

14.2.4 If the Trade Contractor fails to correct defective or non-conforming Work as provided in Subparagraphs 5.5.1, 14.2.1 and 14.2.2, the Owner or Construction Manager may correct it in accordance with Subparagraph 4.3.2.

14.2.5 If the Trade Contractor does not proceed with the correction of such defective or non-conforming Work within a reasonable time fixed by written notice from the Construction Manager, the Owner or Construction Manager may remove it and may store the materials or equipment at the expense of the Trade Contractor. If the Trade Contractor does not pay the cost of such removal and storage within ten days thereafter, the Owner or Construction Manager may upon ten additional days' written notice sell such Work at auction or at private sale and shall account for the net proceeds thereof, after deducting all the costs that should have been borne by the Trade Contractor, including compensation for the Construction Manager's additional services made necessary thereby. If such proceeds of sale do not cover all costs which the Trade Contractor should have borne, the difference shall be charged to the Trade Contractor and an appropriate Change Order shall be issued. If the payments then or thereafter due the Trade Contractor are not sufficient to cover such amount, the Trade Contractor shall pay the difference to the Owner or Construction Manager.

FIGURE A5.6. (*continued*)

14.2.6 The Trade Contractor shall bear the cost of making good all work of the Construction Manager or other contractors destroyed or damaged by such removal or correction.

14.3 ACCEPTANCE OF DEFECTIVE OR NONCONFORMING WORK

14.3.1 If the Owner or Construction Manager prefers to accept defective or non-conforming Work, he may do so instead of requiring its removal and correction, in which case a Change Order will be issued to reflect reduction in the Contract Sum where appropriate and equitable. Such adjustment shall be effected whether or not final payment has been made.

ARTICLE 15

TERMINATION OF THE CONTRACT

15.1 TERMINATION BY THE TRADE CONTRACTOR

15.1.1 If the Work is stopped for a period of thirty days under an order of any court or other public authority having jurisdiction, or as a result of an act of government, such as a declaration of a national emergency making materials unavailable, through no act or fault of the Trade Contractor or a Trade Subcontractor or their agents or employees or any other persons performing any of the Work under a contract with the Trade Contractor, or if the Work should be stopped for a period of thirty days by the Trade Contractor because of a failure to receive payment in accordance with the Contract, then the Trade Contractor may, upon seven additional days' written notice to the Construction Manager, terminate the Contract and recover from the Owner or Construction Manager, as the case may be, payment for all Work executed and for any proven loss sustained upon any materials, equipment, tools, construction equipment and machinery, including reasonable profit and damages.

15.2 TERMINATION BY THE OWNER OR CONSTRUCTION MANAGER

15.2.1 If the Trade Contractor is adjudged a bankrupt, or if he makes a general assignment for the benefit of his creditors, or if a receiver is appointed on account of his insolvency, or if he persistently or repeatedly refuses or fails, except in cases for which extension of time is provided, to supply enough properly skilled workmen or proper materials, or if he fails to make prompt payment to Trade Subcontractors or for materials or labor, or persistently disregards laws, ordinances, rules, regulations or orders of any public authority having jurisdiction, or otherwise is guilty of a substantial violation of a provision of the Contract Documents, then the Owner or Construction Manager may, without prejudice to any right or remedy and after giving the Trade Contractor and his surety, if any, seven days' written notice, terminate the employment of the Trade Contractor and take possession of the site and of all materials, equipment, tools, construction equipment and machinery thereon owned by the Trade Contractor and may finish the Work by whatever method he may deem expedient. In such case the Trade Contractor shall not be entitled to receive any further payment until the Work is finished.

15.2.2 If the unpaid balance of the Contract Sum exceeds the costs of finishing the Work, including compensation for the Construction Manager's additional services made necessary thereby, such excess shall be paid to the Trade Contractor. If such costs exceed the unpaid balance, the Trade Contractor shall pay the difference to the Owner or Construction Manager.

AGC DOCUMENT NO. 8b • GENERAL CONDITIONS FOR TRADE CONTRACTORS UNDER CONSTRUCTION MANAGEMENT AGREEMENTS • JULY 1980

FIGURE A5.6. (*continued*)

Index

Advertising, 209-210

Bid volume, 28
Boycotts, 309-310
Brain picking, 78-81, 322
Bribes and payoffs, 299-304, 307-308
Brochures, 90-128
   categories of, 90-91
   designing, 92-100
   reach the right audience, 91-92
Budget estimate, 207
Business development, 3
Business letters, 129-137

Calls:
   business lunches, 146-147
   follow up, 146
   initial contact, 142-146
   making appointments, 142
   preparation, 148
Career growth, 323-324
Career planning, 314-317
Changed conditions, 265-268
Changes, change orders, and claims, 264-285
Claim letters, 272-276
Claims:
   by avoiding, 284-285
   by change orders, 268
   by impact or ripple, 269
   by inspection, 270-271
   by interpretation of plans and specifications, 271-272
   by owner-furnished items, 269-270
Client involvement, 207-208
Client relationships, 203-204, 236-237

Clients proper, 15-17, 23
Closing sales, 209
Community involvement, 210-211
Competition, 22
   between professional services, 241-242
Competitive pressure, 74-89
Conference problems, 340-346
Conferences:
   minutes, 151-152
   participation, 150-151
   Robert's Rules, 150
   setting up, 149
Construction industry, 1-2
Contracts:
   construction, 245-246
   construction management, 61-69
   consulting, 59-61, 62-63, 243-245
   cost plus fixed fee, 52-57
   day rate, 57-59
   design and construct, 247-248
   lump-sum, 41-47
   lump-sum plus units, 47
   packaging, 70-71
   proper, 11
   size, 7-8
   turnkey, 69
   types and scopes, 40-44
   unit-price, 47-52
Contractual alternatives, 186-187
Contract volume, 28
Creativity, 219

Deductive change orders, 282
Disclaimer clause, 243-244
Diversification, 27
Documentation, 277-279

Emergencies, 321-322
Establishing marketing plans, 39
Estimating, 3
Ethical considerations, 299-310
Equipment, 157-160
Exceptions and stipulations, 184-185
Exercise for sales calls, 347-348
Expense accounts, 205-206
Expression, manner of, 130-132

Feasibility studies, 233-234
Financial data, 160-163
Finding the right job, 317-319
Forecasting, 215-216
Franchise extension, 219-220

Game theory, 28-29

Image through dress, 204-205
Innovation, 212-213
International marketing, 286-298
  long-range business development, 294-298
  marketing techniques, 288-291
  selling to, 286-288
  short-term business development, 291-294

Late payments, 185-186
Legal assistance, 281-282
Location, 9

Market:
  fitting other markets, 22
  fixed practices, 23
  monitoring, 81
  size, 20-21
  sources of data, 24-25
  who serves, 21
Market conditions, reaction to, 86
Market cycles, 13, 21
Marketing aids, 100-127
Market penetration, 26, 225
Market survey, 19-26
Markup, 228-232
Monitoring, market, 81
Moving average, 26
Multiheaded clients, 35-37

Negotiations, 249-263
  appeals to reason, 255
  concluding deal, 260-261
  legal talent, 257-258
  mutual endeavor concept, 254-255
  right opportunity, 249-251
  self-control and temper, 256-257
  setting stage, 252-254
  strategy, 251-252
  subordinates, 258-259
  timing, 259-260
  trades, 259-260

Office politics, 319-321
Opportunism, 85-86
Owner's records, 279-281

Personal contacts, 34, 234-235
Personnel, 157-160
Price chiseling, 86-89
Pricing, 17-18, 23, 30, 221-232
  long-range effects, 227-228
  strategy, 223
Planning, career, 314-317
Preliminary design, 207
Prequalification, 154-181
Presentations, 238-239
Private work, 31-33
Product development, 212-220
Professional liability, 243-248
Professional relations, 237-238
Professional responsibility, 311-312
Professional services, 233-242
Profitability, 29-30
Proposals, 183-201
  legal requirements, 189-198
  for professional services, 239-241
Protecting innovation, 217-219
Protective pricing, 227
Public agencies, 11
Public relations, 237-238
Public work, 31-33

Quality, 13

Reaction to market conditions, 86
References, 163-164
Relations with third parties, 14-15
Risk sharing, 71-73

Sales force, 33-34

Salesmanship and advertising, 202-211
Sample correspondence, 331-339
Sample exercises, 327-330
Schedule, 12
Scope of work, 9-11, 27, 30
Size of contract, 7-8
Subcontractors, 163-164
Suppliers, 163-164

Technical letters, 129-137
Timing, 13-14
Types of work, 8, 19-20

Underpricing, 225-226
Unethical competition, 81-84
Use of allies, 76-77

Value engineering, 30-31, 67-69
Verbal communication, 138-153
Vindictive owners, 283-284
Volume, 15, 26-27

Work:
 scope is, 9-11, 27, 30
 type of, 8, 19-20